FORKLIFT SAFETY

A Practical Guide to Preventing
Powered Industrial Truck
Incidents and Injuries

NEWS / IN BRIEF

Forklift driver, 20, killed in Medford warehouse accident

AN EVERETT man died yesterday in a freak accident at a Medford beer-distributing warehouse when a forklift loaded with cases of beer tipped on top of him. Med Police Lt. Lawrence Groves said Stephen L... died shortly after noon i... Gr... truck distri...

Worker killed in freak accident

A 57-year-old factory worker was fatally strangled at a Pointe Claire warehouse late Sunday after accidentally driving his forklift into a rope.

The accident happened at 11:50 p.m. at the Zellers distribution centre on the Trans-Canada Highway service road.

Police say the man became entwined in the rope. The rope was being used as a barrier to prevent forklift operators from driving on to a freshly-painted part of the warehouse floor, a Zellers spokesman said.

Fellow factory employees tried to administer CPR to the victim, but to no avail. He was pronounced dead on arrival at the Lakeshore General Hospital minutes later.

"It was obviously a horrible accident," said Penny Westman, director of legal services for Zellers Ltd.

Both Zellers officials and the Commission de la santé et sécurité du travail are investigating the accident, she said.

Oshawa youth injured in forklift accident

By Joanne Anderson
Staff Writer

A lesson in lift truck operating ended in serious injury for a 17-year-old Oshawa ... overturned, pinning ...

enough to pull Mr. Harden out from underneath it.

Mrs. White, who has taken a St. John Ambulance course through social services, stayed with the youth until an ambulance arrived.

She said he was bleeding from the mouth, and while he appeared unconscious she thought he could hear her as she tried to reassure him that help was coming.

Had help not been there immediately to get the lift truck off the youth Mrs. White feels the accident might have been far more tragic. "His eyes were already rolling back in his head and he was turning blue," she said.

Randy Henry, part-owner of Henry Buildall, says Mr. Harden was learning how to drive the truck and had been with the company for only two months.

He says the Oshawa store manager was training the youth and the accident is the first the company has ever had while training a new ...

...when this lift truck overturned pinning him underneath ...accident occurred.
PHOTO BY WALTER PA...

Forklift crushes worker to death

A Michigan man was crushed under a forklift when he disconnected a hydraulic ho... while making a repair.

Kent County Sheriff's D... partment spokesman Edwar... Westhouse said 48-year-ol... Arthur Johnson suffered massive chest injuries and died in the accident, which happened while the lift was raised and loaded with wood.

Sgt. Westhouse said the lift mechanism and load collapsed on Johnson when he pulled ... hose, causing a loss of ...

Man critically hurt in forklift accident

A 36-year-old mechanic was critically injured Wednesday when he was pinned under the wheel of an 11-ton forklift, according to Elk Grove Township Fire Department officials.

The accident occurred around 7 a.m. at Wil-Lift, 2405 Hamilton Drive, Elk Grove Township...

A16 THE TORONTO STAR Wednesday, July 13, 1994

Falling crate kills man at Malton warehouse

BY PETER EDWARDS
STAFF REPORTER

A 52-year-old Mississauga father was killed when a 1,800-kilogram crate fell on him from a forklift vehicle at a Malton warehouse, Peel Region police say.

The crate, containing an industrial sewing machine, was being removed from a truck at Thriftcargo Ltd. on Dorman Rd. near Pearson airport at 9:50 a.m. yesterday.

It slipped off the forklift, pinning Express Courier driver Alan To-

Science Centre where he was pronounced dead.

"He was just an all-around family man," said Tovell's brother-in-law, Bob Pottruff. "He was a caring husband and father. F... put his family first."

Tovell had been married to Carol for 23 years, and was proud of her and their daughter, Penny. Pottruff... particularly proud when P... marks in English at sch... Tovell coached Penn... and also coached min... truff said.

The family is await... the accident.

"This is devastati... Pottruff said. "Al... ...one. He would...

Tovell

Man crushed by forklift

A Sacramento man was killed Sunday in West Sacramento when he was crushed by a forklift.

Toby Smedley, 31, was driving a forklift at a food service ware...

SAFETY

OSHA fines firm for lift truck safety

Alleged failures to safely operate and maintain a fleet of industrial trucks and train their operators figure prominently in an enforcement action announced recently by the U.S.'s Occupational Safety and Health Administration (OSHA). OSHA's message to industry: practice safety or pay a high price.

After a "wall to wall" inspection of Boise Cascade's Rumford, Me., pulp and paper mill, OSHA has proposed penalizing the company $245,000 for alleged, willful violations of federal rules for safely operating industrial trucks. The agency also is asking additional penalties for allegedly f...

FORKLIFT SAFETY

A Practical Guide to Preventing Powered Industrial Truck Incidents and Injuries

George Swartz, CSP

GOVERNMENT INSTITUTES
ROCKVILLE, MD

Government Institutes, Inc., 4 Research Place, Rockville, Maryland 20850, USA.

Copyright © 1997 by Government Institutes.
All rights reserved.

01 00 99 98   5 4 3 2

No part of this work may be reproduced or transmitted in any form or by any means, electronic or mechanical, including photocopying, recording, or any information storage and retrieval system, without permission in writing from the publisher. All requests for permission to reproduce material from this work should be directed to Government Institutes, Inc., 4 Research Place, Rockville, Maryland 20850, USA.

The reader should not rely on this publication for identification of specific equipment operations and maintenance repairs, or specific points that apply to a particular manufacturer's equipment or authorized substitution. The author and publisher make no representation or warranty, express or implied, as to the completeness, correctness or utility of the information in this publication. In addition, the author and publisher assume no liability of any kind whatsoever resulting from the use of or reliance upon the contents of this book.

Follow manufacturers' standards for the use, maintenance and repair of equipment. Refer to manufacturers' operating, training and maintenance manuals for guidance on repairs and replacement parts. Call manufacturers for technical assistance on operating or maintenance procedures not provided for in the manufacturers' manuals. The Occupational Safety and Health Administration (OSHA) can also provide information on proper operating methods.

Printed in the United States of America.

Library of Congress Cataloging-in-Publication Data
Forklift safety / by George Swartz.
    p.   cm.
Includes index.
ISBN: 0-86587-559-6
   1. Fork lift trucks--Safety measures 2. Materials handling--Safety measures   I. Title.
TL296.S38   1997
621.8'63--dc21

96-37925
CIP

This book is dedicated

to Frank Cava, former Vice President of Personnel at Pittsburgh Bridge and Iron Works,

for starting me in this career many years ago, and

to my daughter, Laura, and my son, Eric.

TABLE OF CONTENTS

List of Figures and Tables xi
Foreword xv
Preface xvii
About the Author xix
Acknowledgments xx

1. THE NEED FOR OPERATOR TRAINING 1

ACCIDENT REDUCTION THROUGH TRAINING 2
    Employers Are Penalized for Noncompliance 4
COSTS OF ACCIDENTS 7
    Direct/Indirect Costs 10
INITIAL TRAINING RECOMMENDATIONS 10
    Successful Training Programs in Industry 12
THE NEED FOR EMPLOYEE SAFETY 15
    Tip Over 17
    Legal Considerations 20
ANSI B56.1 RECOMMENDATIONS FOR TRAINING 21
    Operator Qualifications 21
    Operator Training 21
OSHA'S RECOMMENDED TRAINING PROGRAM CONTENT 24
    Truck Related Topics 24
    Workplace Related Topics 24
SUMMARY 25
REFERENCES 26

2. HOW LIFT TRUCKS WORK 29

UNDERSTANDING FORK LIFT TRUCKS 29
STABILITY OF POWERED INDUSTRIAL TRUCKS 30
    Load Center 30
    Stability Triangle 31
    Lateral Stability 34
    Dynamic Stability 35
    Longitudinal Stability 35

OVERVIEW OF POWERED INDUSTRIAL TRUCKS	36
Cars/Forklifts: The Difference	38
New Technology	40
SUMMARY	41
REFERENCES	42

3. MODELS OF POWERED INDUSTRIAL TRUCKS 43

LOW LIFT MODELS	48
Platform Trucks	48
Pallet Trucks	48
HIGH LIFT TRUCKS—RIDER TYPE	48
Counterbalance Rider Trucks	48
Narrow-Aisle Trucks	50
Rider Reach Trucks	50
Order Picking Trucks	50
Sideloaders	51
Turret Trucks	52
Hybrid Machines	52
HIGH LIFT TRUCKS—FLOOR OPERATED	52
Manually Propelled Trucks	52
Platform Trucks	53
Walkie Fork Trucks	53
Walkie Straddle Trucks	54
Walkie Reach Trucks	54
Counterbalanced Walkie Trucks	54
SUMMARY	57
REFERENCES	57

4. POWERED WALKIE/RIDER PALLET TRUCKS 59

GUIDELINES FOR THE PROPER USE OF WALKIE RIDERS	62
Safety Features	62
Operating Principles	62
SUMMARY	68
REFERENCES	68

5. PRE-USE INSPECTIONS 69

PRE-USE INSPECTION FORMS	70
Steering and Horn	71
Brakes	71
Lights	72
Controls	72
Coolant Level	72
Fire Extinguisher	72

Forks	73
Tires	73
Gauges	74
Hydraulic Oil Leaks	74
Fuel Level/Oil Level	74
Propane Tank	75
Overhead Guard	75
Additional Items	75
SUMMARY	79
REFERENCES	79

6. GENERAL SAFETY GUIDELINES ... 81

GUIDELINES AND GENERAL SAFETY RULES	81
General Guidelines for Students	81
Traveling with Power Equipment	83
Dock Safety Rules	90
Battery Charging Area	92
Gas Powered Vehicles	94
Gasoline and Diesel Fuel	96
Maintenance	96
Tire Safety	97
General Safety Requirements (1910.178)	98
Operators	99
SUMMARY	99
REFERENCES	100

7. SAFETY AT THE DOCK ... 101

DOCK SAFETY	103
Wheel Chocks/Truck Restraints	103
Automatic Trailer Restraint Devices	104
Dock Plates	107
Dock Inspections	114
SUMMARY	115
REFERENCES	115

8. SAFETY OF PEDESTRIANS ... 117

PHYSICAL HAZARDS	117
TRAINING SAFEGUARDS	122
Operator/Employee Comments on Pedestrians	122
Pedestrian Safety	123
National Safety Council Guidelines	124
SUMMARY	125
REFERENCES	125

9. PERSONAL PROTECTIVE EQUIPMENT (PPE) 127

- THE OSHA PPE STANDARD 127
- EQUIPMENT REQUIRED BY THE PPE STANDARD 134
  - Cold Weather Clothing 134
  - Head Protection 134
  - Eye and Face Protection 135
  - Safety Shoes/Boots 135
  - Respiratory Protection 138
  - Hearing Protection 138
  - Hand Protection 139
  - Hazard Assessment for PPE 139
  - Employee Training 140
- SUMMARY 140
- REFERENCES 140

10. PROTECTING THE HEALTH OF POWERED INDUSTRIAL TRUCK OPERATORS 141

- CARBON MONOXIDE 141
  - Control Methods for CO 143
- NOISE 145
- CHEMICALS IN THE WORKPLACE 146
  - Physical Hazards 147
  - Health Hazards 147
- RADIATION 151
- ERGONOMICS 151
  - Lights, Alarms, Mirrors 155
- SUMMARY 156
- REFERENCES 156

11. USING JOB HAZARD ANALYSIS FOR POWERED EQUIPMENT 159

- WHAT IS JHA—AN OVERVIEW 160
  - Starting the Process 160
  - Proper Forms 162
  - Priority Basis Rationale 169
  - Starting a JHA 169
  - Management Training 171
  - Additional Comments on JHA 173
- SUMMARY 174
- REFERENCES 174

12. INVESTIGATING ACCIDENTS AND INCIDENTS 175

- INCIDENT INVESTIGATION 176
- ACCIDENT INVESTIGATION 179

SUMMARY	185
REFERENCES	186

13. TIPS FOR TRAINING OPERATORS ... **187**

UNDERSTANDING TRAINING	188
Training Tips	189
Use Your Listening Skills	190
Training of Supervisors	191
Developing a Training Plan	191
Group Methods of Training	192
Hands-on Training	195
Quizzes	195
Audiovisuals	195
SUMMARY	198
REFERENCES	198

14. FORKLIFT RODEOS AND FORKLIFT RALLIES ... **199**

WHY HOLD A FORKLIFT COMPETITION?	199
ELEMENTS OF CONSIDERATION FOR COMPETITION	200
Management Commitment	200
Site Location	201
Demonstration of Skills Testing and Judging	201
Selection of Participants	207
Set Up of the Test Site	209
Judging of the Event	213
Determination of Winners	213
Awards and Recognition	215
SUMMARY	217
REFERENCES	217

15. MISCELLANEOUS ISSUES ... **219**

REDUCING FORKLIFT DAMAGE	219
LIFT TRUCK FORKS	220
How Forks Are Inspected	222
PALLET SAFETY	225
USE OF LIFT CAGES/WORKING PLATFORMS	228
ENVIRONMENTAL CONSIDERATIONS	231
Maintaining Water Quality	233
Storing Hazardous Waste	233
Preparing Waste for Shipment	234
Lift Truck Maintenance/Environmental Controls	234
FIRE FIGHTING	236
VOLUNTARY PROTECTION PROGRAMS (VPP)	238

x / Forklift Safety

JOB SAFETY OBSERVATION, (JSO) 240
    What Jobs to Select 242
    Informing Employees of JSO 243
    Scheduling JSO's 243
SUMMARY 245
REFERENCES 246

APPENDIX A: OSHA Proposed Rule for Training Powered Industrial Truck Operators 247
APPENDIX B: Quiz Questions for Forklift Operators 291
APPENDIX C: Sources of Help for Forklift and Powered Truck Training 323
APPENDIX D: OSHA 1910.178 Guidelines for Powered Industrial Trucks 331

GLOSSARY 347

INDEX 355

LIST OF FIGURES AND TABLES

Table 1.1. Percentages of Forklift Accident Causes 1
Table 1.2. Classification of Forklift Fatalities 2
Table 1.3. OSHA's Fatality and Injury Statistics 3
Table 1.4. Causes of Accidents—OSHA Investigation Summaries 4
Figure 1.1. Hazards of Elevated Forks 5
Figure 1.2. Don't Block Your Vision 9
Figure 1.3. Example of Damaged Product 16
Table 1.5. OSHA Study of Tip Over Accidents 17
Figure 1.4. Forklift Falling Off of a Dock 18
Figure 1.5. Tip Over and Seat Belt Decal 20
Figure 1.6. Beware of Dock Edges 25
Figure 2.1. Tipping Fork Lift 31
Figure 2.2. Stability Triangle 32
Figure 2.3. Hazards of Making Sharp Turns 32
Figure 2.4. Oversize Loads Can Tip a Lift Truck 33
Figure 2.5. Stability Triangle and Turns 34
Figure 2.6. Raising a Load and Potential for Tipping 36
Figure 2.7. OSHA's Examples of Tip Over Potential 37
Figure 2.8. Comparing the Turning Radius 39
Figure 2.9. Blind Spot While Driving 40
Figure 3.1. Models of Industrial Trucks 44
Table 3.1. U. S. Shipments of Powered Industrial Trucks 47
Figure 3.2. Maneuverability of a Lift Truck 49
Figure 3.3. Use of a Scissor Lift 51
Table 3.2. Analysis of Lift Truck Classifications and Models 55
Figure 4.1. Steel Toe Boot Damaged by Hand Truck 60
Figure 4.2. Properly Operating a Powered Hand Truck 63
Figure 4.3. Operating Handle Brakes 64
Figure 4.4. Inspection Form for Powered Hand Trucks 66
Figure 4.5. Damage to Racking 67
Figure 5.1. Inspecting a Lift Truck 70
Figure 5.2. Checking Steering and Brakes 71
Figure 5.3. Checking Tires 74
Figure 5.4. Forklift Daily Record Form 76
Figure 5.5. Pre-Use Inspection Form 77
Figure 6.1. Damaged Overhead Fan 83
Figure 6.2. Arrange Loads Properly on Pallets 84

Figure 6.3. Properly Traveling on a Ramp	85
Figure 6.4. Proper Stacking of Product	86
Figure 6.5. Back Rest and Forks	87
Figure 6.6. Operating in Reverse	88
Figure 6.7. Clear and Clean Aisle Ways	89
Figure 6.8. Reduce Speed at Turns	90
Figure 6.9. Manual Use of Wheel Chocks	91
Figure 6.10. Battery Charging with the Proper PPE	93
Figure 6.11. Safe Battery Handling with a Hoist	95
Figure 6.12. Operator Inspecting Lift Truck	97
Figure 7.1. Consequences of Not Chocking Wheels	103
Figure 7.2. Automatic Trailer Restraint	104
Figure 7.3. Trailer Backing and Signage	106
Figure 7.4. Wheel Chock with Flag	106
Figure 7.5. Handling a Portable Dock Plate	108
Figure 7.6. Dock Slope Guidelines	109
Figure 7.7. Dock Safety Checklist	110
Figure 7.8. Dock Safety Awareness	114
Figure 8.1. Safe Entry/Exit Door for Employees	119
Figure 8.2. Look Before You Back	120
Figure 8.3. Fork Lift Counterweight Hazards	121
Figure 8.4. Fork Lift Entry From Out of Doors	123
Figure 8.5. Tight Turning Radius	124
Figure 9.1. PPE Employee Training Form	130
Figure 9.2. PPE Location Survey for Power Equipment Operators	131
Figure 9.3. ASSE/ANSI Eye and Face Protection Guidelines	136
Figure 10.1. Noise Exposure Graph	146
Figure 10.2. NFPA 704 Label	149
Figure 10.3. HMIS Label	150
Figure 10.4. Safe Lifting from a Pallet	154
Figure 11.1. JHA Master List	163
Figure 11.2. JHA Form #2 - Blank	164
Figure 11.3. JHA Form #2 - Completed	166
Figure 11.4. JHA Form #3 - Blank	168
Table 11.1. Step 1 in Recharging a Battery	171
Figure 11.5. JHA Form #3 - Completed	172
Figure 12.1. Accident Pyramid	176
Figure 12.2. Damage to Wall From Forklift	177
Figure 12.3. Battery Charger Incident	178
Figure 12.4. Blank Incident Reporting Form	180
Figure 12.5 Completed Incident Reporting Form	181
Figure 12.6. Hazards of Feet Extending from a Lift Truck	183
Figure 12.7. Accident Investigation Form	184
Figure 13.1. Instructors' Self Check List	193
Figure 13.2. Log for Lift Truck Quizzes	194
Figure 13.3. Interactive Video in Use	197

Figure 14.1. Evaluate Operators in Advance 202
Figure 14.2. Show and Tell Test Form 203
Figure 14.3. Speed Stacking Illustration 205
Figure 14.4. Judges Scoring Sheet - Speed Stacking 206
Figure 14.5. Skills Evaluation Forms 208
Figure 14.6. Maneuvering Skills for a Rodeo 209
Figure 14.7. Judges Scoring Sheet - Maneuvering Skills 210
Figure 14.8. Basket Stacking Illustration 211
Figure 14.9. Judge's Scoring Sheet - Basket Stacking 212
Figure 14.10. Guidelines for Judges 214
Figure 14.11. Penalty Point Code for Judges 216
Figure 15.1. Protective Post for Racking 221
Figure 15.2. Fork Wear Versus Capacity 223
Figure 15.3. Measuring Fork Wear with Calipers 224
Figure 15.4. Proper Pallet Stacking Patterns 226
Figure 15.5. High Stack of Empty Pallets 227
Figure 15.6. Safe Use of a Lifting Cage 229
Figure 15.7. Storing a Lift Cage 231
Table 15.1. Advantages and Disadvantages of Lift Truck Fuel Options 232
Figure 15.8. Spill Clean Up Kit 234
Figure 15.9. Fire Extinguisher Classification 237
Figure 15.10. Fire Extinguisher Location 238
Figure 15.11. Audit Questionnaire - Forklift Training 240
Figure 15.12. Audit Scoring Form 241
Figure 15.13. Job Safety Observation Form 244

FOREWORD

This is the most comprehensive book ever written on this important workplace operation. I wish I had this resource when I was a practicing plant safety director. This book is a must for those safety and health professionals working in manufacturing, distribution, or warehousing. It is also must reading for non-safety professionals, managers, and supervisors responsible for powered industrial truck (PIT) operations. This is not an academic dissertation on the issue, but rather a hands-on pragmatic approach to what needs to be done when operations include PIT.

Mr. Swartz rightly points out in Chapter One that management is responsible and accountable for safe operations. The loss of life, serious injury, and illness caused by operating powered industrial trucks are tragedies because they are preventable. In addition, the risk potential for damage or destroying property, material, and product is enormous—companies have gone out of business as a result of catastrophic losses due to unsafe operation of powered industrial trucks.

It is critical and good business to reduce or eliminate losses. In the event the government regulatory climate heats up and a PIT regulation is promulgated, serious application of this book will provide your PIT operations with a program which will put you in compliance with most, if not all, provisions that would be included in a regulation. Mr. Swartz has provided the reader with a chapter-by-chapter, step-by-step, "cookbook" resource which covers the essential elements of a powered industrial truck program, which, if applied properly, will reduce death, injuries, workers' compensation, and property damage.

Many safety and health professionals are focused on carpal tunnel syndrome and bloodborne pathogens risks, and rightly so. However, we need to balance our approach and time working on various workplace risks. I would suggest that you review the incident data in Chapter One and also review your own workplace data and exposure to PIT risks. Most companies will find a compelling reason to apply the PIT principles and best practices Mr. Swartz has compiled in this valuable resource.

Jerry Scannell
President
National Safety Council

PREFACE

Forklift Safety was prepared to assist in the reduction of incidents and accidents associated with powered industrial trucks. The need for improved forklift safety, and safety with other equipment, has never been greater. OSHA statistics indicate that there are approximately 100 workers killed each year as a result of powered industrial truck accidents. Similar accidents throughout Canada also take their annual toll of life and limb.

It is not unusual to read a local newspaper in which the sad details of an industrial forklift accident are highlighted. A few of these descriptive accidents help tell the story: "Worker Killed in Freak Accident"; "Falling Crate Kills Man in Malton Warehouse"; "Oshawa Youth Injured in Forklift Accident"; "Forklift Truck Skids Off a Loading Dock"; "Maintenance Worker Falls from Forks of Lift Truck". This list grows daily.

Safety experts agree that when proper safety training and enforcement are used, accidents decline. This same principle holds true for lift trucks. Many organizations do not properly train operators. Statistics from the workplace are evidence of this laxity of enforcement—many thousands of injuries and millions of dollars in losses are caused by forklifts and improperly trained operators.

How to safely operate powered equipment is a broad and specialized subject. A skilled operator has to know and understand many rules and guidelines. This book contains a wide range of information, which can be used to improve operator skills and overall forklift safety. *Forklift Safety* was written as a training guide; it provides a checklist approach to powered truck safety. The reader is provided with chapters on why training is important, how lift trucks work, dock safety, PPE, health issues, inspection processes, rodeos and rallies, powered hand trucks, job hazard analysis, accident investigation, and more. The fifteen chapters are well documented and provide a variety of forms, tables, figures, and illustrations to assist the reader. To identify some of the real hazards in the workplace, many chapters contain "Accident Facts," which provide a brief synopsis of actual events.

The appendices contain OSHA's proposed rule on training requirements, quizzes for operator knowledge and procedure, additional sources of information for forklift safety, and the OSHA 1910.178 guidelines for powered equipment.

Because the information is broad and detailed, the book can be used to model a complete powered equipment program, improve an existing program, or provide guidance for the college or university that utilizes material handling or safety courses in its curriculum.

Plant managers, human resource and personnel staff, safety committees, consultants, supervisors, safety directors, risk managers, enforcement officials, manufacturers of powered equipment, and organizations that offer training programs for powered equipment safety could benefit from this material.

The figures and illustrations are extensive and will help the beginning operator, as well as reinforce the subject of industrial truck safety for veteran operators. The various tables, forms, and checklists provide for a wide range of subject matter and will be handy for training and inspection.

It is hoped that readers will fully utilize this publication to improve their safety programs. The needless loss of life, employee injury, production costs, and building damage takes a heavy toll each year. This book will also prove an asset toward the elimination or reduction of OSHA citations. Forklift manufacturers are a valuable resource for product knowledge, purchasing, and training programs, and should be called upon for additional assistance.

George Swartz, CSP
Chicago, Illinois
April 1997

ABOUT THE AUTHOR

George Swartz has been involved in the safety profession for over twenty-five years. His experience includes manufacturing, warehousing and distribution, auto repair shops, and heavy construction. He has been employed as a safety officer by Pittsburgh Bridge and Iron Works, Buell Division of Envirotech Corporation, and his current employer, Midas International Corporation. He also has particular expertise in the area of safety auditing. A frequent lecturer, speaker, author, and trainer, Mr. Swartz has delivered presentations throughout the United States. He has traveled extensively throughout the U.S., Canada, and Europe in his profession.

Mr. Swartz currently serves on the boards of directors for the National Safety Council, the Environmental Resource Institute, and Project Safe Illinois. He is vice president for Business and Industry for the National Safety Council, chairman of the editorial board for *Professional Safety* magazine, and a member to the safety and health advisory committees at the University of Illinois - Chicago and Northern Illinois University. He is also a lecturer for safety courses at Northern Illinois University. He testified before OSHA in Washington, D.C. during hearings regarding the dangers of asbestos exposure for brake mechanics.

Mr. Swartz has been awarded the honor of Fellow by the American Society of Safety Engineers, the Distinguished Service to Safety Award by the National Safety Council, and Outstanding Service Award by the Environmental Resources Institute. He was also honored for service as Chairman of the National Safety Council's Business and Industry Division.

Mr. Swartz holds a B.A. from the University of Pittsburgh, a M.S. in Safety from Northern Illinois University, a C.A.S. in Safety from Northern Illinois University, and an M.S. in Managerial Communications from Northwestern University. Mr. Swartz has been a Certified Safety Professional since 1980.

ACKNOWLEDGMENTS

The author wishes to thank the following organizations for their contribution of materials and information for this publication: Clark Material Handling Company; Rite-Hite Corporation; The National Safety Council; *Material Handling Engineering* magazine; The Hyster Company; J.J. Keller and Associates; *Modern Materials Handling* magazine; The Industrial Truck Manufacturers Association; The American Society of Safety Engineers (ANSI); *Plant Engineering* magazine; Yale Materials Handling Corporation; Atlas Companies; and The Occupational Safety and Health Administration.

In addition, without the able assistance of Laura Davison, Phyllis Luzader, and Linda Pekar, the preparation of this entire manuscript would not have been possible.

The author would also like to thank the following individuals for their assistance and technical review of the manuscript: Ron Koziol and Joe Lasek of the Industrial Department of the National Safety Council, Clark Material Handling Company, and Rita M. Mosley, Industrial Safety Consultant/Supervisor for the Department of Illinois Commerce and Community Affairs.

1

THE NEED FOR OPERATOR TRAINING

It is estimated by OSHA that there are more than 68,400 accidents each year from powered industrial trucks. Approximately 100 workers lose their lives each year as a result of powered industrial equipment accidents. The number of day to day incidents that involve close calls, damage to product, equipment and buildings would be in the many thousands. Each of these incidents provides an opportunity for something more serious to occur.

Statistics show that the causes of lift truck accidents involve many factors. OSHA conducted a study, "First Report of Serious Accidents" 1985-1990, and published the data in Table 1.1.

Table 1.1. Percentages of Forklift Accident Causes

OSHA has identified key factors which have contributed to employee accidents from lift trucks. Tip overs are responsible for over 25% of all accidents.

1.	Tip over	25.3%
2.	Struck by powered industrial truck	18.8%
3.	Struck by falling load	14.4%
4.	Elevated employee on truck	12.2%
5.	Ran off dock or other surface	7.0%
6.	Improper maintenance procedures	6.1%
7.	Lost control of truck	4.4%
8.	Truck struck material	4.4%
9.	Employees overcome by CO or propane fuel	4.4%
10.	Faulty powered industrial truck	3.1%
11.	Unloading unchocked trailer	3.1%
12.	Employee fell from vehicle	3.1%
13.	Improper use of vehicle	2.6%
14.	Electrocutions	1.0%

It should be noted that a small percentage of actual accidents involve tip over. The 25.3% figure quoted by OSHA may be high.

In another study of forklift fatalities published by the Bureau of Labor Statistics, (BLS), information was provided on 170 fatal powered industrial truck accidents. This is illustrated in Table 1.2.

Table 1.2. Classification of Forklift Fatalities, CFOI, 1991-1992

In Table 1.2, OSHA classified forklift fatalities for the years 1991-1992. As in Table 1.1, tip overs lead all other categories.

How Accident Occurred	Number	Percentage
1. Forklift overturned	41	24
2. Worker struck by material	29	17
*3. Worker struck by forklift	24	14
*3. Worker fell from forklift	24	14
5. Worker pinned between objects	19	11
6. Forklift struck something or ran off the dock	13	8
*7. Worker died during forklift repair	10	6
*7. Other accident	10	6
Total	170	100

*tie

ACCIDENT REDUCTION THROUGH TRAINING

Manufacturers of powered industrial equipment and safety professionals feel that accidents and injuries can be reduced if operators are properly trained. In order to improve safety, the following should be a part of the accident prevention overall plan and include:

- Properly designed equipment with safety and ergonomics as prime considerations.
- A committed management.
- In depth training programs for all operators.
- Vigorous enforcement of the safety program.
- Auditing techniques for assurance.
- Re-testing at specified intervals.

Table 1.3 illustrates the estimated number of accidents that could be reduced through proper operator training.

Table 1.3. OSHA's Fatality and Injury Statistics

OSHA has identified current injury and fatality statistics and has projected what the new training guidelines will accomplish.

OSHA states that through the improper operation of powered industrial trucks, there are:

- 85 fatalities each year
- 34,902 serious injuries
- 61,800 non-serious injuries

If an employer were to comply with the newly proposed powered industrial truck training program from OSHA, the risks would fall to:

- 68-63 fatalities - 25% reduction
- 10,898 to 14,118 serious injuries
- 15,450 non-serious injuries

	Current OSHA data	After new program compliance
Fatalities	85	68-63
Serious injuries	34,902	10,898-14,118
Non-serious injuries	61,800	15,450

These numbers are very noteworthy. Effective training programs have been estimated to reduce injuries by as much as 70%. It is apparent that the needless toll of injured, maimed or killed operators takes a serious toll on society. Families and employers are devastated by any loss such as this. It stands to reason that every effort should be made to prevent as many incidents and accidents as possible.

During the OSHA research on the proposed standard they queried the computer for all reports that contained the key word "industrial truck." Table 1.4 illustrates the OSHA findings. There were 4,268 total reports in the system that resulted in 3,038 fatalities, 3,244 serious injuries and 1,413 non-serious injuries. (Many of the reports of the accidents were the result of multiple fatalities and/or injuries).

The study of this information produced the data in Table 1.4.

Table 1.4. Causes of Accidents—OSHA Investigation Summaries

OSHA researched some 4,268 accident reports on lift truck accidents and was able to identify specific accident causes.

	Cause	Number of Reports
1st	Operator inattention	59
2nd	Overturn	53
3rd	Unstable load	45
4th	Operator struck by load	37
5th	Elevated employee	26
6th	No training	19
7th	Overload, improper use	15
8th	Accident during maintenance	14
*9th	Obstructed view	10
*9th	Improper equipment	10
*11th	Falling from platform or curb	9
*11th	Not powered industrial truck accident	9
*13th	Other employee struck by load	8
*13th	Carrying excess passenger	8
*15th	Vehicle left in gear	6
*15th	Falling from trailer	6
17th	Speeding	5
	TOTAL	339

*tie

Figure 1.1 illustrates the hazards of being under or on elevated forks.

Employers Are Penalized for Noncompliance

The need for training has never been greater. An article in the July 10, 1996 BNA Reporter identified an employer in Canada that was cited for $73,000 by the Ontario Occupational Health and Safety Act that resulted in the death of a worker. An employee was crushed by a 1,590 pound beverage dispensing machine that was not secured to a forklift when the machine was loading it at a dock. In another case, an employer in Miami, Florida was cited for not providing sufficient forklift training to his employees.

Figure 1.1. Hazards of Elevated Forks

The danger of being under forks or standing on forks, two areas that contribute to workplace injuries (Courtesy of Clark Material Handling Company).

A worker working off of a work platform and was crushed while helping to install a refrigeration line. The employer was cited for:

- Failure to train operators in safe operations of forklift trucks.
- Failure to secure a safety platform to forks of a truck.
- Failure to use an overhead guard on a forklift truck.

The employer argued that the operator had past forklift experience when he was hired. The trial judge ruled that reliance on the experience of workers gained in their past employment is insufficient to meet the requirements of the cited standard.

An employer in Virginia was fined $112,000 in proposed OSHA penalties following the death of a forklift operator. The operator who was not a qualified electrician, was installing a bracket adjacent to a high voltage switch when he came in contact with live parts.

OSHA issued two willful violations for failure to maintain a minimum safe distance between employees and exposed parts. The employer also allowed an unqualified employee to work near exposed parts.

Another employer was cited for $74,700 in proposed penalties for the death of an employee at a retail food distributorship in Pennsylvania. The operator died after striking an overhead beam.

Surprisingly, studies have shown that many powered equipment operators throughout industry have little or no formal training, prior to the newly proposed OSHA powered industrial truck training rules, (see Appendix A), the agency offered very little in the way of language for training requirements. The original 1970 standard reads "only trained and authorized operators are permitted to operate powered industrial trucks, training methods shall be devised by the employer." In other words employers receive little guidance regarding:

- The method of training to be used.
- What points should be covered in employee training.
- What is used to authorize an operator—what criteria should be used.

The newly proposed OSHA training standard currently asks that:

- Every employer is to develop a training program.
- Every operator will have to qualify to operate the vehicle(s).
- The effectiveness of the program used for training will have to be evaluated.
- Refresher and remedial training must be ongoing.
- The training is to result in operator certification.
- The program will require re-certification and recordkeeping—contents of the program to include core training curriculum, trainee evaluation, the specific vehicle (truck) and the operating environment.
- Operators to be tested and certified for each piece of equipment.

Employers must realize that operator training will pay big dividends. The real and genuine thrust of operator training is teaching all employees to work safely under day-to-day operating conditions. Many of these conditions cannot be duplicated in a training program. Operators, once they have absorbed the new training guidelines, will be able to better protect themselves when these new challenges are encountered.

Benefits of comprehensive operator training for employers and employees include:

- Lower risk to employees—fewer injuries.
- Less damage to equipment.
- Less damage to product.
- Greater compliance with federal, state and local regulations.
- Lower maintenance costs for powered equipment.
- More pride, good will and protection for employees and the company.
- Less likelihood of monetary citations from agencies.
- Less likelihood of litigation.

COSTS OF ACCIDENTS

Powered industrial trucks can and do inflict injury to individuals and damage to buildings, equipment and product. In addition to the need for employee training, there is a need to provide the proper piece of equipment for the environment it will operate in. Consider the need for fire training awareness for the operators and the concern for the building the lift truck will affect.

According to Factory Mutual Engineering and Research statistics, the human element was the prime factor in 220 of the 353 lift truck related losses from 1987-1992. A Lift truck can be the correct type for the building and its contents, be properly maintained and can have ideal operating conditions within, but a poorly trained or careless operator can turn the vehicle into a means of destruction or a means of ignition source on wheels.

Fire was the costliest peril, in the losses from 1987-1992, accounting for more than $146 million of the more than $160 million lift truck related losses.

In addition, improperly operated lift trucks can create dollar losses by damaging in-rack sprinklers. The leakage caused by this damage could cause the fire system to freeze up in the winter. Operators can strike fire doors and damage the operating mechanisms so they do not function. Racking can be struck and damaged, possibly causing collapse and the loss of product and injury.

Some of the most severe losses have occurred while operators were handling flammable liquids. The load can be dropped, the contents of the load spilled, which is then ignited by hot surfaces or sparks from the truck.

ACCIDENT FACT

In 1982, the largest loss involving a lift truck also involved flammable liquids. It took place in a department store warehouse and distribution center. The building was set ablaze when a box of aerosols fell from a 15 foot storage rack. The aerosols ruptured when they struck the floor. Vapors were released and were ignited by a nearby electric forklift. The fire spread rapidly and when the event was over, the 1,200,000 square foot warehouse was leveled at the cost of $148 million.

In another example of a huge loss and the need for training, a lift truck operator in a paint company spilled 10 gallons of a flammable liquid on the floor. A spark from the lift truck ignited the vapors resulting in a fire that destroyed the entire plant. When it was all over, the loss was greater than $50 million.

Factory Mutual went on to state that given the examples of the above losses, operator training is necessary. Operator training should consist of:

- Maneuvering and unloading.
- The types and classes of trucks where they should be used.
- Precautions for working in hazardous materials areas.
- Precautions for handling hazardous materials.
- Loading and transporting goods.
- Basic operating procedures.
- Maintenance, refueling and charging.
- Operator safety tips.

Facility management may wish to supplement the above training items with additional programs. The rear of this chapter has the set of guidelines for operator training from the ANSI B56.1 Standard. When in-house training is used in addition to training that can be provided by an outside source, operators can become more aware of hazards in their facilities. Forklift manufacturers can provide programs for this purpose.

Additional cost considerations for accidents and incidents are associated with lift truck operator training. The majority of reports of losses for lift trucks involve water damage, which occurs when they run, back, load, raise into or sideswipe sprinkler systems, cross mains, branch lines, risers, or sprinkler heads. Operators often fail to consider sprinkler piping while raising stacks, or may fail to leave adequate clearance between a load and overhead piping with the mast raised. This damage not only includes the costs of repairing the sprinklers and water damage to goods and property, it could cause the system to fail if a fire started.

Impact accidents from poorly trained operators could topple storage shelves, buckle roofs if support members are knocked from their base and puncture containers or tanks of chemicals. Some of the chemicals could be flammable or toxic. Environmental considerations must be a part of such exposures. Containing spills and preventing any hazardous or toxic materials from entering sewers, waterways, offices and into products and goods is critical.

To help prevent costly damage as identified above, management can install a few safeguards:

- Place fire extinguishers on trucks, or, have units close to any lift truck traffic for immediate use.
- Install barriers near fire doors to take the impact of poor driving rather than the door.
- Paint overhead obstructions yellow with black stripes to allow for a visual reference for the operator, doorways, pipes, sprinkler systems, and structural members.

- Mark all pipes for contents so they can be easily traced to shut-off valves.

- Install protective railings or barriers by doors, racking, shelving or other machinery. Paint yellow or other bright colors for high visibility.

- Anchor and tie in racking to each other as well as to structural building members with support capability to prevent toppling and buckling of racks. This can also provide help in earthquake zones.

- Use proper signage warning of such hazards.

- Ropes or metal strips which make noise should a raised load be near a point of impact.

- Conduct operator training so that employees can identify the hazards in all of these situations.

OSHA estimates that substantial savings to facilities can take place when operators are properly trained. Reduced losses in property damage and training-related litigation will help reduce property damage by $8 million to $42 million each year. An additional $770,018 will be saved each year in damages and court costs that would have been awarded as a result of injuries caused by deficiencies in industrial truck operator training. Operators can easily run into an individual or into the building structure, machines, equipment or processes. Part of the prevention is the efforts in training. The graphic illustration in Figure 1.2 demonstrates the need for operator awareness.

Figure 1.2. Don't Block Your Vision

A reminder that blocked vision can cause accidents (Courtesy of the National Safety Council).

Direct/Indirect Costs

When any accident takes place and after the injured employee has been provided with medical treatment, the associated costs can be significant. Accidents contain two types of costs:

- Direct (Insured)

- Indirect (Uninsured)

- Insured costs include the full medical costs and the indemnity costs if the employee loses time from work. Uninsured costs are many:

  - Time spent investigating the accident.
  - Other workers stop to look at the accident.
  - Employee is usually paid for the day of injury.
  - Employee is slower when returning from lost time.
  - Employee makes return visits to doctor.
  - Potential for product damage.
  - Potential for equipment damage.
  - Replacement of employee while he is off.
  - Possible OSHA inspection triggered.
  - Permanent disability requires vocational training.
  - Reputation of company tarnished.

More indirect items could be added to the list. Also, most all of the indirect items listed would be associated with each accident.

Safety professionals view accident costs like an iceberg. The direct costs are visible and can be measured. Indirect costs—under the water level of the iceberg, are much larger and hard to measure. It has been estimated that indirect costs are at least a one to one ratio for costs and as high as five to one. For each direct dollar spent, at least five can be spent indirectly.

INITIAL TRAINING RECOMMENDATIONS

OSHA has outlined recommendations for operator training. For *operator selection*, prospective operators of any powered industrial truck should be identified based upon their ability to be trained and accommodated to perform job functions that are essential to the operation and safety of a powered industrial truck.

Management and trainers can determine the capabilities of a prospective operator to fulfill the demands of the job. This determination should be based upon the tasks that the job demands. The Americans with Disabilities Act (ADA) is to be a part of operator consideration. The employer must identify all of the aspects of the job that the employee must meet or perform while accomplishing his or her task. These aspects could include the level at which the employee can see or hear, the physical demands of the job and the environmental extremes of the job.

Some important factors to be considered for the safety of the operator, pedestrian and other workers is the ability of the candidate to see and hear within reasonably acceptable limits. This would include, within the vision requirement, the ability to see and hear both at a distance and peripherally. There could possibly be a requirement for the prospective operator to distinguish between different colors. These would be primarily red, yellow and green. Environmental extremes that might be demanded of the operator being considered could be the ability to work in areas of excessive cold or heat. Lifting manual loads may also be a part of an operator's overall responsibility in addition to driving the vehicle.

Methods of training should include lectures, demonstrations, written testing, oral testing and operator skills evaluations. The combination of these factors have been used in the past to train operators. Overall operator skills and ability are essential to safely perform the job function. The process can be made more understandable and interesting through the use of slides, films, photos, videos and other visual aids. The material has to be understood by the operator.

As an example, OSHA's definitions of lift truck stability and moment appear to be complex. A trainer can easily explain tip-over if a lift truck model or other illustrations are used. The stability triangle is also difficult for some to grasp because of the invisible feature that is used as a reference. Good visual aids and focused training will go a long way to safeguard employees.

Benefits of strong visual aids include:

- Employees being trained remain more attentive during the presentation if strong visuals and graphical information is included. These visuals increase the effectiveness of the training.
- The use of visual presentations allow the trainer to ensure that all of the necessary and required information is being covered during the program.
- Visuals help to break up lengthy discussions which help in employee comprehension.
- Visuals help in the retention of the material being provided.

In addition, employees like being challenged. Where written tests are a part of the program awards such as T-shirts, coffee mugs, baseball hats, etc. should be presented to those individuals that attain special marks, such as perfect test scores. Materials such as these are offered by many employers. Employees are more likely to pay closer attention when their hard work is rewarded with something useful as well as for the gain in safety knowledge.

Training program content must allow for training the operator for each type of powered industrial truck. Makes and models have special differences. Therefore the operator must know these differences if he is to operate different models. The training must also include information on the various work environments the operator will encounter. Overall general safety rules that are applicable to all industrial trucks must be included.

The employer, because of knowledge of the workplace, should be able to determine and design the training programs offered. Employee safety teams and committees can be helpful in program development. After all, who knows more about the hazards of a particular job than those employees who work with them each day. The OSHA Voluntary Protection Program (VPP) has established criteria for safety excellence. For those employers that are interested in this special recognition, refer to Chapter 15, "Miscellaneous Issues" for more information on VPP. The VPP process requires that employees be a big part of an overall safety program in a facility.

Management and trainers are to be reminded that some employees learn more in a classroom while others learn from watching and performing hands-on training. The use of electronic media and hands-on training can be very effective.

Successful Training Programs in Industry

Operator training can vary from employer to employer. Unfortunately, OSHA's original definition of operator training was weak and open to a variety of interpretations. Despite this flaw in the regulation, many employers have adopted operator training programs that are very noteworthy. The programs of these organizations were featured in various issues of *Material Handling Engineering and Modern Materials Handling* magazines.

Kodak began training its lift truck operators in 1955. They recognized the need for operator training long before OSHA came on the scene. Also, the programs utilized at Kodak far and away exceeded the training requirements advocated by the OSHA regulations in 1970.

Through the years the content of this company-created program has been expanded as well as the enhancement of Kodak's commitment to safety. Every lift truck operator at Kodak must successfully complete a 9 hour course, a written test and an operators performance test. The program includes five modules, and covers such topics as safety, vehicle inspection, battery charging and proper operating practices.

A student that passes both the operators performance test and written exam is issued a provisional operators license. At that point they can begin operating one of the company's several thousand fork lift trucks. During the following six weeks to six months (this depends on how often they operate the lift truck) the students supervisor monitors the student's performance. At the end of this probationary period the supervisor may apply for a permanent license for the student.

Kodak feels very strongly about its operator training. No one operates a lift truck there without being licensed. The program includes summer employees and temporary help.

Kodak feels that the program has paid big dividends along the way by reducing injuries, reducing equipment damage and product related losses through handling.

Signode Corporation is a leading manufacturer of steel strapping. At one plant 154 employees participated in a 4 1/2 hour training course. Rather than use an in-house program, Signode chose to use a program from a local lift truck manufacturer's distributor. The distributor provided the instructor for the program as well as the classroom instruction booklets.

One of the key components of the program was a focus on the physics and geometry of lift truck design. This gave operators a better understanding of the practical use and theories of lift trucks. Hands-on driving requirements were included along with a written test.

Students who passed the testing requirements received an operator's license, along with a cloth patch and a sticker which identified them as being certified.

Signode felt that the time spent training the operators more than paid for itself as compared to lost production time from the training program itself.

Champion International, an organization in the pulp and paper industry, embarked on a special operator training program. Even though Champion has had operator training programs in the past, the new program focused on operator training as well as competition. The managers at Champion decided to try an operator program that places special emphasis on measuring operator performance through competition with prizes going to the winners.

A basic operator training program was forwarded to the mills and other facilities by the corporate loss prevention department. Each location was then asked to customize the program. The intent was to have the facilities tailor the operator training to match the needs of the facility. Different mills use different pieces of equipment. Operating conditions are also different in each facility. The objective was to have the operator training as close to the reality of the operations as possible.

The typical program consisted of four hours of classroom training followed by four hours of hands-on driving instruction. The instructors come from the hourly and management ranks. These individuals received special training for their roles as teachers. Supervisors are then asked to monitor post-training driving performance for the certified operators.

As an added program incentive, individual rodeos are sponsored at each mill. Employees receive prizes and have a chance to compete at Champion's "Olympics" in North Carolina.

Alcoa has a reputation in the safety field as being a progressive company with a long history of safety excellence. At the company's plant in Warrick, Indiana, a special technical training center was built to house the lift truck course along with other training programs.

The basic operators course consists of four hours of classroom training and four hours of hands-on driving instruction. Included in the program are pre and post testing segments to measure the effectiveness of the program.

Alcoa feels that safety is the most important part of their business. They feel that the training works and that it has paid big dividends along the way.

14 / Forklift Safety

Xerox is a well-known corporation that produces quality products and has a lift truck training program that is also first class. Xerox operators are required to complete six hours of classroom training and two hours of hands-on training. A written test is also included.

Sources for the written test come from the NIOSH booklet on operator training, questions on Xerox's safety rules and items from its material handling handbook.

Operators that complete the program and pass the exam are required to complete 16 hours of hands-on departmental training. They then have to complete a 30-day probationary period in a department. If the operator passes the probationary period, a supervisor signs a license for him to become a qualified material handler. The license is valid for two years.

After two years, operators must complete a safety renewal course. They must also submit to a sight and hearing exam. If the operators pass these processes, they are issued another license that is good for another two years.

Coors Brewing is another well-known corporation with a quality product. The lift truck training program at Coors is touted as being state of the art. The program includes company-produced video tapes, written exercises, hands on experience, and extensive on-the-job training. Operators are certified annually and a lift truck rally each year reinforces the program.

Coors has made a commitment to teach inexperienced new hires. These trainees must complete an 80 hour program before they are assigned to a job as an operator. The operators then work for a live trainer for another 80 hours on the job.

Live trainers, themselves top notch operators with community college instructor training, evaluate trainees for motor skills, vehicle control and safe operating techniques.

The award winning videotapes are realistic. The intent is for an operator to acquire the skills needed for the job. Scenes on the videotapes are from actual Coors departments. "Accidents" are staged in the videos to add a touch of realism.

Coors has made a major investment in operator training. It has paid off in many ways. A reduction of 50% was reported for warehouse accidents. Handling damage was reduced by more than $80,000 each year.

Raymond Corporation overhauled their operator safety training program several years ago. Lift truck safety has a high priority today and Raymond has developed a state of the art program. It is called "Safety On the Move." The program features a modular approach to training. This design feature allowed the program to be structured to the level and specific needs of different audiences without losing key safety information. The eight modules include:

- Safety First, an overview of the course and general lift truck safety issues (45-60 minutes).

- Operators daily checklist (35-40 minutes).
- Truck travel and proper practices (30-40 minutes).
- Safe load handling concepts (30-40 minutes).
- Ramps and loading docks, and safe maneuvering in these areas (20-30 minutes).
- Battery safety procedures (20-30 minutes).
- Operating the lift truck, including individual hands on instruction—per person (30-60 minutes).
- Operator analysis, a "final" session to follow any hands on training—per person (30-60 minutes).

In addition, Raymond has produced a video tape on safety and how to operate each model they produce. A five day "train the trainers" program assists in the desire to offer quality operator training.

THE NEED FOR EMPLOYEE SAFETY

With the trend of the 1990's being a downsized corporation, many employees are given extra duties and assignments in the workplace. This includes lift truck operators. Some organizations are reluctant to train operators because of the time and expense. In other cases, some organizations fail to train employees because of high turnover. Why invest the money in training when the employee may not be around in the near future, is how some employers think.

Without proper training some organizations may find it difficult to survive in business. The United States now participates in a global economy and for an organization to survive it has to keep pace with competition as well as establish cost-cutting opportunities.

It has been said that what many organizations fail to realize is that they have already paid for elaborate operator training programs through product damage alone. Powered industrial truck operator training is about 80% safety training. Operating safely in the work environment that continually changes presents a big challenge to employees. Employers must know the particular hazards within their plant. Just as no single training program, not even OSHA's, can cover every contingency an operator would face in the workplace, this book cannot identify every single rule or hazard that may be faced by operators.

Employee involvement in developing safe work rules will pay big dividends, this has been proven in VPP sites where these facilities average a 43% better incidence rate than comparable industry. Opportunities abound for employers to take advantage of employee knowledge as one way in developing training programs. A recent survey regarding the levels of training offered by many organizations included:

- Walkie lift truck operators need training as much as rider lift truck operators.
- Management seldom realizes its part in placing the company at risk by neglecting training.
- Operator training is a halfway measure without certification and enforcement.

- Untrained or careless lift truck operators are responsible for almost as much equipment damage as poor maintenance.

- Hidden damage in loads exacts heavy costs in dissatisfied customers, scrap, production delays and lost business.

- The operating environment must be considered when training programs are initiated. A poorly trained operator could easily jeopardize a company's 100% good parts production schedule when he is asked to retrieve the parts and place them in a delivery truck.

- Training the trainer is an important part of the total company's approach because the trainer's job may include enforcement of the rules as well as employer and accident recordkeeping.

Figure 1.3 is an example of damaged product and lost time as a result of an operator not driving properly.

Figure 1.3. Example of Damaged Product

Operators can easily cost an organization many dollars in lost production and damaged product.

Training programs provided by powered equipment manufacturers and distributors are probably the most widely used in industry. Unfortunately, only about 25% or all new truck purchases include the taking advantage of operator training programs. Many of the programs offered are of high quality. It is expected that manufacturers will modify their programs once OSHA approves the new forklift training requirements. It is up to an organization to inquire as to the scope of the training offered. The checklists in the back of this chapter can help in the selection process.

In England, the purchase price of a piece of powered equipment includes primary operator training. A definite opportunity exists for employers to take advantage of manufacturers programs.

Tip Over

The highest percentage of fatalities are caused by lift truck tip overs. One-fourth of all injuries quoted in the OSHA study in Table 1.2 are tip over related. This number may be too high regarding just injuries. One study (Table 1.5) of workplace fatalities lists turnovers as responsible for 42% of the total fatalities.

Table 1.5. OSHA Study of Tip Over Accidents

In the OSHA study of 53 fatalities turnover accidents were attributed to:

- The elevated load vehicle being out of control (speeding, mechanical problems, etc.) 7 @ 13%

- Vehicle being run off/over the edge of the surface (such as a dock area) 4 @ 8%

- Attempting to make too sharp a turn (excessive speed, unbalanced load, etc.) 4 @ 8%

- Employee jumped from overturning vehicle being pulled by another vehicle 2 @ 4%

- Vehicle skidded or slipped on slippery surface 2 @ 4%

- Wheels on one side of vehicle ran over raised surface or object 2 @ 4%

- Vehicle tipped over when struck by another vehicle 1 @ 2%

It is apparent from the data in this chapter that the tip over has to be a priority part of a safety training program. From the various material reviewed, keeping the operator within the confines of a truck during tip over will most likely save his life. Before focusing on the mechanical means to protect the operator, training programs and enforcement of safe driving rules are imperative. Management must train, train, and train some more to educate operators in how to transport a load, the stability triangle and those factors that go into preventing tip overs.

When a lift truck tips over on its side, or, goes forward off a dock, or, falls counterweight first into a dock, as Figure 1.4 illustrates, the operator can be ejected. Many times the operator attempts to jump clear of the lift truck. In some cases the employee escapes with minor injuries. In other cases the overhead guard pins the operator to a stationary object and causes serious injury or death. The causes of tip over are many; this list is not all inclusive:

- Operator drives over a hole, block of wood, rock or obstacle and tips the lift truck.
- Operator drives into a ditch or off a bank or ramp.
- Operator fails to ensure wheels are chocked on a trailer, trailer creeps or moves forward, operator goes forward and falls into the dock area. Another scenario; operator falls into dock area when counterweight goes down first.
- Operator is traveling at a high rate of speed and makes a sharp turn.
- Operator is traveling with an elevated load—makes a turn or runs over a rut, block or wood, etc.
- Operator is traveling with only the mast elevated and makes a turn.
- Operator decides to turn on a ramp.
- Operator fails to ensure tires are inflated or sound.

Figure 1.4. Forklift Falling Off of a Dock

A fall from the dock can be catastrophic. Operators must chock the wheels of trailers and take personal responsibility for this process (Courtesy of Clark Material Handling Company).

The center of gravity is a major problem with forklifts. Even though all four wheels are touching the floor, the stability of the vehicle is governed by the two front wheels and the center of the rear axle. The narrow wheel base, load, overhead guard, speed, operating surface, tire pressure and operator control help effect tip overs. Speeds as low as 5 miles per hour can tip over a lift truck.

At present, there is no OSHA rule or requirement that mandates lift truck seat belts. During the hearings held by OSHA to develop a required seat belt rule for employees while operating any company vehicle, it was suggested that slow moving vehicles be excluded from the proposed rule. A slow moving vehicle could be a lift truck. Most operate at speeds of 10 miles per hour or less. Yet, as the data indicate, the tip over causes the highest percentage of injuries and fatalities.

Can anything be done to eliminate tip over through engineering controls? Could a device or alarm alert the operator that a tip over was imminent? Can a safer seat and restraint be devised for operator protection? Some safety literature for counter-balanced trucks suggests the following for employee tip over safety:

- Don't jump from the lift truck.
- Hold on tight.
- Brace your feet.
- Lean away from the side that is lowest during a tip.

On a stand-up or narrow aisle lift truck, the manufacturers recommend that the operator leave the vehicle in the event of a tip over. Operators can step out the rear opening to the truck and get clear of the moving vehicle.

Another piece of safety literature recommends that the operator brace his feet, reach up and grasp the overhead guard with both hands and brace his body while the lift truck tips over. Being that hands-on operator training cannot simulate a tip over it seems reasonable to say that based on these recommendations, only someone that is athletic, quick to react and perhaps physically fit should operate a lift truck. A typical lift truck operator may not be able to duplicate the physical demands as recommended above. In any event, this simulated safety hazard cannot be practiced so that all operators could help protect themselves during a tip over.

The wing seat, hip restraint or "mid-torso" restraint seat and seat belt appear to be the most viable means of operator protection for a sit down lift truck. The wing seat allows for a padded wing to protrude out of both sides of the operator's seat to keep him in the seat during a tip over. The seat belt also ensures that he will stay in the confines of the seat and overhead guard. It must be noted that there has been a concern by some that the impact on a surface while the operator is in the wing seat can possibly injure the neck because of the momentum of the head to the side when the vehicle makes impact.

The winged or modified seat helps prevent the employee from jumping out also. An incident at a Des Moines, Iowa, warehouse of a major company gives evidence of wing seat effectiveness. As the operator was exiting a trailer with a load, the trailer pulled away from the dock. The rear of the lift truck fell into the dock area. The operator was about to jump out but the wing seat prevented this. He bounced on the dock area and the lift truck tipped sideways but did not tip over. His seat belt was not fastened because of his frequent getting off and on the lift truck, the operator said. The only injury to the operator was a sore neck. Figure 1.5 illustrates a graphic that can be posted on the lift truck as a reminder to operators.

Based on the data and potential for injury or worse, employers should insist on the new-improved seats from the various manufacturers, and, the wearing of a seat belt on all sit down pieces of equipment. Hip restraints, wing seats and mid-torso restraints are not substitutes for a seat belt but must be used in conjunction with a seat belt.

Figure 1.5. Tip Over and Seat Belt Decal

Decals can be placed on the lift truck to serve as reminders that tip overs can injure or kill, and, that operators are to buckle up and learn the applicable safety rules (Courtesy of Clark Material Handling Company).

Legal Considerations

The employer has to be concerned with legal ramifications regarding lift truck safety. OSHA, state safety codes and the EPA must always be complied with. This is a big burden for all of industry. Another consideration is the ever-present subject of litigation.

A recent article presented a very compelling reason to provide proper operator training. A lift truck operator was seriously injured at work. The operator evidently committed a serious error which led him to his injury. He was thoroughly trained and authorized to operate a lift truck. He had been trained to identify the hazards in his work environment and warned about the danger he had disregarded. That disregard caused the injury.

The operator had passed all of the fork lift tests with high scores. His trainer's evaluation was positive.

When the employee's lawyer was confronted with the test and the course book the trainer used, the lawyer did not press for a law suit.

This instance had spared the employer untold dollars and legal fees and perhaps a large judgment. The employers program began some time before the accident when they decided to provide operator training and create a permanent paper trail. There are no guarantees or insurance against lawsuits. Proper training that provides for all of the requirements of the manufacturers and proposed OSHA guidelines will help to prevent many injuries from ever occurring.

ANSI B56.1 RECOMMENDATIONS FOR TRAINING

Note: The following is from the original ANSI B56.1 standard. OSHA has modified some of these provisions in their proposed standard.

Operator Qualifications

Only trained and authorized persons shall be permitted to operate a powered industrial truck. Operators of powered industrial trucks shall be qualified as to visual, auditory, physical and mental ability to operate the equipment safely according to the data below.

Operator Training

Personnel who have not been trained to operate powered industrial trucks may operate a truck for the purposes of training only and only under the direct supervision of the trainer. This training should be conducted in an area away from other trucks, obstacles and pedestrians.

The operator training program should include the users policies for the site where the trainee will operate the truck, the operating conditions for that location, and the specific truck the trainee will operate. The training program shall be presented to all new operators regardless of previous experiences.

The training program shall inform the trainee that:

- The primary responsibility of the operator is to use the powered industrial truck safely following the instructions given in the training program.
- Unsafe or improper operation of the powered industrial truck can result in death or serious injury to the operator or others, damage to the powered industrial truck or other property.

The training program shall emphasize safe and proper operation to avoid injury to the operators and others and prevent property damage and shall cover the following areas:

- Fundamentals of the powered industrial truck(s) the trainee will operate, including:
  - Characteristics of the powered industrial truck(s), including variations between trucks in the workplace.
  - Similarities to and differences from automobiles.
  - Significance of nameplate data, including rated capacity, warnings and instructions affixed to the truck.
  - Operating instructions and warnings in the operating manual for the truck, and instructions for inspections and maintenance to be performed by the operator.
  - Type of motive power and its characteristics.
  - Method of steering.
  - Braking method and characteristics with and without loads.
  - Visibility with and without load, forward and reverse.
  - Load handling capacity, weight and load center.
  - Stability characteristics with and without load and without attachments.
  - Controls, locations, function, method of operation, identification of symbols.
  - Load handling capabilities, forks, attachments.
  - Fueling and battery charging.
  - Guards and protective devices for the specific type of truck.
  - Other characteristics of the specific industrial truck.

- Operating environment and its effect on truck operations which includes:
  - Floor or ground conditions including temporary conditions.
  - Ramps and inclines with and without load.
  - Trailers, railcars and dockboards (including the use of wheel chocks, jacks, and other securing devices).
  - Fueling and battery charging facilities.
  - The use of "classified" trucks in areas classified as hazardous due to risk of fire or explosion as defined in ANSI/NFPA 505, Powered Industrial Trucks.
  - Narrow aisles, doorways, overhead piping and other areas of limited clearance.
  - Areas where the truck may be operated near other powered industrial trucks, other vehicles or pedestrians.
  - Use and capacity of elevators.
  - Operating near edge of dock or edge of improved surface.
  - Other special operating conditions and hazards which may be encountered.

- Operation of the powered industrial truck including:
  - Proper pre-shift inspection and approved method for removing from service a truck which is in need of repair.
  - Load handling techniques, lifting, lowering, picking up, placing and tilting.
  - Traveling with and without loads, turning corners.
  - Parking and shutdown procedures.
  - Other special operating conditions for the specific application.

- Operating safety rules and practices, including:
  - Provisions of the consensus standard regarding safety rules and practices.
  - Provisions of the consensus standard addressing care of the truck.
  - Other rules, regulations or practices specified by the employer at the location where the powered industrial truck will be used.

- Operational training practices, including:
  - If feasible, practice in the operation of powered industrial trucks shall be conducted in an area separate from other workplace activities and personnel.
  - Training practices shall be conducted under the supervision of the trainer.
  - Training practices shall include the actual operation or simulated performance of all operating tasks such as loading, maneuvering, traveling, stopping, starting and other activities under the conditions which will be encountered in the use of the truck.

- Testing, retraining and enforcement:
  - During training, performance and oral and/or written tests shall be given by the employer to assure the skill and knowledge of the operator in meeting the requirements of the Standard. Employers shall establish a pass/fail requirement for such tests. Employers may delegate such testing to others but shall remain responsible for the testing. Appropriate records shall be kept.
  - Operators shall be retrained when new equipment is introduced, existing equipment is modified, operating conditions change or an operator's performance is unsatisfactory.
  - The user shall be responsible for enforcing the safe use of the powered industrial truck according to the provisions of the Standard.

As noted in the proposed standard:

Note: Information on operator training is available from such sources as

- Powered industrial truck manufacturers.
- Government agencies dealing with employee safety.
- Trade organizations or users of powered industrial trucks.
- Public and private organizations.
- Safety consultants.
- The National Safety Council.

OSHA'S RECOMMENDED TRAINING PROGRAM CONTENT

Powered industrial truck operator trainees shall be trained in the following topics unless the employer can demonstrate that some of the topics are not needed for safe operation.

Truck Related Topics

- All operating instructions, warnings and precautions for all types of trucks the operator will be authorized to operate.
- Similarities to and differences from the automobile.
- Controls and instrumentation, location, what they do and how they work.
- Power plant operation and maintenance.
- Steering and maneuvering.
- Visibility (including restrictions due to loading)
- Fork and attachment adaptation, operation and limitations of their utilization.
- Vehicle capacity.
- Vehicle stability.
- Vehicle inspection and maintenance.
- Refueling or charging, recharging batteries.
- Operating limitations.
- Any other operating instruction, warning or precaution listed in the operators manual for the type of vehicle which the employee is being trained to operate.

Workplace Related Topics

- Surface conditions where the vehicle will be operated.
- Composition of probable loads and load stability.
- Load manipulation, stacking and unstacking.
- Pedestrian traffic.
- Narrow aisles and other restricted places of operation.
- Operating in hazardous classified locations.
- Operating the truck on ramps and other sloped surfaces that could affect the stability of the vehicle.

The Need for Operator Training / 25

- Other unique or potentially hazardous environmental conditions that exist or may exist in the workplace.
- Operating the vehicle in closed environments and other areas where insufficient ventilation could cause a buildup of carbon monoxide or diesel exhaust.

SUMMARY

With over 800,000 pieces of power equipment in the workplace and over one million operators, the opportunity for injury or something more serious is ever-present. Training of operators is paramount wherever a piece of power equipment exists.

The tip over causes more fatalities than any other forklift accident. The prevention of tip over must come from proper training and the enforcement of the appropriate safeguards. Manufacturers feel that the winged seat, hip restraint seat or mid-torso seat and seat belt will provide improved safety for the operator. Overall prevention is the key. This part of operator training must be included. The graphic illustration in Figure 1.6 should remind readers that tip overs can be deadly.

Figure 1.6. Beware of Dock Edges

A graphic reminder that driving off of a dock edge can be deadly (Courtesy of the National Safety Council).

Economic losses take place because of poor operating training and control. Millions of dollars have been lost as a result of fires, property damage, damage to equipment and, of course, to individuals.

Until all employers are required to maintain a comprehensive operator training program, the economic losses will continue. Those employers that have maintained comprehensive programs boast of reduced accidents and economic savings. The value of employee training cannot be over emphasized.

REFERENCES

BNA Reporter. "Canadian Firms Fined in Two Worker Fatalities." Washington, DC: Bureau of National Affairs, Inc., July 10, 1996, p. 160.

Entwisle, F. "Preventing Lift Truck Overturns," *Modern Materials Handling*, January 1995, p. 19.

Factory Mutual. "Lift Trucks," *Record*, March/April 1993, pp. 3-8.

Feare, T. "Lift Truck Training: Readers React," *Modern Materials Handling*, March 1995, pp. 57-58.

Feare, T. "Workplace Safety: Lift Truck Safety Belts—A Good Idea," *Modern Materials Handling*, February 1992, p.17.

Gould, L. "Part II, Lift Truck Training," *Modern Materials Handling*, September 1989, pp. 72-76.

Krill, B. "OSHA Cites Employer for Lack of Operator Training," *Materials Handling Engineering*, February 1991.

Materials Handling Engineering, "Lift Truck Injuries, Can We Reduce Them?" February 1996.

Modern Materials Handling, "Riding the Forks; A Thrill That Can Kill," June 1994.

News and Trends. "Raymond Overhauls Operator Training," *Modern Materials Handling*, August 1991, pp. 12-13.

Occupational Hazards Magazine, "Hazards of Forklifts," August 1996, p. 13.

Operator Restraint for the Lift Truck—A Compendium. Battle Creek, MI: Clark Equipment Company, Industrial Truck Division, September 19, 1983.

OSHA Proposed Rule for Training Powered Industrial Truck Operators in General, Maritime Industries, 60 FR 13782, March 14, 1995.

Schwind, G.F. "Lift Trucks Making Changes for 2000," *Material Handling Engineering*, October 1995, pp. 88-110.

Schwind, G.F. "No Proof of Driver Training Could Be Hazardous to Your Company," *Material Handling Engineering*, October 1994, p. 18.

Schwind, G.F. "Operator Training: More than Just Driving a Lift Truck," *Material Handling Engineering*, June 1990, pp. 33-42.

Torok, D.B. "Lift Truck Safety: It Pays To Train," *Materials Handling Engineering*, August 1990, pp. 52-28.

Weinstock, M.P., "Raising the Standard for Lift Truck Operators," *Occupational Hazards*, January 1994.

Workplace Safety. "Wing Seat Deters Injury," *Modern Materials Handling*, April 1991, p. 17.

2

HOW LIFT TRUCKS WORK

There are many that would argue that operating a car is very similar to that of operating a fork lift; if you can drive a car you can drive a fork lift. Driving both is simple, so they say. This is not true; a car and a forklift are different. Operating the lift truck or other piece of power equipment takes a high degree of skill.

To be able to safely operate a lift truck the operator must have an in-depth understanding of the vehicle. Note the particulars of forklift trucks and some important operating guidelines.

UNDERSTANDING FORK LIFT TRUCKS

Powered industrial trucks may be powered by propane (LPG), gasoline, diesel or liquified natural gas (LNG) engines or by electric motors. Each motor or engine and associated components may be upgraded. The entire vehicle may be upgraded and the entire truck may be approved by a nationally recognized testing laboratory for operation in certain classified hazardous areas. This would not apply to propane but would include mufflers on internal combustion engines, switches and wiring on electric trucks.

These hazardous areas are those parts of a plant, factory or other workplace where there exists or may exist concentrations of flammable vapors, combustible dust or easily ignited fibers or flyings. These items would increase the risk of fire or explosion.

The current OSHA regulation, 1910.178, for powered industrial trucks contains descriptions of the various divisions, classes and groups of classified hazardous areas and some of the materials whose presence would cause those areas to be classified. (The current OSHA regulations for reference are in Appendix D). Since OSHA was promulgated, the number of substances whose presence causes the hazards of fire and/or explosion have greatly increased.

Readers should study the National Fire Protection Associations NFPA 505-1992, Fire Safety Standard for Powered Industrial Trucks, Including Type Designation, Areas of Use, Maintenance and Operation.

STABILITY OF POWERED INDUSTRIAL TRUCKS

This section will provide specific information from the OSHA research regarding the principles of stability. Operators are to be trained in the following definitions:

- Load center.
- Stability triangle.
- Lateral stability.
- Dynamic stability.
- Longitudinal stability.

For additional information on these definitions see Appendix C.

Load Center

Operators of powered industrial trucks must understand truck stability and load carrying ability. Every load that has to be lifted and moved may not be perfectly level, straight or secured to keep from spilling.

Operators must be trained in how to handle asymmetrical loads when their work includes this activity. Unusual loads must be safely moved. Operators must know and understand the principles of moment and stability such as the load center of the vehicle. The load center is the distance from the vertical face of the forks to the center of gravity of the load. Capacity of the forklift is on the nameplate which is on the vehicle, usually this distance from the mast is 24 inches.

The distance of 24 inches assumes that the load is symmetrical and the operator can pick it up correctly. The load could be within the limits of the lift truck but if the load is out on the edge of the forks, as in Figure 2.1, the truck cannot lift it safely. If the load is not symmetrical, the load center may change. When load centers are understood, loads that are overweight should not be picked up. Also, this overload condition can damage the chains, components, other parts of the truck, drop a load and damage product. Overload can cause the truck to tip forward, with the counterweight raising off the floor, resulting in property damage and potential harm to the operator.

Operators can be taught the principles of load stability. Truck stability contains a few basic principles. Operators have to understand that there are many factors that influence vehicle stability; vehicle wheelbase, track, height and weight distribution of the load, condition of the tires, and the location of the counterweight of the vehicle.

Figure 2.1. Tipping Fork Lift

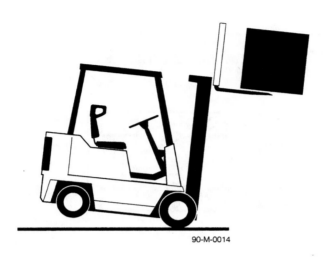

Even if a load that is within the capacity of the lift truck is on the edge of the forks and elevated, the lift truck can tip forward (Courtesy of Clark Material Handling Company).

Stability Triangle

One of the most important items for operators to understand about a piece of powered equipment is the stability triangle. The determination of whether an object is stable is dependent on the moment of an object at one end of a system being greater than, equal to, or smaller than the moment of the object at the other end of that system. Operators can easily understand this concept when it is compared to a see-saw or teeter-totter. If one end has a heavier load than the other, the heavy load will go down. In technical terms, if the product of the load and distance from the fulcrum(the moment) is equal to the moment at the other end of the device, the device is balanced and it will not move.

However, if there is a greater moment at one end of the device, the device will try to move downward at the end with the greater moment.

Figure 2.2 illustrates this invisible stability triangle. When the machine is unloaded the black dot is directly in the center of the triangle. When an operator adheres to specific guidelines, and follows safe procedures and manufacturers guidelines, the black dot will remain within the triangle.

32 / Forklift Safety

However, if an operator is traveling too fast with a load and makes a sharp turn, the invisible black dot can easily leave this triangle thus making the truck unstable. The result could easily be a tip-over; the one item that causes the most fatalities. Figure 2.3 serves as a reminder that sharp turns can tip a lift truck.

Figure 2.2. Stability Triangle

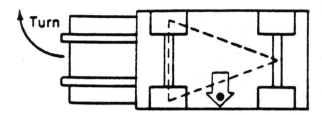

The invisible dot under the lift truck carriage can easily move out of the safe zone within the stability triangle when a sharp turn is made.

Figure 2.3. Hazards of Making Sharp Turns

Sharp turns can easily spill product and tip the lift truck (Courtesy of the National Safety Council).

Here are a few more situations that can affect the stability triangle:

- An operator lifts an over-capacity load with his forks, the rear of the lift truck (counterweight) goes up as in Figure 2.4. This could easily damage the lift mechanism, spill the load, shake up the operator when the load falls off, or cause personal injury. When the heavy load was lifted, the black dot moved outside of the stability triangle.

Figure 2.4. Oversize Loads Can Tip a Lift Truck

Notice the lift truck on the left, the load weighs 4500 pounds with a 24" load center. Take this same weight and extend its load center and the rear wheels will come up off the floor (Courtesy of the National Safety Council).

- An operator is driving up an incline with a heavy load and decides to turn around. The load's weight and the momentum of the truck in the turn causes the truck to tip over. Figure 2.5 illustrates how the stability triangle can shift during a sharp turn.
- An operator has just placed a pallet on a high rack in a warehouse. He backs away and continues to operate the lift truck without lowering the mast. He continues down a long aisle with the mast elevated and makes a sharp turn at the end of the aisle. The lift truck tips over even though it is unloaded.

More examples could be used but the message should be obvious with these examples. Speed, heavy loads, tire inflation, condition of tires, loads carried too high, the weight of the mast and work/floor surfaces can tip a lift truck. The consequences can be deadly. In each case the stability triangle was violated.

Almost all counterbalanced power industrial trucks have a three point suspension system. The lift truck is supported at three points even though most have four wheels. The two front wheels provide two points of the triangle. The rear axle, the steer axle, has a pivot pin in its center. Lines from the two front

34 / Forklift Safety

wheels to the rear axle pivot point, the triangle, (invisible of course), can be drawn. Note the drawing below which identifies how the triangle works.

Figure 2.5. Stability Triangle and Turns

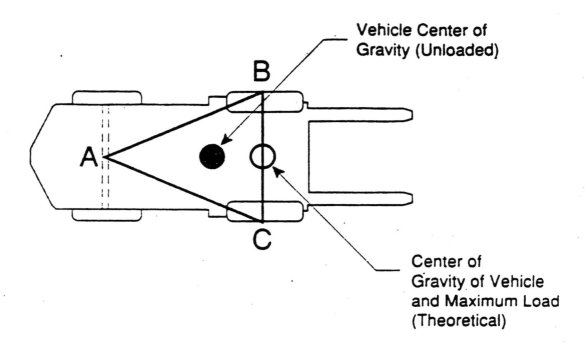

When the vehicle is loaded, the combined center of gravity shifts toward line B-C. Theoretically, the maximum load will result in the center of gravity at the line B-C. In actual practice, the combined center of gravity should never be at line B-C. The addition of additional counterweight will cause the truck center of gravity to shift toward point A and result in a truck that is less stable laterally.

Lateral Stability

The lateral stability of a powered industrial truck is determined by the position of action (a vertical line that passes through the combined center of gravity of the vehicle and the load) in relation to the stability triangle. When the vehicle is *not* loaded, the location of the center of gravity of the truck is the only factor to be considered in determining the stability of the truck.

As long as the line of action of the combined center of gravity of the vehicle and the load falls within the stability triangle, the truck is stable and will not tip over. However, if the line of action falls outside the stability triangle, the truck is not stable and may tip over.

Factors that will have an effect on the lateral stability of the truck include the method of how the operator placed the load on the truck. Also, the height of the load above the surface on which the vehicle is operating, the operating surface and the degree of lean of vehicle must be considered.

Dynamic Stability

Dynamic stability results from the vehicle and load when they are put into motion. Stability considerations such as braking, cornering, lifting, tilting, moving and lowering loads affect truck safety. The transfer of weight and the resultant shift in the center of gravity is due to the dynamic forces created by the machine movements.

Operators must exercise extra caution for personal safety when handling loads that cause a vehicle to approach its maximum design characteristics. The operator has to have the ability to safely size up a load that is about to be lifted. This judgment does not just come from experience but is also learned in the classroom. All loads must be carried as low as possible with the forks raised only high enough to clear the road or floor surface. Levels of speed must be made gradually, turns or corners are to be taken with a low load, slowly and cautiously. Tilting must be made with caution. Even with all of these safeguards in place, the condition of the floor, operating surface, tires, weather and truck condition must be considered. OSHA feels that no precise rules can be formulated to cover every eventuality. When operators are properly trained the probability of injury or incident is reduced.

Longitudinal Stability

Longitudinal stability of a counterbalanced powered industrial truck is dependent on the moment of the vehicle and that of the load. As mentioned earlier, a load heavier than the rated load capacity of the truck will cause the vehicle to tip forward. The distance from the backrest of the forks to the center of the load determines the load being lifted safely. Forks must be placed under the load as far as possible and mast tilted back to stabilize the load. When the system is balanced the vehicle will not tip forward. When the load moment is greater than the vehicle moment, the greater load moment will force the truck to tip forward. Figure 2.6 demonstrates that a load at the end of the forks and lifted high can tip the lift truck forward.

For a better understanding of stability, OSHA published several illustrations of a stable versus unstable vehicle. Note the drawing on the left of Figure 2.7; even though the load is high and the truck base is tilted, the combined center of gravity is not in the tip over mode. It is important to point out that this drawing can be misleading to operators. (OSHA did not intend for the illustration to be used as an example of when a danger point is reached as much as trying to illustrate how a tip over occurs.) With a tilted base a load should never be raised because this would endanger the operator. The indication by OSHA is that "the vehicle is stable." Even though the vertical stability line is on the inside of the wheels, this is a dangerous move. If an experienced operator were lifting a load this high on a tilted base, he would not consider the vehicle as being stable.

36 / Forklift Safety

Figure 2.6. Raising a Load and Potential for Tipping

A truck becomes less stable when a load is lifted. A heavy load can cause the truck to tip and the load can be dropped (Courtesy of Clark Material Handling Company).

In the drawing on the right of Figure 2.7 the load is much higher, the vertical stability line is on the outside of the wheel and the same base tilt is in place. The combined center of gravity provides for an unstable lift and a tip over.

OVERVIEW OF POWERED INDUSTRIAL TRUCKS

Each powered industrial truck type has its own operating characteristics as well as inherent hazards. When training operators in overall safety, it is important to focus on vehicles and vehicle types that the operator will use. Safety features may not be the same for each of the manufacturers on the same piece of equipment. In other words, just because an operator has come to expect certain controls to be in a certain place in a certain fashion, the operator shouldn't be surprised to find those same controls placed differently or functioning differently on other models.

How Lift Trucks Work / 37

Figure 2.7. OSHA's Examples of Tip Over Potential

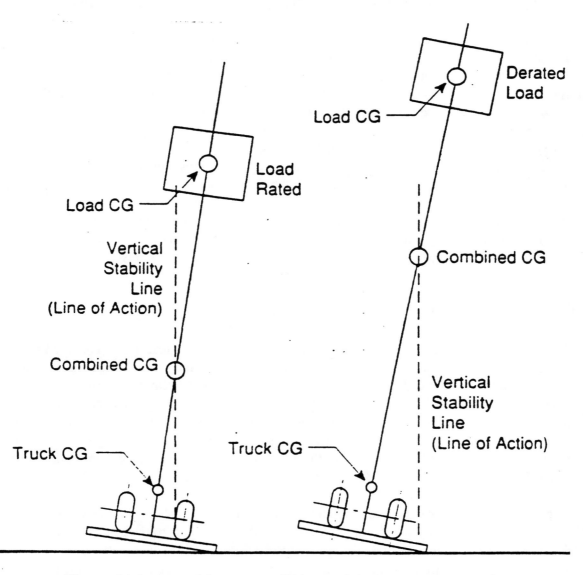

The vehicle is stable   This vehicle is unstable and will continue to tip over

OSHA illustrated two situations where tip over can easily take place. The higher the load the greater the potential.

Depending on the type of powered industrial truck, it can be described as a machine that can carry, push, pull, lift, stack or tier product and material. All of this while the machine is mobile and under power. Is it any wonder that it takes a great deal of skill and dedication to operate a lift truck safely?

Some models allow for the operator to be elevated along with the load such as an order picker model. Order pickers are used to store or retrieve materials from racks, shelves, or bins that may be high stacked. Platforms on order pickers may or may not be fully enclosed to protect the operator from falling. For personal protection the operator must wear a body harness or belt which is attached to the overhead guard or mast. The lanyard must be capable of safely absorbing the shock of a fall. The lanyard must be short enough to limit the fall. The anchor point on the lift truck, approved by the manufacturer, must be capable of sustaining the impact of the fall. Operators must be thoroughly taught how to wear, utilize and maintain these belts or harnesses. Strict disciplinary procedures should be in place to ensure compliance for this potentially life threatening safety infraction.

Research has shown that safety belts which are just worn about the waist can be harmful if a fall occurs. Operators should wear only safety harnesses for maximum safety when elevated. Manufacturers of these devices are to be contacted for proper care, use and maintenance procedures.

Cars/Forklifts: The Difference

Forklifts are much different than cars. Operators are frequently told that if they can operate a car they can operate a lift truck. Forklifts have a narrower wheelbase and a much higher center of gravity. Forklifts operate on a teeter-totter principle. A Counterweight on the rear of the truck helps balance the load on the forks. When a lift truck is empty there is a significant weight imbalance. An empty forklift does not imply that the forklift is safe. Operators have to be reminded that a car carries its load on the inside center of the vehicle. A lift truck carries its load outside of its supporting base. Cars have four point suspensions, fork lifts have three. With this in mind the forklift can become unstable rather easily. Figure 2.8 illustrates the turning radius when comparing an automobile and forklift.

A car can turn over under certain circumstances; a hole in the road, a sharp turn on a grade or ramp and speeds while driving on dry pavement and/or ice. A lift truck can turn over much easier and at much lower speeds. Speeds as low as five miles per hour can tip over a lift truck. Even empty lift trucks can turn over. Operating environments that include railroad tracks, rough pavement, holes in the roadway, uneven surfaces, ramps, grades or other less than desirable features can be dangerous.

Lift trucks have unique operating capabilities which include braking. A fully loaded lift truck, even though it isn't moving fast, is not easily stopped. When the reaction time of the operator is combined with stopping distance it is very possible that even an experienced operator would not be able to stop in time if a pedestrian suddenly appeared in his path of travel.

Figure 2.8. Comparing the Turning Radius

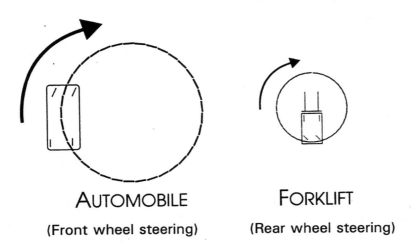

Operators, especially those that are new, are to be reminded that the turning radius of a lift truck is much smaller than that of an automobile (Courtesy of J.J. Keller and Associates).

ACCIDENT FACT

An experienced operator had just set a heavy load down by an operator's machine and was driving back to pick up another load. He was moving very fast. His forks were down and while making a very sharp turn the front wheel went into a large hole in the concrete floor. His lift truck tipped over. Luckily, he did not attempt to jump clear and was only shaken up as a result of this life-threatening incident.

The small wheels on a forklift combined with only two braking wheels doesn't always allow for a sudden stop. There may not be a backup system to stop the vehicle if these systems fail. Maintenance on the vehicle could be lacking which could allow braking systems to fail when most needed. Operators could easily fail to properly inspect the vehicle at the start of a shift and not report defects to anyone.

Operators have to rely on the ability to see over their loads if they have to brake for any reason. Employees have been known to be behind product or structures only to be struck by a load or the lift truck. Figure 2.9 illustrates the hazards of employees not being seen.

40 / Forklift Safety

Figure 2.9. Blind Spot While Driving

Note: *Never raise the load in an attempt to see under it.*

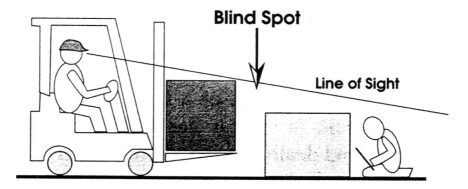

Pedestrians can easily be struck by a load or a lift truck especially if the operator cannot see them (Courtesy of J.J. Keller and Associates).

New Technology

Manufacturers continue to improve the quality and features of powered equipment. Some new changes, which are being discussed by manufacturers and those that have been implemented in industry, are a boost to operator safety.

Employee training and knowledge cannot be replaced by devices but new technology can enhance safety. The operator will still have to be relied upon for sound judgment, thinking and decision making for overall safety.

Because the lift truck tip over causes the greatest number of fatalities, perhaps built-in sensors could ensure that the operator is safely buckled into a wing, hip restraint or mid-torso restraint seat. Also, a device could alert him that a part of his body is outside of the confines of the truck. A device may be able to warn operators of a pending tip over. If a lift truck is too close to a dock edge, an alarm may sound. A common accident is a mast or high load making contact with the building; a device may warn the operator of this problem.

Currently, some systems provide data that records each hour of operation and what maintenance costs are involved. Not only does this method improve maintenance tracking, it alerts the employer and manufacturer of parts breakdown which could influence safety.

Lift trucks have historically lasted for 7,000 hours before an engine needed an overhaul. Today's engines can get up to 30,000 hours. Quieter engines and quieter hydraulic systems are another improvement. This improvement has helped in reducing some noise exposures in the workplace. Electronic control of hydraulics has helped reduce the number of fittings and possible leak points. It is not unusual to see hydraulic fluid on the floor when a lift truck is parked. This improvement could definitely be an asset in reducing slip and fall accidents and improving housekeeping practices.

In Europe, the powered equipment uses a much smaller battery than those in the United States. Battery chargers are continually being improved. A battery's life can be extended by one-third if a quality battery is purchased. OSHA and EPA regulations on battery service areas and chargers may become stronger.

Some additional features regarding powered equipment may become part of everyday operations. Additional proposals by manufacturers are:

- A key reader that would allow only qualified operators to operate specific trucks.
- Cards to read a truck's capability so a shipping supervisor can plan his loading of product.
- Catalytic converters that can bring emissions down to zero.
- Better ergonomic designs that allow for more seat movement, inflatable lumbar seats, less vibration, smaller mast and tilt cylinders and simpler hand and finger switches.
- Plastic pallets could provide greater longevity of pallets.
- Microprocessors that can control the pressures on clamps to reduce damage.

SUMMARY

The information in this chapter was somewhat technical in sections due to the complexity of some of the lift truck designs and handling features. Understanding how a piece of powered equipment operates is essential for all employees. Understanding the stability triangle and those forces that can cause a lift truck to tip are extremely important. Operators have to understand how a lift truck can tip forward or to the side; their lives may depend on this knowledge.

It is also important for operators and prospective operators to understand that the car they may drive to work has only a few similarities to that of a lift truck. One of the problems throughout industry is the improper belief that if you can drive a car you can drive a forklift. There are significant differences in not only the models, but the manufacturers and the type of product being handled. Time in the classroom for these topics will pay big dividends.

REFERENCES

National Safety Council. "Powered Industrial Trucks," *Accident Prevention Manual*, 11th edition. Itasca, IL, 1997, pp. 505-528.

OSHA Proposed Rule For Training Powered Industrial Truck Operators in General, Maritime Industries, 60FR 13782, March 14, 1995

Schwind, G.F. "Lift Trucks Making Changes for 2000," *Material Handling Engineering*, October 1995, pp. 88-110

3

MODELS OF POWERED INDUSTRIAL TRUCKS

Every operator of a powered industrial truck should have a thorough understanding of his machine. For the trucks to be driven safely, the operator must have the ability to operate them with and without loads, at docks, at various industrial settings, at various projects, and in warehouses. The grasp for safety of the machine comes from acquired knowledge and experience. The better trained the operator, the less likelihood of accidents. Not all machines are alike. However, operators must be trained properly to handle each piece of equipment they are authorized to use. The Industrial Truck Association has placed this array of powered industrial trucks into seven classes. Figure 3.1 identifies the various pieces of powered equipment.

Class 1—Sit-down rider, electric, counter balanced trucks (with solid and pneumatic tires).
Class 2—Electric motor narrow aisle trucks (with solid tires).
Class 3—Electric motor hand trucks or hand/rider trucks (with solid tires).
Class 4—Internal combustion engine trucks (with solid tires).
Class 5—Internal combustion engine trucks (with pneumatic tires).
Class 6—Electric and internal combustion engine tractors (with solid and pneumatic tires).
Class 7—Rough terrain fork lift trucks (with pneumatic tires).

The occupational Safety and Health Administration (OSHA) estimates that there are more than 822,000 powered industrial trucks currently in use in the United States. Table 3.1 identifies the U. S. shipments of powered industrial trucks. There are approximately 1.5 operators for each of these vehicles. This may be a conservative estimate of how many operators there really are. Regardless, with this many pieces of powered equipment in the workplace and so many estimated operators as driving these units, opportunity abounds for accidents.

With the variety of powered industrial trucks in use, operators must understand each of the truck's operating characteristics and operator hazards. Lift trucks are divided into two broad categories:

- **Low Lift.** These units are used for transporting products, the load is kept high enough just to clear the floor.

Figure 3.1. Models of Industrial Trucks

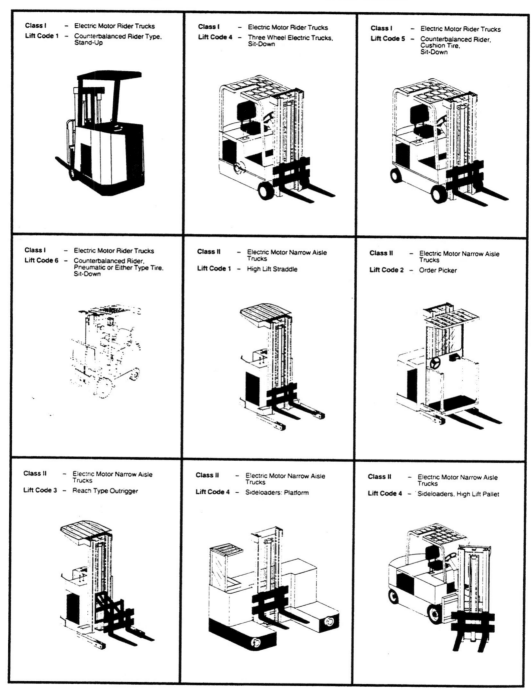

There are many different models of powered equipment as these three pages illustrate (Courtesy of the Industrial Truck Association).

Models of Powered Industrial Trucks / 45

Figure 3.1 (continued)

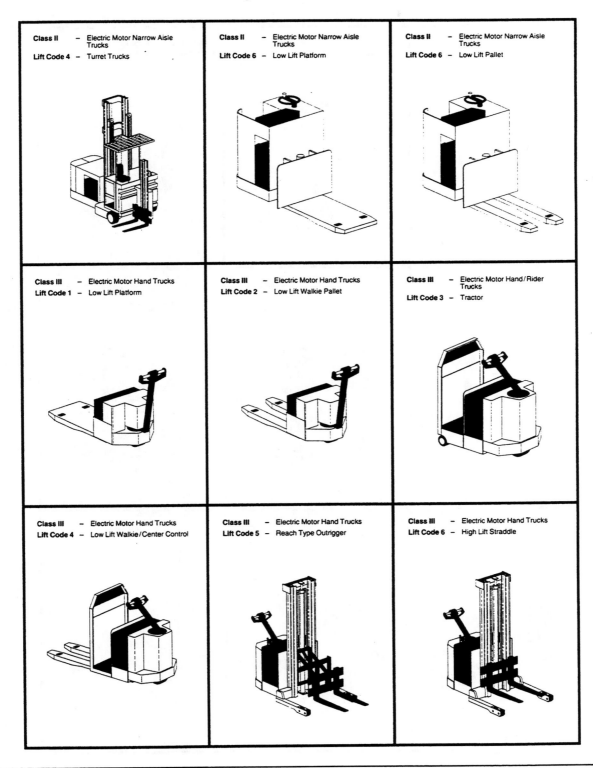

Figure 3.1 *(continued)*

Class III — Electric Motor Hand Trucks **Lift Code 6** — Single Face Pallet	**Class III** — Electric Motor Hand Trucks **Lift Code 6** — High Lift Platform	**Class III** — Electric Motor Hand Trucks **Lift Code 7** — High Lift Counterbalanced
Class III — Electric Motor Hand Trucks **Lift Code 8** — Low Lift Walkie/Rider Pallet and End Control	**Class IV** — Internal Combustion Engine Trucks **Lift Code 3** — Fork, Counterbalanced, Cushion Tire	**Class V** — Internal Combustion Engine Trucks—Pneumatic Tires Only **Lift Code 4** — Fork, Counterbalanced, Pneumatic Tire
Class VI — Electric and Internal Combustion Engine Tractors **Lift Code 1** — Sit Down Rider	**Class VII** — Rough Terrain Fork Lift Truck **Lift Code 1** — All Rough Terrain Fork Lift Trucks	Hand Pallet Trucks

Table 3.1. United States Shipments of Powered Industrial Trucks

YEAR	ELECTRIC RIDER	MOTORIZED HAND	INTERNAL COMBUSTION ENGINE
1982	14,859	13,988	18,553
1983	16,227	16,237	26,245
1984	23,983	21,958	45,338
1985	26,400	22,493	47,844
1986	25,818	24,589	46,195
1987	24,928	24,857	47,945
1988	29,202	27,205	48,535
1989	31,010	29,838	55,104
1990	27,877	26,941	47,702
1991	24,565	23,599	38,406
1992	28,277	27,700	46,183
1993	29,210	28,492	48,947
1994	36,747	34,127	65,027
1995	44,087	37,746	72,685

(Courtesy of the Industrial Truck Association)

- **High Lift.** These units are used for lifting, stacking and retrieving product. High lifts come in a wide variety of designs, models and configurations.

LOW LIFT MODELS

Low lift model trucks transfer and transport product and material while traveling very close to the floor. Usually the load is only 4 to 6 inches from the floor for proper clearance. The name low lift implies that the load is not raised for any stacking. Pallet trucks and platform trucks are the two most common types. Some specialized versions of these models are used in locations that handle dies.

Platform Trucks

Platform trucks are used to transport pallets of materials and dies. Rather than forks being used to drive under the load and lift it, a solid metal platform is used. When a control is activated on the truck, a hydraulic mechanism lifts the load just high enough for good floor clearance. Most models are powered by battery. Some of the models allow for a rider to stand on the equipment. Some models allow for walking with the unit. Capacities can range up to 10,000 pounds for a standard unit.

Pallet Trucks

Pallet trucks contain retractable wheels on rollers on the tips of the forks. The wheels lower when a control is activated so the truck can be placed into a pallet or under a load. After the forks are properly placed under the load, a control is activated which lifts the load high enough for good floor clearance. Steering or drive wheels are at the rear of the truck by the controls and operating handle. Many are powered by electric battery. Capacities range up to 6,000 pounds.

HIGH LIFT TRUCKS—RIDER TYPE

Counterbalance Rider Trucks

Counterbalance rider trucks are one of the most common types of powered equipment used in industry. These models contain adjustable forks, a tilt mast, overhead guard and a counterweight. The load is lifted while it is against the mast and the counterweight at the back of the truck acts as a see-saw. Loads can be lifted very high depending on the model selected. Rear wheels steer the vehicle. The trucks are capable of turning in a small radius; much smaller than that of automobiles. They can be powered by electricity, gasoline, natural gas (LNG), liquid propane (LPG) or diesel fuel. Tires can be pneumatic or solid. Figure 3.2 illustrates the ability of the lift truck to maneuver.

Figure 3.2. Maneuverability of a Lift Truck

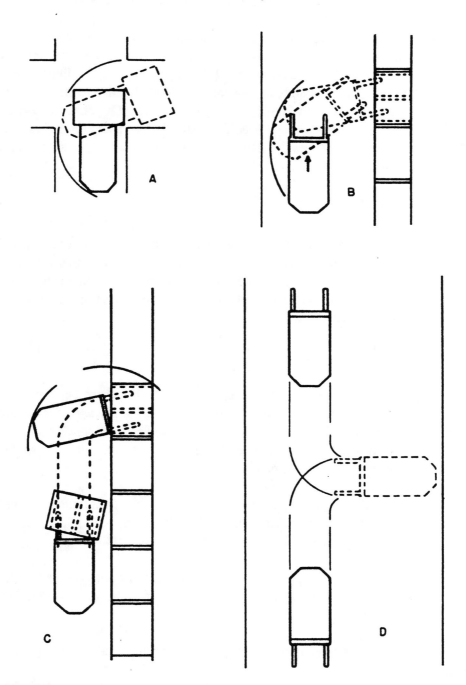

Lift trucks have the ability to maneuver into tight spots as the illustration demonstrates. In each use, an employee could be struck by the unit (Courtesy of the National Safety Council).

Narrow-Aisle Trucks

Narrow-aisle trucks are designed to take advantage of material handling in narrow aisles which are 5-10 feet wide. A standard aisle can be up to 12 feet wide or more. Narrow aisle trucks have two designs; the straddle truck and the reach truck. Both designs allow the operator to be standing while at the controls. The straddle truck has base legs that straddle the pallet or load. Loads are stacked directly over each other. Capacities of straddle trucks range from 2,000 to 4,000 pounds. The distance between the two legs of the lift must be wider than the pallet or load width. The straddle needs only 5 inches or so on each side of the load or rack to maneuver. Straddles do, however, come in contact with racking on a regular basis which can damage and weaken the rack.

Rider Reach Trucks

Rider reach trucks are very similar to walkie type units. Forks extend the load beyond the base legs. Loads can be lifted and deposited without straddling the load. Some feel that this feature makes this model more versatile than the straddle truck. This model requires 6 to 9 inches on both sides of the pallet or load. Capacities of these units range from 2,000 to 6,000 pounds. A special deep reach version is available which permits two-deep stacking.

Order Picking Trucks

Order picking trucks allow the operator to ride up with the forks where he regulates travel, speed, elevation and direction with onboard controls. These trucks are mainly used for assembling less than pallet load quantities. The order picker is a narrow-aisle straddle truck which contains the operator's platform. This model can usually lift at heights up to 15 to 20 feet with a load capacity of up to 2,500 pounds. Some heavy duty units can lift as high as 30 feet.

In higher volume or higher lift systems, some types of guidance is oftentimes provided so that the operator is relieved of the responsibility of steering. Guiderails or an electronic guidewire embedded in the floor may be used. In some cases an operator cab is used.

This unit can be fitted with a level of sophistication which includes a minicomputer control with a video display and keyboard console. Inventory and picking information is provided by remote radio transmission.

Note: some of the order picker/rider picker units are not high lift models and operate like a low lift platform truck.

Another type of unit similar to an order picker is the scissor lift. These models are mostly used for maintenance work or for taking inventory. It is important to block off areas where these units will operate to keep from being struck. They can be top heavy and must not make turns while elevated. While traveling the unit can be hazardous. The operator has the controls with him. Figure 3.3 illustrates the use of a scissor lift.

Figure 3.3. Use of a Scissor Lift

Scissor lifts provide a useful means for maintenance or projects involving inventory. These units can easily topple over or be struck by another fork lift. Operators must be properly trained. The aisle that this unit is operating in has been blocked off.

Sideloaders

Sideloaders are four wheeled vehicles used for transporting and stacking long bulky and difficult to handle loads. The side loader truck loads and carries loads from the side. This width allows the truck to operate in narrow aisles as small as 5 feet. The forks are mounted on an elevating mast in the center of the truck. The mast can be tilted slightly and moved back and forth laterally. For greater stability the load can be lowered with the load resting on two platforms—one on either side of the mast. This also helps to transport the load in a safer manner.

These models can be used indoors or outdoors. Capacities can range from 15,000 to 100,000 pounds with the loads being lifted up to 40 feet. They can be powered by gasoline, LPG or diesel fuel. Pneumatic tires allow for outdoor use and tiering of the load.

Turret Trucks

Turret trucks have high lift capabilities ranging from 30 to 40 feet. They also contain a device which allows the forks or mast to rotate which permits stacking at right angles to the forward direction of the truck. This feature allows the vehicle to operate in narrow aisles. The truck can stack to one side, or to both sides depending on who manufactured it.

Some of these units are essentially counterbalanced trucks with rotating fork attachments which have been added. Other models are specifically designed for narrow aisles, under rail or wire guidance. All models have telescoping masts.

Most are powered by electricity, some of the counterbalanced units may use an internal combustion engine. Load capacities can range from 2,000 to 6,000 pounds. In some models the operator travels with the forks, which provides for high-rise order picking capability. Other models allow for the operator to be at a stationary position at floor level. Some type of location system is often provided to help the operator position the forks at the correct rack opening height.

Hybrid Machines

Hybrid machines are a special class of vehicle which are developed for high-rise, high-volume operations. The need is for the operator to ride up with the load. Unlike turret trucks, the hybrid model has a rigid mast with a lift capability of 40 to 60 feet. This model has been characterized as a cross between a lift truck and an automated storage/retrieval system.

When operating in an aisle, the electric vehicle draws power from the building through overhead connectors. It operates on its own battery power when out of the storage area or when it moves from one aisle to another.

The typical hybrid unit can serve 4 or 5 aisles within a large storage system. Usually, high-volume warehouses use these models.

HIGH LIFT TRUCKS—FLOOR OPERATED

Manually Propelled Trucks

Manually propelled trucks are manually pushed to the desired area. Steel wheels sized for the floor, type of load and working conditions are normally used. Lifting is accomplished through hydraulics or by mechanical means.

The mechanical system can be a hand crank that activates a lifting hoist cable or a lever system that operates in a manner analogous to an automobile jack. The load is held at the desired level with a safety ratchet and pawl. These units usually have a limit of up to 750 pounds.

Hydraulic pump units can be activated by foot pedal, hand lever, or by a battery-powered electric motor. A chain or cable lift is used. In a typical system, the power mechanism activates a hydraulic ram, which raises a yoke carrying one or two sprocketed pulleys. The pulleys push upward against the sprocket chains looped over them. One end of a chain is attached to the fork lifting assembly. Hydraulic lift units typically have capacities ranging to 2,500 pounds. Lift height range can be up to about 7 feet. Telescoping masts on these units which travel from 12 to 15 feet are available.

Platform Trucks

Platform trucks are used for stacking or positioning pallets, dies or loads. Standard high lift platform trucks can lift loads to about 5 feet. Some telescoping models are available and can be used to stack at higher levels.

These trucks are available in walkie and rider designs. Most models are battery powered. Standard units have capacities up to 6,000 pounds but specialized heavy-duty rider versions can go up to 125,000 pounds.

Walkie Fork Trucks

Walkie fork trucks (hand trucks) are basic units which are used throughout industry. Walkies cost about half of what a rider truck costs and are also more economical to operate. Manufacturers state that any operator can be readily trained on these units with the assumption that handling the machine is a part time function. A note of caution here: statistics show that walkie trucks contribute to many industrial injuries as well as damage to racking, product, etc.

Walkie trucks allow for quicker loading and unloading at docks. Where there is a limited floor capacity, such as an elevator, walkies offer an alternative to a rider truck. Walkies can lift as high as 16 feet and can handle loads in the 2,000 to 4,000 pound range. These units can be powered by electricity.

Note: Some walkie units are designed to allow the operator to ride on the truck.

These models run on hard rubber or polyurethane wheels. Most times the steering handle contains most, if not all, of the controls. The steering handle may contain a safety switch that protects the operator from being accidentally pinned while operating the truck in reverse. Some refer to this safety switch as a "belly button."

> **ACCIDENT FACT**
>
> A walkie truck operator had just placed a pallet load of product under a rack and was backing up the unit. Upon turning the control handle he did not allow for a tight turn and pinned his hand between the walkie handle and a rack. He fractured two fingers on his right hand as a result of this serious pinch point.

The braking system can be activated by raising or lowering the handle or spring-loaded steering arm beyond a set angle. These units also contain a switch that engages the brake and shuts off the power system when the steering arm is released. (Refer to Figure 4.3.)

Walkie Straddle Trucks

Walkie straddle trucks have extended outrigger base legs which are mounted parallel to and outside of the lifting forks. This design allows for the trucks center of gravity to be directly beneath the load which helps to keep the truck from tipping forward when the lift is raised. This unit does not have a counterweight because of this.

The truck is designed for narrow aisles and the straddle legs fit a 36 inch wide pallet rather than the 40 inch wide pallet. This unit would thus require a standard or uniform size of pallet.

Walkie Reach Trucks

Walkie reach trucks have an outrigger base similar to the straddle truck. The outriggers are for stability and are not influenced by pallet width because they do not straddle the load. The pantograph mechanism allows the forks to be extended forward beyond the base legs to reach in under a load. Special deep-reach models permit two deep stacking.

Counterbalanced Walkie Trucks

Counterbalanced walkie trucks handle the load in front of it, in a cantilever fashion. The weight on the forks is counterbalanced by the weight of the truck. The load wheels of the vehicle act as a fulcrum. The carriage can travel up and down while carrying the forks. The mast can be tilted forward and back. This model can lift loads of various widths and does not have to utilize a set size pallet width. The long body shape does not make it practical for narrow aisle use.

Table 3.2 provides detailed information on the various classifications and lift code types for all powered industrial trucks.

Table 3.2. Analysis of Lift Truck Classifications and Models

MEMBER	CLASS 1 LIFT CODE 1	4	5	6	CLASS 2 LIFT CODE 1	2	3	4	6	CLASS 3 LIFT CODE 1	2	3	4	5	6	7	8	CLASS 4 LIFT CODE 3	CLASS 5 LIFT CODE 4	CLASS 6 LIFT CODE 1	CLASS 7 LIFT CODE 1
Barrett	•																				
Big Joe					•	•	•			•	•										
Blue Giant						•				•											
BT/Prime Mover	•				•	•			•		•	•	•	•	•	•	•				
Clark	•	•	•	•	•	•	•				•	•	•	•	•	•	•	•			
Crown	•	•	•	•	•	•	•				•	•	•	•	•	•	•				
Dew Engineering	•							•			•	•	•	•	•	•	•				
Drexel			•																		
E-Z-GO Textron																					
Forano																	•				
Hyster	•	•	•	•	•	•	•				•	•	•					•	•		
K-D Manitou																					•
Kalmar AC	•	•	•			•	•				•	•			•		•	•	• •		
Komatsu		•				•	•				•	•	•		•	•	•	•	•		
Linde/Baker	•	•	•		•	•	•			•	•	•			•	•	•	•	•		
MCFA/Caterpillar	•	•	•		•	•	•				•	•	•	•	•	•	•	• •	•		
MCFA/Mitsubishi	•	•	•		•	•					•	•	•	•	•	•	•	•	•		
Multiton						•	•														
Nissan US/CAN	•	•	•	•		•					•	•	•	•	•	•	•	•	•	•	
Nissan Mexico		•	•	•																	
Pettibone Tiffin																	•				•
Raymond	•				•	•	•		•												
Simpson Machine																				•	
TCM/CIM		•	•								•	•	•	•		•	•	• •	• •		
TCM US/CAN		•	•					•			•	•	•	•		•	•	• •	• •		
Teledyne Specialty																					•
Toyota US/CAN		•	•			•	•	•			•	•		•	•	•	•	• •	• • •	• •	
Toyota Mexico		•	•			•	•														
Wiggins Lift																	•		•		•
Yale	•	•	•	•	•	•	•			•	•	•	•	•	•	•	•	• •	• • •		

(Courtesy of the Industrial Truck Association)

Table 3.2. *(continued)*

ANALYSIS

CLASS 1 – ELECTRIC MOTOR RIDER TRUCKS
- LIFT CODE 1 – COUNTERBALANCED RIDER TYPE, STAND UP
- LIFT CODE 4 – THREE WHEEL ELECTRIC TRUCKS, SIT-DOWN
- LIFT CODE 5 – COUNTERBALANCED RIDER TYPE, CUSHION TIRES, SIT-DOWN (INCLUDES HIGH AND LOW PLATFORM)
- LIFT CODE 6 – COUNTERBALANCED RIDER, PNEUMATIC OR EITHER TYPE TIRE, SIT-DOWN (INCLUDES HIGH & LOW PLATFORM)

CLASS 2 – ELECTRIC MOTOR NARROW AISLE TRUCKS
- LIFT CODE 1 – HIGH LIFT STRADDLE
- LIFT CODE 2 – ORDER PICKER
- LIFT CODE 3 – REACH TYPE OUTRIGGER
- LIFT CODE 4 – SIDE LOADERS, TURRET TRUCKS, SWING MAST AND CONVERTIBLE TURRET/STOCK PICKERS
- LIFT CODE 6 – LOW LIFT PALLET AND PLATFORM (RIDER)

CLASS 3 – ELECTRIC MOTOR HAND TRUCKS
- LIFT CODE 1 – LOW LIFT PLATFORM
- LIFT CODE 2 – LOW LIFT WALKIE PALLET
- LIFT CODE 3 – TRACTORS (DRAW BAR PULL UNDER 999 LBS.)
- LIFT CODE 4 – LOW LIFT WALKIE/CENTER CONTROL
- LIFT CODE 5 – REACH TYPE OUTRIGGER
- LIFT CODE 6 – HIGH LIFT STRADDLE
- LIFT CODE 7 – HIGH LIFT COUNTERBALANCED
- LIFT CODE 8 – LOW LIFT WALKIE/RIDER PALLET

CLASS 4 – INTERNAL COMBUSTION ENGINE TRUCKS – CUSHION TIRES ONLY
- LIFT CODE 3 – FORK, COUNTERBALANCED (CUSHION TIRE)

CLASS 5 – INTERNAL COMBUSTION ENGINE TRUCKS – PNEUMATIC TIRES ONLY
- LIFT CODE 4 – FORK, COUNTERBALANCED (PNEUMATIC TIRE)

CLASS 6 – ELECTRIC AND INTERNAL COMBUSTION ENGINE TRACTORS
- LIFT CODE 1 – SIT-DOWN RIDER (DRAW BAR PULL OVER 999 LBS.)

CLASS 7 – ROUGH TERRAIN FORK LIFT TRUCKS
- LIFT CODE 1 – ALL ROUGH TERRAIN FORK LIFT TRUCKS

(Courtesy of the Industrial Truck Association)

SUMMARY

Each piece of power equipment is unique. The models may look the same, but the controls, pedals, etc., may be different depending on the manufacturer. The proper piece of equipment should be purchased or leased depending on the needs of the job. The manufacturers of the equipment can help provide the expertise and specific training for the equipment.

Employees are to be properly trained for each piece of equipment they must operate. Any operator that has not been properly instructed on a particular piece of equipment must not be allowed to operate the powered industrial truck. Management must exercise a strong sense of enforcement here.

Selection of the proper size and capacity of the truck is very important. Any attachments or modifications to any vehicle must be in writing from the manufacturer. Professional help is needed to assist in vehicle and parts selection.

REFERENCES

Industrial Truck Association, Washington, DC, 1996.

Kelly, M. "Understanding Lift Trucks," *Engineering Digest*, January 1994.

National Safety Council. "Powered Industrial Trucks," *Accident Prevention Manual*, Itasca, IL, 1992, pp. 155-176.

OSHA Proposed Rule for Training Industrial Truck Operators, U.S. Department of Labor, 60FR13782, March 14, 1995.

Plant Engineering Magazine, Plant Engineering Directory, pp. F20-24.

4

POWERED WALKIE/RIDER PALLET TRUCKS

Almost any workplace that requires the movement of product, will have a hand jack, hand truck walkie or pallet jack; these units are called several different names. It could be battery-powered or a manual model. It might be designed to only lift the product a few inches off the floor before transporting the load. It might be equipped with provisions to raise the load many feet into the air. Some models allow the operator to ride on the unit.

These models described above are very common throughout industry. Just as common are the injuries that these piece of power equipment can cause. Many operators are not trained in how to safely operate these units.

It looks simple—just walk up to the machine, turn a knob or control, and move it forward, or backwards. Push another button and raise the load. Now, push another button or turn a knob and the entire machine is moving a pallet load of product. Simple, yes. Dangerous, absolutely. This chapter will provide safety rules and operating tips for powered walkie trucks, as they are most commonly called. More walkie lift trucks are produced each year than any other type of powered equipment. Despite the fact that they cost less than a forklift and are versatile, these machines are responsible for their share of injuries each year.

It's important to note that walkie type injuries occur to operators and non-operators in the industrial setting. Members of management have commented that they didn't realize that the machine could be so dangerous; this being said after an accident. "I only wanted to move that heavy box on the pallet. The machine looked so innocent; I've watched employees handle similar loads many times."

Make no mistake about it, anyone can be injured while operating a walkie truck. The most common types of injuries are:

- ■ A foot being caught between the floor and the bottom structure of the machine. This usually occurs when walking forward with the load behind while holding the hand on the controls. The person is walking slower than the machine or they stop. The machine continues to move and pins the foot, causing injury. Note Figure 4.1; an employee's foot was saved from injury by steel-toe boots; when a walkie truck ran over his foot.

Figure 4.1. Steel Toe Boot Damaged by Hand Truck

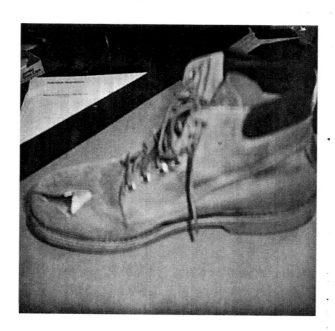

A powered hand truck ran over the operator's foot. The steel toe saved the employee from injury. Note the damaged boot.

- Hands can be pinched between the handle and a stationary object, such as racking. This usually occurs when the operator doesn't allow for sufficient clearance when moving the load while the hands are on the control handle. As the machine moves, the hands are pinned against something causing a caught-between type injury.

- A device referred to as a "belly-button" is a safety feature on the handle. The operator can push on this button, which is the center of the control handle, with their stomach and the machine will reverse its travel.

- Another common hand injury occurs from reaching under the unit. Never reach under a vehicle to remove an obstruction without first removing the key. Be sure you're not in the path of travel of other vehicles when looking under a pallet truck.

- Employees or pieces of equipment or product can be struck by the machine or load. While moving forward or backward, anyone or anything can be struck by the machine.

- The operator can be struck by a falling load. It is not uncommon for an operators to lift a load and have part of it fall back onto them (or someone else). Also, operators can strike racking and knock down items onto themselves or others.

- Battery acid on the unit can cause harm. Many operators service their walkie trucks and it is not uncommon for battery acid to harm the skin or eyes.

- The walkie truck could be driven off of the dock. The danger to the operator or to someone else at the dock is apparent because of the nature of dock areas. The edge of the dock could be misjudged or the operator could fail to chock the wheels of a trailer being used. If the truck falls, there is property damage and a work stoppage. If the employee falls, there could be serious consequences.

Powered hand trucks are similar to lift trucks in many ways. In the case of powered walkie trucks, the operators must protect themselves from the dangers associated with the vehicle. With large lift trucks, the operators usually are not in danger as a result of being an operator. The operators are in a seat inside the confines of the vehicle, or, are standing up and surrounded by metal on three sides.

A hazard assessment for the selection of personal protective equipment is essential when operating a walkie type truck. Operators as well as fellow workers may be required to wear:

- ANSI approved hard hats. Bump caps must not be allowed because they offer little or no protection from falling objects.

- ANSI approved steel-toe shoes or boots. The majority of personal injuries that occur to walkie truck operators are to the feet. Metatarsal shoes take foot protection a step further than steel-toe shoes. This type of footwear should be seriously considered.

- Gloves are now a part of OSHA's Personal Protective Equipment (PPE) standard. To protect from cuts, splinters, sharp edges, and minor pinch points, gloves are important. The heavier the glove, the better the protection they provide.

- ANSI approved eye and face protection where the job or rules warrant such equipment. Standard safety glasses with side shields may be a part of everyday required equipment on the job. Safety goggles and face shields may be required for certain jobs, such as battery care or charging.

- Hearing protection, such as ear muffs or ear plugs, may also be a requirement of the job or to be used for specific assignments or hazards.

GUIDELINES FOR THE PROPER USE OF WALKIE RIDERS

Safety Features

These guidelines apply to most walkie type power equipment. There may be older models that have been modified, therefore, some of these rules may not apply.

- Hand controls on the handle allow for rotation to propel the vehicle forward or in reverse. Once the grips are released, the vehicle stops.

- The handle acts as a brake when it is in the vertical or horizontal position.

- Walkies are three wheeled vehicles. Two wheels are mounted at each side of the front end of the unit; and one wheel is at the rear of the unit, which is at the same end as the steering handle. The single wheel acts as a swivel for steering and maneuvering.

- Walkie trucks travel slower than the average forklift. Walkies usually travel about three miles per hour. Walkie riders may have more speed. Walkies are now being built with programmable electronic speed controls. This feature allows for an increase in battery life. Operators can drive a rider too fast and tip it over while turning.

Some European design rider units offer side gates that keep the operator within the width of the walkie rider truck platform. A hand-hold is also provided for greater security.

Newer vehicles have small solid-state drive and lift controls which offer variable speed control. These new controls avoid the forward-reverse jerking which can spill or damage loads. Also, these quick movements have been known to dump operators off of the walkie-rider or to cause other accidents for walkie operators.

Explosion-proof walkie trucks are available where such conditions exist. Processes such as ink, paint, chemical, or fiber manufacturing contain hazardous situations.

Operating Principles

If traveling with a load on a walkie or powered walkie, and the load is in front of the operator, one or both hands are to be on the controls. If the load is behind the operator, one hand should be on the control handle while the operator is facing the direction of travel. The operator must stand far enough to the side to prevent being struck by the equipment, as well as to concentrate on obstacles along the way. Speed control while walking is essential for operator safety.

Figure 4.2 illustrates the proper method of operating a powered hand truck.

Figure 4.2. Properly Operating a Powered Hand Truck

Operators are to stand to the left or right of the operating handle of the powered unit (Courtesy of Rite-Hite Corporation).

If a load is too large the operator, will have to travel forward with the load behind. Operators must always face the direction of travel.

Loads should be approached with the forks square. Forks should be lowered and the load entered evenly. Be sure the forks are wide enough to support the load properly. Go under the load as far as possible before lifting.

- Before lifting any load, ensure that the load is centered and stable. If possible, loads should be shrink-wrapped for stability in handling, as well as for safer storage. Straighten loads and make them more secure before moving.

- Raise the load high enough to make safe floor clearance. Avoid ruts, bumps or other obstacles in the path of travel.

- Approach the load deposit area cautiously, then raise the load after the vehicle is stopped. Only raise a load when ready to set it down. The walkie could easily tip over and spill the load if it is high enough. This, of course, could be very hazardous to the operator or anyone else in the area. Being that the walkie truck will not have an overhead guard, operators must exercise caution at all times to guard against falling objects.

- Do not raise or lower the load while the machine is in motion.

- New operators should be thoroughly trained before actually working with the unit. Practice should be in a remote area where the operator can practice turning, handling, lifting, stacking, and maneuvering without danger to others as well as product and property. Figure 4.3 illustrates the testing of the brakes on a powered hand truck. New operators must know the braking system.

Figure 4.3. Operating Handle Brakes

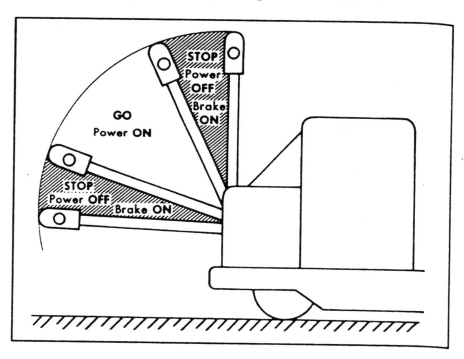

Note the shaded section of the operating handle. When the handle is in the proper position the unit will either stop or move (Courtesy of the National Safety Council).

- Only one person should operate the walkie at one time. "No riders" must be the rule.

- Loads should be kept as low as possible when stacked and transported. If a load is too high when entering a trailer, parts of the load can strike the top of the trailer and fall back onto the operator. The heights of the dock plate and entry to the trailer can be deceiving.

- Ensure trailers are properly chocked and the brakes are set before entering them.

- Thoroughly inspect the pallet jack on each shift before use. Document the findings on the appropriate form. Figure 4.4 contains an inspection form for powered hand trucks.

- Pedestrians always have the right of way.

- Read the literature from the manufacturer for proper operating guidelines and instructions. Each manufacturer may have different operating rules.

- Before starting, check:
  - Load wheels.
  - Steering.
  - Horn.
  - Operating handle and controls.
  - Gauges.
  - Directional and speed controls.
  - Load data plate.
  - Safety signs and labels.
  - Braking mechanism.
  - Any other safety devices.

- Never exceed the operating capacity of the vehicle. If the load can't be lifted, do not force it.

- Avoid sudden starts and stops. It is possible to be thrown from a rider-type pallet jack during a sudden stop.

- When on a rider model, always come to a complete stop before getting off of the vehicle.

- Remove any vehicle from service if a defect or missing safeguard is discovered.

- Keep employees away from areas where a load has to be deposited or lifted—warn others of your intentions.

- Go slow on dockboards or bridgeplates; be sure they are secure before crossing them.

- Do not push or pull loads or other vehicles with the walkie. Do not allow others to push or pull your vehicle.

- Allow for wide swings into aisle ways and other areas. Sound your horn to alert others. Figure 4.5 shows damaged racking as a result of a powered hand truck.

- When parked, shut off the unit, lower the load, and be sure it is not blocking a doorway, aisle, or emergency equipment, remove the key.

- Never speed. Even an empty walkie can tip over.

Figure 4.4. Inspection Form for Powered Hand Trucks

POWERED PALLET JACK INSPECTION FORM

Operators, inspect your walkie or rider pallet jacks at the start of each shift. If any problems are discovered, notify your supervisor. *Never operate a vehicle that is not safe!*

ITEM	STATUS YES	NO
1. Is the horn functioning?	☐	☐
2. Are the gauge and meters functioning?	☐	☐
3. Are there any cracks in the forks?	☐	☐
4. Are wheels functioning and not broken?	☐	☐
5. Are the tires safe?	☐	☐
6. Do forward and reverse controls work?	☐	☐
7. Are any leaks detected; battery/hydraulic?	☐	☐
8. Is there any external damage?	☐	☐
9. Are labels and signs in place?	☐	☐
10. Is the load backrest in place?	☐	☐
11. Does the steering arm move freely?	☐	☐
12. Does the lift control work?	☐	☐
13. Does the braking mechanism work?	☐	☐
14. Is the standing platform on a rider a non-skid surface?	☐	☐
15. Is the cover for the electrical controls in place?	☐	☐
16. Does the battery show any signs of corrosion?	☐	☐
17. Other _____	☐	☐

Comments: _____

Operator Name: _____ Date: _____

Vehicle Name or Number: _____

A safety check list for powered hand trucks.

- When traveling up or down a ramp, keep the load or load end on the down side. Never turn on a ramp.

- Use sound judgment, do not take chances or short cuts.

- Inspect flooring in trailers or railway cars before entering. Support the nose end of trailers to avoid tipping.

Figure 4.5. Damage to Racking

Powered hand trucks contribute their share of damage to buildings.

- Be sure the hands and soles of shoes are free of grease or oil. On rider-type walkies, a non-skid paint can help keep the operator from slipping on the surface.

- Keep feet, hands, head, and other body parts confined to the running lines of the rider walkie.

Notify your supervisor if any problems arise during the operation or inspection of your pallet truck.

SUMMARY

Powered pallet jacks are very useful, but can also be very dangerous. Many individuals are lulled into a false sense of personal safety because these units appear to be so simple in design and control handling. Operators of these machines should go through the same intense training as forklift operators. Written tests, daily checking of the machine, skills driving, personal protective equipment, and overall job knowledge must be a part of the program. To aid in skills evaluation, refer to appendix B for the power hand truck quiz.

It should be noted that powered hand walkie truck operators must be trained and authorized to operate the equipment just as forklift operators.

REFERENCES

Coastal Video Booklet, "Industrial Low-Lift Trucks," Virginia Beach, VA, 1993.

Dessoff, A. "OSHA Looks at Powered Hand Truck Safety" *Safety & Health Magazine*, October 1994, pp. 72-76.

Schwind, G.F. "Powered Walkie Trucks: New Designs Stretch Applications," *Material Handling Engineering Magazine*, December 1995, pp. 47-55.

"Selection Guides for Walkie Hand Trucks," *Plant Engineering Magazine*, December 11, 1995, pp. 77-78.

"Using Motor Hand Trucks Safely," *Plant Engineering Magazine*, July 10, 1995, p. 21.

Yale Industrial Trucks. "Handle with Care," *Yale Instructors' Manual*, Portland, OR, 1972.

5

PRE-USE INSPECTIONS

A safe piece of power equipment depends on the operators' skills, training, management concern and enforcement, and intrinsic safety. The piece of equipment must be given constant attention to ensure it is safe. As with an automobile, longevity and performance have a lot to do with the care it is given. If someone is driving their family car and the handling or brakes don't feel right, the car is checked. Powered equipment operators also have a "feel" for their vehicle and can usually determine if maintenance is necessary. One sure way of regularly checking the condition of equipment is to check all of the necessary parts of the machine before putting it to use. After all, a typical day in an industrial setting puts a lot of wear on the brakes, steering, tires, suspension, and drivetrain.

Operators must be trained to perform pre-use inspections. Prior to each shift, each operator must perform a thorough, documented evaluation. To do this, a form has to be devised for each different piece of power equipment. As an example, a counterbalance lift truck and a walkie pallet truck would require two separate forms. A few of the items checked may be similar, but the evaluation may be a little different.

Once the lists have been developed, each operator should be thoroughly trained by using the forms and evaluating each piece of equipment. Operators should be given a clipboard to hold the form in place. Clipboards should be placed on a large wall or board so that the forms are readily available. In addition, quick availability allows for an inspection each shift by management to ensure that:

- Each piece of equipment is being thoroughly inspected.
- Each form is properly completed as required.
- Any need for maintenance can be detected so proper repairs can be made.
- Vital parts of equipment can be replaced. If necessary, replace the piece of power equipment.
- OSHA requirements will be met regarding pre-use inspections.
- Other facility employees, as well as visitors, would be further protected as a result of safer equipment.
- Maintenance can properly schedule repairs as a result of ongoing inspections.

70 / Forklift Safety

Figure 5.1 illustrates an employee performing an in-depth inspection of a lift truck.

Figure 5.1. Inspecting a Lift Truck

Operators must use a comprehensive inspection form when inspecting the lift truck (Courtesy of Rite-Hite Corporation).

PRE-USE INSPECTION FORMS

One of the most important elements is the form itself. Power equipment operators are to focus their inspection efforts on:

- Steering and horn.
- Brakes.
- Lights and alarm.
- Tilt.
- Mast.
- Raising and lowering forks.
- Coolant level.
- Fire extinguisher.
- Brakes.
- Tires.
- Gauges.
- Leaks.
- Fuel level.
- Oil level.
- Propane Tank.
- Overhead guard.

Pre-Use Inspections / 71

First, the purpose of an inspection is to ensure that the unit is safe. There must be a standing rule that if any defects are discovered, the piece of equipment is to be taken out of service.

Individually, detailed information of these major items are as follows:

Steering and Horn

- With power on, completely turn the steering wheel to the left and right. There should be no squealing from any steering pump. The wheel should feel normal and not loose to the operator. The horn should be tested at this time.

Brakes

- Test *service brakes* by pushing in the pedal. Pedals should be solid, not spongy. There should be no fade or drift. When pushing on the pedal, hold for at least 10 seconds to ensure a solid brake.

Figure 5.2. Checking Steering and Brakes

Operators must manually check the brakes and steering prior to putting the unit in operation. Note the seat belt (Courtesy of Clark Material Handling Company).

- For *seat brakes* on electric powered lift trucks, the operator must get onto the seat. On some of the older models, as the operator pulls forward and lifts up off the seat, the lift should come to a stop. This procedure can be dangerous and is not advisable for obvious reasons. In addition, a seat belt cannot be worn if using this method. A built-in seat switch acts as a safeguard on newer models. Carefully listen to the sounds of the machine; there should be silence when the operator is off the seat.

- The *parking brake* should be activated and tested to ensure it functions properly. Figure 5.2 illustrates an operator testing the steering and brakes.

Lights

- Where lights are needed for driving into dark trailers, in a facility or out of doors, lights should always be functional. Emergency lights must also be fully functional.

- Flashing lights on the top or side of the power equipment provide for increased plant safety; these lights should always be visible and functional.

- In addition to the horn being an alarm to warn others, the back-up alarm on sit-down models may help increase plant safety. Some models of alarms can be adjusted and some are self-adjusting for desired sound levels. The maintenance department should make this adjustment.

Controls

- Lift the forks while seated or standing in the unit; be sure to be at the controls—not off of the equipment. Raise the forks as high was they will go. Tilt the mast forward and backward. Ensure everything is working as it should.

- Check the chains while the forks are flat on the floor and the chains are loose. Use a rag or gloves to keep grease off your hands.

- Knobs, handles, and controls should all be functional and in place.

Coolant Level

- Always check coolant levels when the liquid is cold. Hot coolant can cause severe burns. When power equipment changes operators during shift changes, the coolant is sure to be hot and dangerous. Do not attempt to check the level at that time.

Fire Extinguisher

- An ABC fire extinguisher is necessary on power equipment that may be operating in remote areas of the plant or facility grounds. A fire extinguisher that would be readily available inside a plant (every 50 - 75 feet of travel), could be very close to a forklift that is on fire.

- Check the tag to ensure it has been serviced by a professional during the past year. The gauge arrow should read full, the pin and seal should be intact.
- Be sure the bracket supporting the extinguisher is in place.
- Of course, proper training in the use of fire extinguishers is necessary.

Forks

- Forks should not be bent or distorted in any way. Look for cracks, also. Periodically, the maintenance department should conduct mag particle testing or die penetrate testing on forks to ensure there are no hidden breaks or defects.
- Never weld or straighten forks—replace them.
- When a lift truck is parked, the forks should always be flat on the floor for maximum safety. See Chapter 15 for additional information regarding forks.

Tires

Tires and wheels on powered equipment are very important. As the old saying goes, the brakes stop the tire, the tires stop the vehicle.

- Look for cuts, breaks, missing chambers of rubber, and signs of wear. Bald tires can be dangerous to everyone. Check for imbedded material in the tires. The wearing of gloves can be important here.
- Where tires are pneumatic, ensure tire pressure is appropriate. Low air pressure can have an effect on the vehicle's stability, steering, and stopping distance. If more air pressure is needed, check with maintenance or a mechanic. Be sure to mark this problem on the inspection form.
- If tires contain multi-piece or split rims, use extreme caution. Do not attempt to fix or repair this tire. A trained mechanic is necessary. Safeguards must be taken to prevent serious injury. These tires must be completely deflated before removal. Serious injury or death could be the result. In addition, never weld on this piece of equipment.

Follow manufacture' guidelines for tire safety. Figure 5.3 provides a graphic illustration regarding the proper evaluation of tires.

Figure 5.3. Checking Tires

Employees that are trained and knowledgeable in tire maintenance are to service them. Tire pressure is important to safety (Courtesy of Clark Material Handling Company).

Gauges

- Check all gauges to ensure they are functioning properly. Hour meter and amp meter are important. Log the proper information on the inspection form.

Hydraulic Oil Leaks

- Are hoses leaking? Are there oil or hydraulic spots under the vehicle after it is moved? Be sure to alert your supervisor to this problem. Not only can a leak cause damage to the vehicle; it could greatly effect steering, lifting, maneuvering or braking. Also, leaks pose a high risk for slip and fall hazards. Is any visible damage apparent?

Fuel Level/Oil Level

- The gauges will identify fuel or power levels. A fuel gauge where gasoline or diesel are needed will have a gauge such as that of a car. Liquid propane, (LPG), will have a gauge on the tank, which is strapped onto the back of the vehicle.

- If connecting or disconnecting a LPG hose, and wear gloves and eye or face protection, and do not smoke.

- If checking a small battery, be sure to wear protective equipment. A gauge could indicate if enough power was in the battery.

- Check oil just as you would in a automobile by using the dipstick. Be careful with the hot oil. Use a rag to wipe the oil off, do not use a part of the lift truck or your shirt.

Propane Tank

- Never smoke near a propane tank when handling a tank, hose, or connection. Wear gloves and eye/face protection. A propane tank weighs at least 40 pounds and is clumsy in addition to being heavy. Lift with the legs and arms, not with the back. Get help if possible to lift tanks. Ensure the tank is strapped in place.

Overhead Guard

- Check the welds and bolted areas of an overhead guard. Notify maintenance if any cracks or loose bolts are observed.

Additional Items

- Check for any additional items such as side markings, load limit plate, seat belts, condition of seat, condition of pedals, etc.

Many lift truck training programs have a full videotape or long segment dedicated to pre-use inspections. It's a must for operators to view these tapes. It may take patience to view the tape and understand the inspection process. Experienced operators will grasp the depth and necessity of the training information more quickly than beginners.

The list of items to check, which are identified in this chapter, seem rather long. Like the video on inspection, both are time-consuming. This analogy is used because an experienced operator requires only a few moments to properly follow a fork lift checklist. It takes much longer to read from this chapter or watch a videotape on the subject.

The following pages contain several examples of inspection forms from various sources. An organization can choose any specific form they wish or modify it to fit their particular situation.

The most important part of this process is to have each operator do a thorough job of documenting each inspection before each piece of power equipment is used. Management must follow-up to ensure everyone is trained, the forms are being completed, and proper repairs are being made on a timely basis.

Two sample equipment inspection forms are illustrated in Figures 5.4 and 5.5.

Figure 5.4. Forklift Daily Record Form

FORK LIFT DAILY RECORD
(ALL TRUCKS)

Truck Asset Number:	Date (MMYY):
Make and Model	Warehouse Number:
Fleet Number:	**Exception Codes:** 1. B-Meter 2. I-Available Equipment Idle 3. T-Meter Turned Over or Replaced
Dept:	**Safety Checks:** 1. ✓ - OK 2. X - Needs Service

Date		Equipment Meter Readings: In Service Hours		Downtime Hours	Exception Codes	Equipment Safety Checks - Operator												Maintenance Checks			
Wk End	Day	START	END	TOTAL	B-I-T	Brakes	Horn	Lift/Tilt	Lights	Oil Leaks	Reach&Forks	Steering	Tires	Engine Oil	Backup Alarm	Overhead Guard	Lift Chains	Battery	Coolent		Operator
	1																				
	2																				
	3																				
	4																				
	5																				
	6																				
	7																				
	8																				
	9																				
	10																				
	11																				
	12																				
	13																				
	14																				
	15																				
	16																				
	17																				
	18																				
	19																				
	20																				
	21																				
	22																				
	23																				
	24																				
	25																				
	26																				
	27																				
	28																				
	29																				
	30																				
	31																				

Use a form such as this for daily operator checks.

Figure 5.5. Pre-Use Inspection Form

This form provides another choice for daily operator checks.

78 / Forklift Safety

Figure 5.5 *(continued)*

EQUIPMENT HOURS				VISUAL CHECKS						OPERATIONAL CHECKS						SERVICE	
Date	METER READING		Total Service	Out of Service	Battery Fluid / Cable	Tires	LIGHTS		GAUGES		Hydr. Leaks	Horn	Steering	BRAKES		Lift / Tilt	Forward / Reverse	Opr. Int.	Job Code	Work Order / Invoice #
	Start	End					head / tail	warning	hour meter	discharge indicator				service	deadman					
14																				
15																				
16																				
17																				
18																				
19																				
20																				
21																				
22																				
23																				
24																				
25																				
26																				
27																				
28																				
29																				
30																				
31																				

SUMMARY

It is very important that all operators perform an inspection of their piece of power equipment prior to the start of a shift. Appropriate forms should be developed and used. Equipment checks must take place each day and the appropriate follow-up must take place for necessary repair.

Management must be involved in this process. Employees are to be taught how to inspect and use the form(s). There has to be a procedure that allows for disconnecting the battery, removing the key, chocking the wheels, tagging, and removing equipment for servicing. Not only is this a necessity because of personal safety, it is an excellent method to track maintenance costs and vehicle replacement. If the equipment cannot be maintained by staff, manufacturers should be contacted for proper servicing.

REFERENCES

Clark Material Handling Company. *Operator Training Manual*, Lexington, KY, 1997.

Clark Material Handling Company. *Employer's Guide to Material Handling Safety*, Lexington, KY, 1990.

National Safety Council. *Powered Industrial Trucks*, Accident & Prevention Manual, Itasca, IL, 1992, pp. 155-175.

NIOSH. *Outline for Training Powered Industrial Truck Operators*, DHEW Publication No. 78-199, Cincinnati, Ohio, October 1978.

U.S. Dept. of Labor, OSHA, Proposed Rule for Training Powered Industrial Truck Operators in General, Maritime Industries, 60 FR 13782, March 14, 1995.

6

GENERAL SAFETY GUIDELINES

Powered equipment operators must know, understand, and be able to put into practice the myriad of safety rules and guidelines that apply to their jobs. When looking at other occupations in a work setting and comparing them to a powered industrial truck operator, the powered equipment operator must have a much greater grasp of their job hazards. Forklift operators are required to know all of the rules that affect both themselves and their fellow workers. This requirement would also include the safety of visitors, as well as the care and handling of product, and protection of property.

As with many learning experiences that you confront on a daily basis, you gain more knowledge when the message is repeated over and over again. The overall knowledge gained thus increases. In regard to power equipment safety rules and guidelines, employees who operate various forms of power equipment must be given the applicable rules on a regular basis to reinforce job knowledge. The process of transferring this necessary knowledge to the operator should come in small enough doses (sessions), so that learning properly takes place. It is recommended that all operators experience training each year. This annual exposure to the rules will help to reinforce what is required of the operators to make them safer operators. The information that management wishes the operator to process and retain must be offered in a manner that is easily understood and absorbed. Above all, any safety regulation should have a purpose and this purpose should be knowledge so that proper safeguards can be met.

In other words, just because a safety rule exists on paper and operators are told to obey and follow it, that doesn't mean the rule will be followed. The information must make sense to the operator. Why does such a rule exist? What are the consequences for the individuals if they don't abide by the rule itself? Management or an operator trainer will not be successful if they don't offer explanations of why there is such a rule. In fact, using real-life experiences to demonstrate why a specific rule is essential will go a long way to help reinforce the rule. Operators will more readily remember the discussion on how fellow employees or other individuals experienced a personal injury because they failed to abide by the rule. Make the rules believable and applicable.

GUIDELINES AND GENERAL SAFETY RULES

The various safety handbooks and manuals produced by the power equipment manufacturers offer a wide variety of "do's" and "don'ts" for operating the equipment. The National Safety Council and other safety organizations also offer a wide variety of safety practices that are a part of their publications for operator training. OSHA regulations contain a wide variety of operating procedures for the protection of the operator. This publication contains many of these guidelines, which can be of great importance to the operator.

To maximize powered equipment safety, readers, trainers, manufacturers, operators and purchasers should be knowledgeable in:

NFPA #30	Flammable and Combustible Liquids Code
NFPA #58	Storage and Handling of Liquified Petroleum Gases
ANSI/UL 583	Safety Standard for Electric, Battery, Powered Industrial Trucks
ANSI B 56.1	Powered Industrial Trucks, Part II
ANSI/NFPA 505	Powered Industrial Trucks
OSHA 1910.30	Other Working Surfaces (Dockboards/Bridge Plates)
OSHA 1910.37	Means of Egress
OSHA 1910.38	Employee Emergency Plans and Fire Prevention Plans
OSHA 1910.95	Hearing Protection
OSHA 1910.106	Flammable and Combustible Liquids
OSHA 1910.132-138	Personal Protective Equipment
OSHA 1910.134	Respiratory Protection
OSHA 1910.157	Portable Fire Extinguishers
OSHA 1910.178	Powered Industrial Truck Standards
OSHA 1910.1000	Exposures to Carbon Monoxide Gas and Other Air Containments

Additional parts of the OSHA regulations, fire codes, and various standards may affect powered equipment operators. It is important to inquire with the appropriate agency for assistance.

General Guidelines for Students

To serve as a more convenient resource to the reader, the various rules and guidelines have been broken down into several categories. For some of the rules, there may be an overlap. The operators, as well as those who are instructing operators, should have an easier time in assisting the training efforts by having a ready reference point. For more guidelines, refer to OSHA 1910.178 in Appendix D. This appendix will identify the various hazardous atmospheres and the types of powered equipment that can be used in such atmospheres.

Traveling with Power Equipment

- Operating surfaces should be smooth and free of holes, bumps, or any other restriction to safe operation.

- Paths of travel should also be free of oil, grease, water, or any other element that could affect turning, starting, traveling, or stopping.

- Pedestrians should have designated walking areas that do not create an exposure to moving equipment.

- Overhead hazards should be relocated if possible. Items such as sprinkler pipes or gas pipes could be moved to prevent being struck by the lift truck load or mast. Other hazards that cannot be moved—such as doorways, racking, beams and other obstacles—should be highlighted with black and yellow stripes. Proper lighting and added visual awareness signals will also assist in greater operator safety. Figure 6.1 identifies an overhead fan that was struck by a forklift mast.

Figure 6.1. Damaged Overhead Fan

While a load was being raised the load struck this fan and caused hundreds of dollars in damage.

- Flashing lights, convex or panoramic mirrors, stop signs, speed limit signs, painted walkways, intersections, and machine and storage areas should be highlighted as much as possible to prevent collisions or damage to property.

- Before lifting any load, properly adjust and lock the forks in place to prevent movement. Wear gloves to protect the hands.

- Keep loads centered on the forks so that there is a balance of load weight. Keep the heavier part of the load towards or against the mast as illustrated in Figure 6.2. If the heavier part of the load is facing away from the operator, turn the load around by approaching it from the opposite side. This may require lifting, moving it to a spot that allows the lift truck to turn around, and approaching it with the heavy side toward the forks.

Figure 6.2. Arrange Loads Properly on Pallets

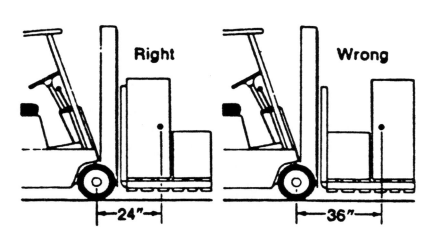

When lifting an unevenly loaded pallet, keep the heavy load toward the mast (Courtesy of the National Safety Council).

- Always allow for proper clearance when lifting, traveling or setting down a load. Pre-plan the route of travel before picking up an odd-shaped load. If a load is too long or wide, such as rolls of carpeting, make provisions with the supervisor ahead of time.

- Always be on the alert for employees on foot, visitors, other pieces of power equipment, emergency equipment, and tight quarters. Give other pieces of power equipment as much room as possible.

General Safety Guidelines / 85

- Do not tailgate other equipment. A good rule of thumb is to keep three to five lengths behind the lift truck in front of you. Keep alert to their stopping or turning. Brake lights on all vehicles improve operator safety.

- Travel up or down any grade slowly. Any grade can be hazardous.

- Never indulge in stunt driving or horseplay.

- When pulling into an elevator, approach and enter slowly with the forks first. The lift truck should keep centered on the elevator. Once in place, place the lift truck controls in neutral, lower the load to the floor, turn off the power, and set the brake. (Elevators should be inspected on a regular basis depending on federal, state, or local requirements. It is recommended that at least an annual inspection and certification take place through an agency that is licensed to do so.)

- When approaching a ramp, keep the load upgrade. This keeps it against the mast and backrest for stability.

- When traveling down a ramp, keep the load upgrade for the same purpose—for greater stability. It is recommended that empty lift trucks be operated with the forks downgrade; even for those pieces of power equipment that contain attachments. Figure 6.3 illustrates the safe means of ramp use.

Figure 6.3. Properly Traveling on a Ramp

Use the proper procedures for traveling on a ramp—with or without a load. These guidelines affect both safety and the lift truck (Courtesy of Clark Material Handling Company).

86 / Forklift Safety

- Never operate a lift truck if a leak in the fuel system exists. Have the leak fixed.

- In the event of a chemical spill, know where the appropriate cleanup kit is stored. Handle all cleanup situations per safety guidelines. Wear the appropriate PPE. Properly dispose of waste per EPA guidelines.

- Ensure that the forks of the lift truck are long enough to safely handle the load. Forks should be at least 2/3 the length of the load being moved.

- Loads should be made as stable as possible before moving bags, boxes and other individual items; they should be cross-stacked for greater stability. If possible, shrink-wrap pallets to allow for greater safety when transporting, handling, or storing. Figure 6.4 reminds operators to keep materials properly stacked.

Figure 6.4. Proper Stacking of Product

Safe stacking of product will increase safety and reduce the time lost from re-stacking (Courtesy of the National Safety Council).

- Never turn on a ramp or slope; always travel straight up or down the incline. Once you are on level surfaces, turns become much safer.

- Never make turns while a load is elevated because the lift truck can tip over. The higher the load is raised, the greater probability that the stability of the lift truck is effected.

General Safety Guidelines / 87

- When lifting a load, tilt it back for greater stability. Loads should always be against the back rest. Tilting the load allows the load to be cradled. It's more difficult to spill a load after it has been properly tilted. Ensure that the back rest is in place. Figure 6.5 provides a graphic of the required back rest.

Figure 6.5. Back Rest and Forks

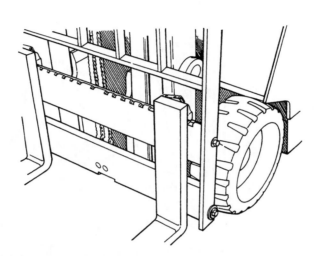

Operators are to inspect for the conditions of forks and the operator's protective back rest. The back rest keeps product from falling on the operator (Courtesy of Clark Material Handling Company).

- Never drive up to anyone who is standing in front of a machine or other object. Should the operator make an error, the person would easily be pinned by the lift truck. Alert everyone near the load that it is about to be spotted; have them stand clear.

- Always keep your arms, hands, legs, feet, and head within the confines of the lift truck. It is not unusual for an operator to be moving a lift truck and have a foot, arm, or head make contact with another object.

- Obey traffic rules as required by your employer and as identified by regulations. Be sure to yield the right of way to other vehicles where required.

- Be sure your lift truck is properly rated for the areas in which it may be traveling. Guidelines for fire and explosion are outlined in the OSHA 1910.178 rules, shown in Appendix D.

- When coming to a crosswalk, slow down and be prepared to stop. Using the horn to alert other operators and pedestrians adds to improved safety. Use caution when approaching or proceeding through any blind spot.

- Always look in the direction of travel. There may be times when the operator won't be able to look over the top of the load for safe travel; therefore travel in reverse. Figure 6.6 illustrates the proper method of traveling with the load in reverse.

Figure 6.6. Operating in Reverse

When an operator cannot see over a load, travel in reverse (Courtesy of Clark Material Handling Company).

- When traveling, always operate at a speed that allows you to have full control of the equipment. Bring the lift truck to a safe stop in a safe manner. Stop without damaging or spilling the product being handled.

- Slippery floors are invitations to accidents if the operator does not slow down.

- If, while traveling through a facility, you see a block of wood or other obstacle or hazard in the path of travel, stop, get off, and remove the hazard. Just steering around a hazard allows it to remain a hazard for the next operator. Figure 6.7 is an example of a clear and clean aisle way with materials safely stacked.

Figure 6.7. Clear and Clean Aisle Ways

Product should be safely stacked, aisles should be clean and clear of obstructions.

- Reduce speed when making turns. Anticipate obstructions or personnel when making blind corners. Keep loads as low as possible when transporting. Figure 6.8 illustrates the hazards in not reducing speed at turns.

- Accept no riders on your power equipment. Never allow anyone to ride or be lifted on your forks.

- Never use your lift truck to lift another lift truck. Never tow trailers, railroad cars, or other power equipment.

- Powered equipment should be equipped with an all-purpose (ABC) fire extinguisher if the equipment travels where there are no extinguishers available. There should be only 50 to 75 feet of travel to an extinguisher if an operator needs one. The nature of the hazards in the facility will determine the distance to an extinguisher. Local fire codes can help determine this.

- Great care must be exercised when lifting a load to deposit it on racking. Be sure the lift truck is not in motion when any load is being raised or lowered. Once the load has been made level up to the spot where it will be deposited, tilt forward, then lower the forks to deposit the load.

Figure 6.8. Reduce Speed at Turns

Reducing speed during turns will prevent many accidents (Courtesy of the National Safety Council).

- A lift truck can tip over when empty. Keep the mast low when traveling. Avoid ruts, holes, or obstacles. Make turns at slow speeds. Never turn on a ramp.

- Check with the manufacturer regarding the safety and application of any add-on devices, such as clamps or hoists.

- If a unit has to operate on a roadway, be sure to add flashing lights, reflectors, and a slow-moving vehicle emblem to the powered equipment. The operator and any other employee on the roadway must wear a reflective vest. Someone to direct traffic may be necessary.

Dock Safety Rules

- Before entering any truck or trailer, ensure that the wheels have been chocked to prevent movement.

- Check the condition of an empty trailer before entering. Do not enter a trailer without checking for holes or other conditions that could endanger the operator; the floor and landing wheels are necessary items to inspect.

- Ensure that the landing gear on trailers is capable of supporting the load. Place support jacks under the nose of the trailer for additional support.

- Automatic trailer locking devices are an excellent method for restraining a trailer. Operators must be sure to activate the device that locks onto the ICC bar before entering a trailer. Ensure wheels are chocked. Figure 6.9 illustrates manual chocking.

Figure 6.9. Manual Use of Wheel Chocks

Wheel chocks are an absolute for every trailer or truck being entered.

- Ensure there is sufficient light at the dock area for productivity, as well as safety.
- Edges of docks should be painted yellow to serve as a reminder to operators.
- Properly maintain dock systems such as levelers, dock plates, ramps, and weather shelters.
- Dock areas should be kept dry, clean, and clear of debris. Non-skid paint will help prevent skidding and provide more traction.
- Keep clear of dock edges and always remember the rear-end swing of the lift truck.

- Positive protection shall be provided to prevent railroad cars from being moved while dockloads or bridgeplates are in position.

- Lift trucks should be equipped with auxiliary lights to help in overall safety. Dock safety lights also help in trailer access/exit safety.

- Don't allow a gas or diesel lift truck to run in a trailer or confined space because of carbon monoxide exposure; always shut off engines in the trailer.

- Ensure a dock plate has been properly set between the dock and trailer. Portable dock plates should be handled by at least two people or with the aid of a lift truck.

- When cleaning under dock plates, be sure to properly rope off and identify the area. Place braces under the elevated dock plate so it won't fall down on anyone.

- Allow for stairs with approved railings, or a fixed ladder for access to dock wells. Do not jump off of docks, rather, use the stairs or ladder.

- Dock plates and ramps should be of the proper rated capacity; check manufacturer guidelines.

- Ensure the trailer is properly docked and aligned before placing wheel chocks and entering it. Stand clear of backing trailers. Stand clear of trailer opening doors, you could be struck by falling product.

- At docks, as well as other parts of a facility, be sure you know what to do in the event of an emergency. Know where to sound an alarm or summon help for medical, fire, or chemical emergency.

Battery Charging Area

- Powered industrial trucks need power from batteries, gasoline, diesel fuel, or propane. Each of these pose a specific hazard to the operator or to other employees. The battery charging installation shall be located in areas designated for that purpose. Batteries pose many dangers to all employees. Batteries are used in approximately 65% of all lift trucks.

- All batteries are heavy. Even the typical 6 or 12 volt battery for the family car is difficult to handle. Batteries for powered equipment weigh hundreds of pounds. When pushing or pulling a battery in or out of a vehicle, it is best to use a mechanical device. When a manual method of handling is used, back injuries and pinched fingers are usually the result.

- Battery acid and corrosion on the battery are harmful to the skin and eyes. Rubber gloves, rubber aprons, face shields, and goggles are a must for personal protection. Rubber boots with built-in steel-toes and metatarsal protection can be used for added protection.

- Emergency spill units should be nearby and all employees should be trained in the use of these units. Where acid is spilled, a neutralizer should be sprinkled on the acid. The neutralizer acts chemically to render the acid harmless. In some cases, the neutralizer changes color to alert the employee that the spill is safe to clean up. Simply place all of the neutralizer into a scoop and place it in a container for proper disposal following EPA guidelines. Of course, all of the required PPE is to be used.

- Where batteries must be periodically filled with water, an automatic fill device is the safest method. Where manual means are necessary, the wearing of PPE is important. Water should not be added until after the battery has been charged. If water and acid spill or overflows, proper cleanup is necessary. Figure 6.10 illustrates the proper use of PPE at a battery charging station.

Figure 6.10. Battery Charging with the Proper PPE

Note that the operator is wearing all of the appropriate personal protective equipment (PPE).

- Keep tools and other metallic objects away from the top of uncovered batteries.
- A "No Smoking" rule must prevail. Batteries contain hydrogen gas, which is explosive. Open flames, sparks, or any other ignition source must be avoided.
- When batteries need maintenance, many organizations use an outside service for this purpose. Batteries may have to be cleaned and serviced and the cells may have to be replaced from time to time.

- Should a fire occur in a battery, the best extinguisher is a carbon dioxide (CO_2) or a multi-purpose unit (ABC).

- Battery chargers should be numbered or labeled to correspond with the appropriate disconnect switches. In event of an emergency, the switch can be moved to the "off" position to stop the electrical current.

- Eye wash units are a must by battery chargers. The best units allow for continuous flushing of the eyes for at least 15 minutes. Employees should be trained in the use of eyewash units. A job hazard analysis on how to properly use eyewash units should be developed, reviewed periodically, and placed by the unit.

- The battery charging machines should be maintained per manufacturers' guidelines. Batteries should be maintained per manufacturers' guidelines.

- The disconnect charger plugs should be on elastic cord or take up reel. When operators disconnect the male plug on the charger from the female plug on the lift truck battery, the electrical cord could be damaged. It's not unusual for the cord to be left on the floor only to have a lift truck run over the cord and damage it.

- Place "No Smoking" signs by all battery changing areas.

- When handling electrolyte, provide a carboy tilter or siphon. Do not attempt to add sulferic acid to a battery.

- A placard or poster from the battery company should be posted by battery changing areas. The battery care and charging rules should be on the poster.

- Areas where charging and maintenance is to take place should be specifically designed for the purpose. The floor should be non-skid, because of the potential of water and battery acid/electrolyte being on the floor.

- Where hoists or cranes are used to lift the batteries, the lifting device should be designed for the job. Hard hats may be necessary when working around the crane. Do not work under a crane. A job hazard analysis should be completed for this job. In addition, pinch points exist that could easily crush a hands or fingers. Wear the proper PPE and use safe hoist handling procedures when handling batteries, as illustrated in Figure 6.11.

- If battery retainer plates are used to secure a battery, they must always be replaced properly after charging. Also, it's possible that the retainer plates may have acid or corrosion on them. PPE is necessary for protection.

- Batteries are to be disconnected before making any repairs on the electrical system.

Gas Powered Vehicles

- Liquid petroleum gas (LPG) is used in approximately 25% of powered equipment in industry. Safety concerns for LPG are to be taken seriously. See NFPA #58 on the safe storage and handling of liquified petroleum gases.

Figure 6.11. Safe Battery Handling with a Hoist

Batteries are very heavy and must be safely handled by a crane or hoist. PPE is very important (Courtesy of Clark Material Handling Company).

- LPG is heavier than air and will settle in low areas. The gas is flammable and could be explosive. A "No Smoking" rule must be strictly enforced during refilling.

- Place "No Smoking" signs where all LPG is stored and handled.

- A job hazard analysis should be performed on refilling and handling LPG.

- A CO^2 fire extinguisher or a multi-purpose ABC extinguisher should be available at all storage and filling areas.

- Wearing the proper PPE is essential when handling LPG. The gas can be very cold and cause damage to the skin and eyes. Heavy duty gloves, face shields or goggles are needed.

- A full LPG tank weighs almost 40 pounds and can be very awkward and clumsy to lift. Use proper lifting techniques or get help in handling cylinders.

- Ensure LPG tanks are properly anchored to the lift truck. Have defective hold-down straps repaired as needed.

- When changing LPG tanks always check the "O" ring on the LPG tank to ensure it is not damaged and meets manufacturer's guidelines. The "O" ring must be seated properly so the fuel doesn't leak.

Gasoline and Diesel Fuel

- Gasoline is highly flammable; it can cause fire or explosion if not handled properly. The "no smoking" rule is paramount when pumping or handling gasoline. The storage and handling of liquid fuels shall be in accordance with NFPA #30.

- Place proper "No Smoking" signs where gasoline is handled or stored.

- Ensure the proper grounding and bonding of all storage tanks and containers to prevent sparking or ignition of fuel vapors.

- If portable containers are used for refueling, be sure they are UL/FM approved. Ensure flame arrestors are intact within the container.

- Properly identify all containers for content.

- Diesel fuel has a higher flash point than gasoline and is more difficult to ignite. However, when heated or pressurized the liquid burns as readily as gasoline. Proper storage is outside and as far away from any building as possible. Follow all fire and OSHA regulations for storage.

Maintenance

- Operators are to check their pieces of equipment before each shift and report any defect to the supervisor or the maintenance department.

- A detailed checklist is to be used by operators before each shift starts. A system must be developed that allows for the daily evaluation of the checksheets so that proper repair can take place. Management must design a system that ensures that the forms are being used by operators and followed up by maintenance. Note the use of a clipboard and inspection form in Figure 6.12.

- Where employees have to be lifted to repair sprinklers, change lights, or perform any other maintenance work, the worker must be in an approved lifting cage. The cage must have a proper handrail, toe board, and a secure means of locking the gate. The lift cage must safely attach to the mast of the lift truck. Forks on the lift truck must be as wide as possible. The person operating the lift truck must not leave the lift truck; this ensures that the person in the lift cage can give immediate instructions to the operator.

- Never drive the lift truck with the safety platform elevated. The safest method is to drive to the selected locations and then allow the person to enter the cage.

- The person in the cage must be wearing the proper PPE.

- The employee in the lift cage should be wearing an approved body harness. Training for the use of the equipment should be provided by the manufacturer's representative.

- All repairs are to be made by authorized personnel.

Figure 6.12. Operator Inspecting Lift Truck

Operators must perform a thorough inspection of their equipment.

- No repairs are to be made in Class I, II, and III locations. Any repairs in fuel systems which involve fire hazards shall be made in locations designated for such repairs.

Tire Safety

Modern Materials Handling Management magazine lists six steps to eliminate any tire-related safety problems:

- For *traction-treaded cushion tires*, replace the tire when the tread pattern is worn to the point of disappearing. For *plain-tread cushion tires*, replace the tire when wear reaches the wear indicator (this is often the top of the brand identification lettering on the side of the tire). If a cushion tire is damaged to the extent that it causes excessive vibration, replace it.
- For pneumatic tires, replace the tire when the tread pattern disappears.

- To operate safely, pneumatic tires must be kept properly inflated. Because pneumatic lift truck tires are rated for high loads, they usually require high air pressure—typically 125 psi. If your maintenance department cannot keep tires inflated at these pressures, consider contracting the job out to a tire service company. High air pressure also makes tire mounting a dangerous job. Again, your facility may wish to contract this out to tire professionals. Use an approved safety cage for any tire work.

- If lift truck tires go through oil or chemicals, switch to oil-resistant tire compounds in order to ensure safe operating characteristics and increase tire service life.

- Make tire inspection an integral part of the preventive maintenance plan.

- Select the best tire for each application. The right tire will be determined by a combination of tread design, tire compounds, tire construction, and the particular application.

All parts which require replacement shall be replaced only by those parts that have equivalent safety as those used in the original design.

General Safety Requirements (1910.178)

- All new powered industrial trucks acquired shall meet the design and construction requirements established by Part II, ANSI, B56.1, except for vehicles intended primarily for earth-moving or over-the-road hauling.

- Approved trucks shall bear a label or other identifying mark indicating approval by the testing laboratory.

- Modifications and additions which affect lift truck or powered industrial truck capacity and safe operation shall not be performed by the user without the prior written approval of the manufacturer. Capacity, operation, and maintenance instruction plates, tags, or decals shall be changed accordingly.

- All name plates and markings must be in place, visible to the operator, and maintained in a legible condition.

- If the truck is equipped with front-end attachments other than those installed by a factory, the user shall request that the truck be marked to identify the attachments, it should also show the approximate weight of the truck along with the attachment combination at maximum elevation with the load laterally centered. Check with the manufacturer for assistance.

- High lift rider trucks shall be equipped with an overhead guard manufactured per the manufacturer's written guidelines. A backrest extension, which prevents product or loads from falling back onto the operator must be in place where loads are lifted high enough to pose a hazard. An overhead guard is not designed to stop a falling capacity load but to protect the operator from small packages, boxes, bagged material, etc.

Operators

- Only trained and authorized operators shall be permitted to operate a powered industrial truck. Methods (such as those in this book) shall be devised to train operators in the safe operation of all powered industrial trucks.

- When a powered industrial truck is left unattended, the forks are to be lowered, controls neutralized, power shut off, brakes set, and key removed.

- If stopped or parked on an incline, the lift truck's wheels are to be chocked.

- OSHA considers a truck unattended when the operator is 25 or more feet from the vehicle or the vehicle is not within their vision.

- No one should be allowed to work or walk under elevated forks; this rules applies whether the forks are empty or loaded.

- Operators must not allow anyone to ride on their powered industrial truck. One's arms, legs, and head must not be placed between the uprights of the mast or outside the running lines of the truck.

- If an operator discovers that a piece of equipment is in need of repair or defective in any way, it shall be taken out of service and management is to be notified.

- Operators shall complete a physical examination of the truck before putting it into service at the start of each shift.

- Always come to a complete stop before getting off of the piece of power equipment.

- When parked by a railroad track keep at least 9 feet from the track.

SUMMARY

It should be apparent that this chapter contains an abundance of operator driving tips, rules, and regulations. This is a significant amount of material to learn and absorb by any student or operator. It is not possible to identify every single rule pertaining to lift trucks or other powered equipment. Suffice it to say that some of the material contained in this chapter may go beyond that offered by the OSHA 1910.178 standard.

A complete printout of the standard, including those references to the types of trucks to use in various hazardous atmospheres and locations, is included in Appendix D.

A new lift truck operator may struggle with many of the rules, not necessarily because of their complexity, but because there are so many. This is another example of the fallacy of comparing a lift truck to a car. The powered equipment operator must know and understand much more information than the operator of a car.

REFERENCES

Clark Equipment Corporation. *Employers' Guide to Material Handling Safety,* Lexington, KY, 1990.

Fernold, D. "Propane Gas: Handle with Care," *Plant Engineering Magazine,* April 1996, pp. 79-82.

Gould, L. "Are Your Lift Truck Tires Safe?" *Modern Materials Handling,* August 1990, p. 69.

National Safety Council. "Powered Industrial Trucks," *Accident Prevention Manual,* Itasca, IL, 1992, pp. 155-176.

Occupational Safety and Health Administration. OSHA 1910.178, Powered Industrial Trucks.

Occupational Safety and Health Administration. *OSHA Outline for Training Powered Industrial Truck Operators,* DHEW Publication No. 78, 1990.

Schneider, D. "Evaluating Propane-Powered Lift Trucks," *Plant Engineering Magazine,* July 10, 1995, pp. 66-68.

Schwind, G. "LP Gas Requires Common-Sense Training," *Material Handling Engineering Magazine,* June 1996, p.18.

"Smart Company Makes Move to CNG," *Natural Gas Applications Industry Magazine,* Fall 1995, pp. A11-A12.

Upp, P. "Lifting the Illusions," *Warehousing Management Magazine,* May/June 1996, pp. 43-45.

Van Doren, J. "Comparing Motorized Hand Trucks," *Plant Engineering Magazine,* June 5, 1995, pp. 62-64.

7

SAFETY AT THE DOCK

Docks can be very dangerous. It's been estimated that between 10% to 25% of all industrial accidents occur at shipping or receiving docks. Docks are most times congested with people, product and machines. Weather conditions can also be a big factor.

With all of these variables in mind, it's important to plan and design for a safe dock, as well as one that is productive. If the design and building of a new dock are being considered, there are specific plans that should be made on size, shape of driveway, weather conditions, and other safeguards.

For those docks that currently exist, and there are many thousands of them, changes can be made to improve safety and efficiency. This one area of a plant or warehouse that may be considered, dangerous can be made safer; but this will take effort and changes will have to be made.

When one considers that the function of a dock is to provide an area where trailers or other vehicles will be backed in to receive or discharge freight, this can have significant implications. Employees can fall off docks, they can be struck by a piece of power equipment, they can drive their forklift off of the dock. A trailer can be pulled away by a truck when it isn't supposed to be moved. A trailer could drift or creep away from the dock as it is being loaded or unloaded. Pedestrians can be harmed at the dock. Loads could fall on customers who are there to pick up products. A piece of power equipment could strike a customer or truck driver visiting the plant.

To begin, a large area in front of a receiving or shipping dock is important so that trailers have the proper space to maneuver into the docking zone. An area at least three feet on each side of the trailer dock should be available. Being that most over-the-road semi-trailer trucks have a length of 65 feet, the area in front of the dock has to have at least this much space for backing and pulling out. Consideration also has to be made for right-angle turns. The outside turn radius must be at least 55 feet with an inside turn radius of at least 25 feet. These dimensions should properly accommodate 65 foot trailers.

In April 1983, the U.S. government began allowing larger trailers on the highways. Trailers as wide as 102 inches are now allowed. Many trailer manufacturers are building these larger units. With the expansion of the trailer, dock doors, seals, and shelters now have to be enlarged to accommodate these larger trailers. Dock safety access must be considered for these larger units. If a dock project is being planned, a contact should be made with a reputable manufacturer of dock equipment so the proper specifications are adopted for maximum dock utilization as well as safety.

It stands to reason that shorter trucks would require less room than longer trucks when backing or turning into a dock. It's important to allow as much room as possible to maneuver the large trailers. Larger turn areas allow for greater safety.

The slope of a ramp is also a very important consideration. A slight sloping driveway is ideal, this allows for the dock height to be about two inches below the average height of a truck bed. Being that a full trailer will be lower when parked at the dock and higher when empty, dock height is important. The design of a ramp slope allows for a 3% grade to be ideal. Slopes of 3% to 6% are considered practical. A slope that goes up to a 10% grade is near the limit. Backing in or pulling out of steep slopes have a bearing on wheel traction.

Consideration for Weather Conditions at Dock Areas Is Very Important

Weather is a serious consideration at the dock because it can effect productivity and the safety of the workers. Weather seals around dock doors keep elements out of the building. Many of these shelters extend around the two sides and top of a dock area. Where a lower section of the shelter meets the dock area, signs will have to be posted to alert employees of the hazards of stepping onto these devices. They are not intended to carry the weight of a person, thus, someone could easily fall through and become injured. There is also the danger of being in the area of the shelter when a trailer is being backed into the building. A fall at this time could be life-threatening.

Weather is a factor to consider on the dock as well. Employees must usually get down from the dock to the spotted trailer to check for wheel chocks, the condition of the trailer, and landing wheels. Ice, snow, or rain can pose risks that are unexpected. Safe access from the dock should be provided by an approved ladder—one that meets OSHA specifications. Stairs with handrail and non-skid steps would also be needed. Jumping off of the dock must be discouraged for obvious reasons. Proper signage, alerting employees to the hazards of jumping from the dock, would be beneficial. Have a de-icer available to use on walking surfaces.

Safety at the Dock / 103

DOCK SAFETY

Wheel Chocks/Truck Restraints

Dock safety involves a safe and secure trailer. OSHA regulations require that the trailer be properly chocked or restrained while being used to accommodate employees and equipment. Through the years, there have been many serious accidents and fatalities which were the result of trailers being pulled from the dock or slowly creeping from the dock. The unwary and unsuspecting lift truck operator suddenly finds himself falling from the dock. The end result can be a badly shaken up operator, a damaged lift truck, damaged product, or a serious injury or death. Figure 7.1 illustrates the consequences of an unchocked trailer.

Figure 7.1. Consequences of Not Chocking Wheels

The consequences of not chocking trailer wheels can be devastating (Courtesy of Rite-Hite Corporation).

104 / Forklift Safety

All of this can be prevented by keeping the trailer in place. At times, this is not an easy thing to do. The responsibility of who shall chock the wheels can result in no one chocking the wheels. Is the semi-driver responsible or is the lift truck operator responsible? Some trucking firms have skilled operators who know that when they spot a trailer, the wheels are to be chocked. When a trucker frequents the same site on a regular basis, this habit of wheel chocking can be reinforced each time. When different semi-drivers frequent the same site, the probability of 100% wheel chock use will be questionable.

The forklift operator is required to ensure the trailer is chocked before entering it. This rule allows for anyone to chock the wheels, but the operators must personally inspect the situation because their lives are on the line.

The ladder and the steps to the dock well area are important, because access to these points must be available so wheels can be chocked. Employees may balk at going outdoors in inclement weather to chock the wheels. The dock may be covered with snow or ice. The dock well area, which is usually sloped, may have poor drainage. The employee might look down at six inches of standing water and decide not to walk in ankle-deep water to place a wheel chock. Also, wheel chocks on snow or ice have reduced or no traction. All of these are considerations to be made regarding manual wheel chocking.

Automatic Trailer Restraint Devices

Several years ago, the age of technology introduced truck restraints to dock safety. A truck restraint is a mechanical device that, when activated, allows for a bar or similar device to be placed against the ICC bar (Interstate Commerce Commission). Almost all of the over-the road-trailers are equipped with this bar. When the trailer is backed up to the dock, the device is self-activated with a sensor or a button is pushed inside the building. Figure 7.2 illustrates the functioning of an automatic trailer restraint.

Figure 7.2. Automatic Trailer Restraint

Automatic trailer restraints are an excellent means of safeguarding operators (Courtesy of Rite-Hite Corporation).

ACCIDENT FACT

A lift truck operator attempted to enter a trailer at a dock while the red light was flashing for the automatic truck restraint. At the moment he was entering the trailer, it pulled away from the dock. His lift truck fell into the hole created between the dock and trailer. He jumped clear and avoided serious injury. He had been trained to respect the green or red lights which give proper warning on when to enter the trailer. Safety at the dock must be followed even when modern devices, such as truck restraints, are used.

The left side of the dock on the inside of the building is equipped with a control panel for activation as well as a red, green, and amber light. A flashing green light reminds the operator that the device is engaged and the trailer is secure. A flashing red light indicates that the device is not activated and the trailer must not be entered. Signage is also present, which allows for increased safety awareness.

To the semi-driver who has backed in the trailer, the rear view mirror is used for guidance. When looking into the mirror, the wheel chock signs and restraint device signs are printed in reverse. In the mirror, the signs are now legible. When backing in to remove a trailer, the flashing green light would be on outside the building to alert the driver that it is okay to pull the trailer away after hooking on to it. A flashing red light indicates that the device is hooked onto the ICC bar; and the operator will be unable to pull the trailer away without causing some damage to the trailer. The restraint devices have a give-point, but not until significant pull has been exerted on them to cause this. Figure 7.3 provides an illustration of a backing trailer and the appropriate signs.

The device is always activated from the inside of the building and this is the responsibility of the operator. Training should take place to allow employees to properly use the device, as well as to alert them to their personal safety. Manufacturers of these devices can provide training to all employees. This would include proper care, maintenance, use, and all safeguards for the equipment.

Regarding the inclement weather for wheel chocking, trailer restraints help the employee or semi-driver avoid the bad weather. The device would hold the trailer in place despite the slippery surface or ankle-deep water in the dock well. For maximum safety, the trailer restraints provide a very big margin of safety for lift truck operators. It is not unusual, however, to hear about an organization which uses both trailer restraints and wheel chocks.

Manual wheel chocks can be enhanced by attaching a rod on top of the wheel chock. The rod bends up at 90° and has a small square attached to it. It gives the appearance of a flag sticking out of the wheel chock. The benefit of this feature is that when painted a fluorescent red or orange, it provides for greater visibility to the semi-driver when backing a truck in, or to the employee at the dock who is looking down at the wheels to see if they are chocked. Note the chock, flag, and chain illustrated in Figure 7.4.

106 / Forklift Safety

Figure 7.3. Trailer Backing and Signage

When appropriate signs are mounted outside the building, the trailer truck operator can read the reverse-printed signs while backing (Courtesy of Rite-Hite Corporation).

Figure 7.4. Wheel Chock with Flag

A metal flag welded onto a metal chock allows for greater visibility and easier handling.

Another new safety device on the market is the wheel restraint. When the trailer can't be restrained with a hook-type ICC bar restraint, or when a trailer doesn't have an ICC bar, the new wheel restraint can be used. The equipment works this way:

- A truck backs up to a loading dock that is equipped with a new wheel restraint.
- The truck backs over the low-profile "sled" in its stored position.
- When the sled makes contact with the tires, it is activated. Dual, steel barriers rise up like an open clam shell; these then position themselves against the truck's rear wheels.
- When they are fully engaged, these barriers can secure a trailer—even when empty, they can secure against a 32,000 pound pull.

Because of its low profile, the device will not interfere with low hanging parts of the trailer. The unit is powered by a screw-drive mechanism.

Dock Plates

Inside the building, the dock plates provide the bridge or ramp to the trailer bed once the doors have been opened and the trailer is in place. The dock plate is also classified as a dock leveler. The device offers the bridge between the trailer floor and the dock. Levelers come in several sizes, but most are 6 to 10 feet wide and 8 to 12 feet in length. The need for specific sizes are best determined by consulting the manufacturers who install them. Devices are also made to spring up at the end of a dock plate to keep a lift truck from driving off the dock plate into the dock well. Yellow or yellow with black stripes around the perimeter of a dock plate gives the powered equipment operator an advantage because of the added visual perception. When the dock edge is painted yellow, this also helps in alerting operators to the hazard.

In the past, many dock plates were of the portable design and were put in place on the back of the trailer either by manually lifting them in place or allowing for a forklift to maneuver them in place. Manual dock plates are still being used in many locations. With this in mind, care must be exercised when manually lifting a dock plate. The device is clumsy and can be heavy. At least two people must use proper lifting techniques and gloves when lifting. If possible, a lift truck should first lift the dock plate and get it as close to the dock edge before it has to be manually placed. If the lift truck can lift the dock plate and set it in place without any manual assistance, this is obviously an ideal method to avoid injury. Keep in mind that portable dock plates can easily slip as a result of the lift truck movement, if not anchored properly between the dock and trailer. Dock plate surfaces should be of a non-skid design. Figure 7.5 illustrates the proper handling of a portable dock plate.

The portable dock plate has almost been replaced by the dock leveler. Capacity of the leveler should be checked before installation to ensure that lift truck, operator, and the load being carried are safe. The lip on the front of the leveler must be capable of safeguarding the operator during lift truck operation. The length of the leveler will determine the ramp grade on which the load will be carried. This grade should be as small as possible to allow for safety.

108 / Forklift Safety

Figure 7.5. Handling a Portable Dock Plate

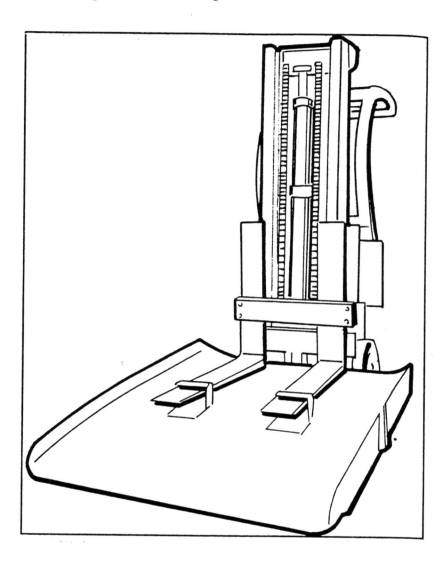

Use a lift truck to place a portable dock plate into place rather than manually handling it (Courtesy of the National Safety Council).

According to the Rite-Hite Corporation, electric pallet trucks being used on dock plates should not exceed a 7 percent dock plate slope. Electric lift trucks should not exceed a 10 percent slope. Gas-powered lift trucks should not exceed a 15 percent slope. The proper leveler for the job depends on the height difference of the trailer bed and dock; this is coupled with the length of the leveler itself.

Figure 7.6. Dock Slope Guidelines

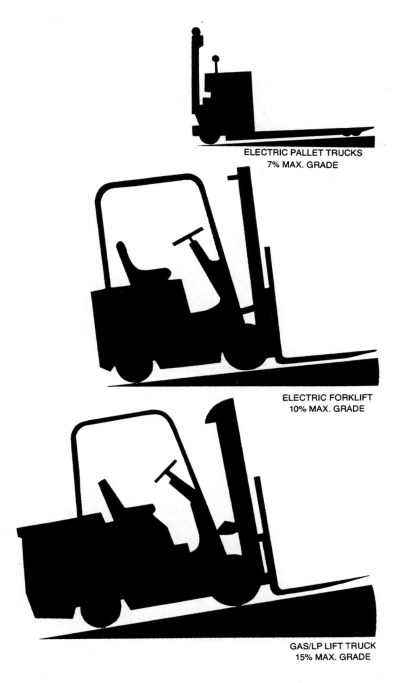

ELECTRIC PALLET TRUCKS
7% MAX. GRADE

ELECTRIC FORKLIFT
10% MAX. GRADE

GAS/LP LIFT TRUCK
15% MAX. GRADE

There are appropriate guidelines for dock plate slope; note the maximum grades (Courtesy of Rite-Hite Corporation).

Figure 7.7. Dock Safety Checklist

Date _____

Company Name _____

Plant Name _____ Plant Location _____

Dock examined/Door numbers _____

Company representative completing checklist:

Name _____ Title _____

A. Vehicles/Traffic Control

1. Do forklifts have the following safety equipment?
   - ☐ Seat belt   ☐ Load backrest
   - ☐ Headlight ☐ Backup alarm
   - ☐ Horn   ☐ Overhead guard
   - ☐ Tilt indicator
   - ☐ On-board fire extinguisher
   - ☐ Other _____

2. Are the following in use? (Yes/No)
   - Driver candidate screening ☐ ☐
   - Driver training/licensing ☐ ☐
   - Periodic driver retraining ☐ ☐
   - Vehicle maintenance records ☐ ☐
   - Written vehicle safety rules ☐ ☐

3. Is the dock kept clear of loads of materials? ☐ ☐

4. Are there convex mirrors at blind corners? ☐ ☐

5. Is forklift cross traffic over dock levelers restricted? ☐ ☐

6. Is pedestrian traffic restricted in the dock area? ☐ ☐

7. Is there a clearly marked pedestrian walkway? ☐ ☐

8. Are guardrails used to define the pedestrian walkway? ☐ ☐

9. Comments/recommendations
   _____
   _____
   _____
   _____
   _____

B. Vehicle restraining

1. If vehicle restraints are used: (Yes/No)
   a. Are all dock workers trained in the use of the restraints? ☐ ☐
   b. Are the restraints used consistently? ☐ ☐
   c. Are there warning signs and lights inside and out to tell when a trailer is secured and when it is not? ☐ ☐
   d. Are the outdoor signal lights clearly visible even in fog or bright sunlight? ☐ ☐
   e. Can the restraints secure trailers regardless of the height of their ICC bars? ☐ ☐
   f. Can the restraints secure all trailers with I-beam, round, or other common ICC bar shapes? ☐ ☐
   g. Does the restraint sound an alarm when a trailer cannot be secured because its ICC bar is missing or out of place? ☐ ☐

A comprehensive inspection of the dock area is important for operator and pedestrian safety (Courtesy of the National Safety Council and Rite-Hite Corporation).

Figure 7.7 (continued)

h. Are dock personnel specifically trained to watch for trailers with unusual rear-end assemblies (e.g. sloping steel back plates and hydraulic tailgates) that can cause a restraint to give a signal that the trailer is engaged even when it is not safely engaged? ☐ ☐
i. Are dock personnel specifically trained to observe the safe engagement of all unusual rear-end assemblies? ☐ ☐
j. Do the restraints receive regular planned maintenance? ☐ ☐
k. Do restraints need repairs or replacement? (List door numbers.)

l. Comments/recommendations

2. If wheel chocks are used: Yes No
   a. Are dock workers, rather than truckers, responsible for placing chocks? ☐ ☐
   b. Are all dock workers trained in proper chocking procedures? ☐ ☐
   c. Are chocks of suitable design and construction? ☐ ☐
   d. Are there two chocks for each position? ☐ ☐
   e. Are all trailers chocked on both sides? ☐ ☐
   f. Are chocks chained to the building? ☐ ☐
   g. Are warning signs in use? ☐ ☐
   h. Are driveways kept clear of ice and snow to help keep chocks from slipping? ☐ ☐
   i. Comments/recommendations

C. Dock levelers
   Yes No
1. Are the dock levelers working properly? ☐ ☐

2. Are levelers long enough to provide a gentle grade into trailers of all heights? ☐ ☐
3. Is leveler width adequate when servicing wider trailers? ☐ ☐
4. Do platform width and configuration allow safe handling of end loads? ☐ ☐
5. Is leveler capacity adequate given typical load weights, lift truck speeds, ramp inclines and frequency of use? ☐ ☐
6. Do levelers have the following safety features?
   Working-range toe guards ☐ ☐
   Full-range toe guards ☐ ☐
   Ramp free-fall protection ☐ ☐
   Automatic recycling ☐ ☐
7. Do levelers receive regular planned maintenance? ☐ ☐
8. Do levelers need repair or replacement? (List door numbers.)

9. Comments/recommendations

D. Portable dock plates
   Yes No
1. Is plate length adequate? ☐ ☐
2. Is plate capacity adequate? ☐ ☐
3. Are plates of suitable design and materials? ☐ ☐
4. Do plates have curbed sides? ☐ ☐
5. Do plates have suitable anchor stops? ☐ ☐
6. Are plates moved by lift trucks rather than by hand? ☐ ☐
7. Are plates stored away from traffic? ☐ ☐
8. Are plates inspected regularly? ☐ ☐
9. Do plates need repair or replacement? (List door numbers.)

10. Comments/recommendations

Figure 7.7 (continued)

E. Dock Doors

                                              Yes   No

1. Are doors large enough to admit all loads without obstruction? ☐ ☐
2. Are door rails protected by bumper posts? ☐ ☐
3. Do doors receive regular planned maintenance? ☐ ☐
4. Which doors (if any) need repair or replacement? _____
5. Comments/recommendations _____

F. Traffic Doors

                                                Yes   No

1. Are doors wide enough to handle all loads and minimize damage. ☐ ☐
2. Does door arrangement allow safe lift truck and pedestrian traffic? ☐ ☐
3. Are visibility and lighting adequate on both sides of all doors? ☐ ☐
4. Which doors (if any) need repair or replacement? _____
5. Comments/recommendations _____

G. Weather sealing

                                                Yes   No

1. Are the seals or shelters effective in excluding moisture and debris from the dock? ☐ ☐
2. Are seals or shelters sized to provide an effective seal against all types of trailers? ☐ ☐
3. Are seals or shelters designed so that they will not obstruct loading and unloading? ☐ ☐
4. Are dock levelers weather sealed along the sides and back? ☐ ☐
5. In addition to seals or shelters, would an air curtain solve a problem? ☐ ☐
6. Do seals or shelters need repair or replacement? (List door numbers.) _____
7. Comments/recommendations _____

H. Trailer Lifting

1. How are low-bed trailers elevated for loading/unloading?
   ☐ Wheel risers ☐ Concrete ramps
   ☐ Trailer-mounted jacks
   ☐ Truck levelers

                                                Yes   No

2. Do lifting devices provide adequate stability? ☐ ☐
3. Are trailers secured with vehicle restraints when elevated? ☐ ☐
4. Do lifting devices need repair or replacement? (List door numbers.) _____
5. Comments/recommendations _____

I. Other Considerations

                                                Yes   No

1. Dock lights
   a. Is lighting adequate inside trailers? ☐ ☐
   b. Is the lift mechanism properly shielded? ☐ ☐
2. Scissors lifts
   a. Are all appropriate workers trained in safe operating procedures? ☐ ☐
   b. Is the lift mechanism properly shielded? ☐ ☐
   c. Are guardrails and chock ramps in place and in good repair? ☐ ☐

Figure 7.7 (continued)

3. Conveyors
   a. Are all appropriate workers trained in safe operating procedures? ☐ ☐
   b. Are necessary safeguards in place to protect against pinch points, jam-ups and runaway material? ☐ ☐
   c. Are crossovers provided? ☐ ☐
   d. Are emergency stop buttons in place and properly located? ☐ ☐

4. Strapping
   a. Are proper tools available for applying strapping? ☐ ☐
   b. Do workers cut strapping using only cutters equipped with a holddown device? ☐ ☐
   c. Do workers wear hand, foot and face protection when applying and cutting strapping? ☐ ☐
   d. Are all appropriate workers trained in safe strapping techniques? ☐ ☐

5. Manual handling
   a. Is the dock designed so as to minimize manual lifting and carrying? ☐ ☐
   b. Are dock workers trained in safe lifting and manual handling techniques? ☐ ☐

6. Miscellaneous   Yes   No
   a. Are pallets regularly inspected? ☐ ☐
   b. Are dock bumpers in good repair? ☐ ☐
   c. Is the dock kept clean and free of clutter? ☐ ☐
   d. Are housekeeping inspections performed periodically? ☐ ☐
   e. Are anti-skid floor surfaces, mats or runners used where appropriate? ☐ ☐

   f. Are stairways or ladders provided for access to ground level from the dock? ☐ ☐
   g. Is the trailer landing strip in good condition? ☐ ☐
   h. Are dock approaches free of potholes or deteriorated pavement? ☐ ☐
   i. Are dock approaches and outdoor stairs kept clear of ice and snow? ☐ ☐
   j. Are dock positions marked with lines or lights for accurate trailer spotting? ☐ ☐
   k. Do all dock workers wear personal protective equipment as required by company policy? ☐ ☐
   l. Is safety training provided for all dock employees? ☐ ☐
   m. Are periodic safety refresher courses offered? ☐ ☐

J. General comments/recommendations

_____
_____
_____
_____
_____

For additional copies of this Loading Dock Safety Checklist, write to Rite-Hite Corporation, 9019 North Deerwood Drive, P.O. Box 23043, Milwaukee, WI 53223-0043.
For more information on loading dock safety, write for a complimentary copy of Rite-Hite's Dock Safety Guide.

This loading dock safety checklist is provided as a service by Rite-Hite Corporation, Milwaukee, Wis. It is intended as an aid to safety evaluation of loading dock equipment and operations. However, it is not intended as a complete guide to loading dock hazard identification. Therefore, Rite-Hite Corporation makes no guarantees as to nor assumes any liability for the sufficiency or completeness of this document. It may be necessary under particular circumstances to evaluate other dock equipment and procedures in addition to those included in the checklist.
For information on U.S. loading dock safety requirements, consult OSHA Safety and Health Standards (29 CFR 1910). In other countries, consult the applicable national or provincial occupational health and safety codes.

114 / Forklift Safety

Proper care, use, and maintenance of these devices are very important. Devices should be inspected regularly. Also, it's important to keep dock areas, dock wells, and areas below dock plates as clean as possible. A casually tossed cigarette has been known to start a fire in a dock well or under a dock plate. Smoking should obviously be controlled in these areas. Fire extinguishers should be available.

If employees have to go under a dock plate to clean it, proper safe guards must be in place. A proper job hazard analysis would be very helpful for employee safety. The area around the dock plate on the inside and outside of the building should be properly identified with yellow hazard tape, signage, and flashing lights. The dock plate should be properly secured in the upright position so that it doesn't fall on the employee. NIOSH-approved dust masks may also be required to prevent inhalation of harmful dusts when sweeping. All employees should recognize this work zone to allow for maximum employee safety.

Dock Inspections

To ensure that the dock area is safe, management should conduct comprehensive surveys of the dock. Those items that are in need of attention should be corrected as soon as possible. Figure 7.7 provides a complete dock safety check list.

Figure 7.8. Dock Safety Awareness

This graphic provides awareness of safety at the dock (Courtesy of the National Safety Council).

SUMMARY

Dock safety is a critical part of a facility's overall safety program. Statistics indicate that up to one of four injuries in industry occur at the dock. OSHA statistics indicate that fatalities, along with numerous injuries and incidents, occur at the dock.

Physical changes would have to be made to a facility to improve dock safety. Wheel chocks or trailer restraints are necessary and will help save lives.

The manufacturers of dock plates, trailer restraints, and dock enclosures, can serve a big need toward a safer dock. These organizations are the experts. Figure 7.8 provides a proper poster setting for dock safety awareness.

REFERENCES

"Creating a Safe Loading Dock," *Plant Engineering Magazine*, April 1994, pp. 86-90.

"Dock Designer Guide," Rite-Hite Corporation, Milwaukee, WI, 1986, 1987-1989.

"Dock Safety," *Materials Management Distribution Magazine*, April 1994, pp. 70-71.

Feare, T. " New Wheel Restraints Advance Goal of Accident Free Docks," *Modern Materials Handling Magazine*, February 1996, pp. 52-53.

Feare, T. "Reduce the Risks in Your Dock's Danger Zones," *Modern Materials Handling Magazine*, November 1995, pp. 44-46.

Ketchpaw, B. "The Loading Dock: Combating New Hazards," *Risk Management Magazine*, July 1986, pp. 44-48.

_____. "Making Your 'Danger Zone' Safer," *Modern Materials Handling Magazine*, 1988 Planning Guidebook, pp. 14-17.

"Safety Consideration for Loading Docks," *Plant Services Magazine*, October 1989, pp. 100-102.

Schwind, G. "Docks: Opening the Door to a Productive Plant," *Materials Handling Engineering*, November 1994, pp. 51-58.

8

SAFETY OF PEDESTRIANS

One of the most overlooked features of a plant and forklift safety program is that of pedestrian safety. Serious consideration has to be given to this part of the safety program, because of the supporting data regarding injuries to employees. OSHA has estimated that some 19% of power equipment fatalities occur to pedestrians as a result of lift truck accidents each year. This is just behind the number one cause of lift truck accidents—tip over at 25%. When a four or five ton lift truck strikes a person, there is no way that the injury will be a minor one.

By observing some basic safety practices and the aid of management involvement, cooperation, and enforcement, the safety of those on foot will be enhanced. Unfortunately, this phase of a safety program has been overlooked in the development of many lift truck safety programs. As a result, management doesn't apply the focus on the safety issue of pedestrians as the OSHA statistics would indicate.

A review of the various safeguards and regulations governing pedestrian safety is sketchy. There isn't much training or injury data to support this subject; perhaps this is one reason that the accident statistics are so high. To help ensure pedestrian safety, program items such as physical hazards, machine operating hazards, and employee awareness are necessary. The training of all employees is essential.

PHYSICAL HAZARDS

As employees or visitors are walking through a facility, physical safeguards must be in place to aid in their safety. Consider the following physical safeguards:

- Walkways which are identified with the proper yellow or white walkway lines are a must. Where powered equipment is used along with a pedestrian walkway, enough room must be provided to allow for both the power equipment and the person on foot. Even though the OSHA guidelines for the aisle width is three feet for each side of the load, more space should be considered. The width of the load being carried, as well as overall safety for this passageway, must be taken into consideration.

- Signs and warnings are another important consideration. Both lift truck operators and pedestrians must comply with stop signs that are mounted where necessity dictates. The signs should be of standard design and be mounted where they are the most obvious. Bright warning signs can be painted onto the floor at key locations. Peel-off signs or labels that depict forklift safety can also be mounted on the floor or wall. These signs can be purchased from sign manufacturers. Keep signs clean and legible.

- Flashing lights allow for a more active visual alarm for everyone. The lights should not be overdone, but be strategically placed in the facility for the greatest effect. Replace defective bulbs as soon as possible.

- Special convex or panoramic mirrors will highlight those areas that may present a blind corner or dangerous intersections. The mirrors must be strategically placed for the greatest effect. Mirrors are to be used by both the person on foot and the powered equipment operator.

- Where doors open to an aisle or area that allows powered equipment to pass, it's important to provide warning lights. A light, amber or red in color, should be installed by the doorway and connected to a trip switch that allows for the light to flash when the door is opened. This will alert operators that someone is about to enter this danger zone. An audible alarm can augment safety when installed with the flashing light. When doors are of solid construction, add a heavy-duty plastic window so anyone can look through before entering. Be sure the modification to the door meets fire codes.

- In addition to the visual and audible alarm at doors that enter into an aisle, sections of hand-rail should be erected for additional safety. As in all situations such as this, hand-rails should be painted yellow for greater visibility. A section of hand-rail directly in front of the door would require pedestrians to go left or right after exiting, rather than walking directly into the path of the vehicle traffic. Warning signs that read "caution" or contain other visual warnings would provide for greater safety.

- Devices mounted onto the floor can activate when the lift truck passes over them; they can also provide a warning with a flashing light or audible alarm. In this case, the alarm and light are activated by the operator's equipment as a means of a pre-warning for employees that there is equipment approaching.

- Hand-railings or barriers adjacent to an aisle that allows power equipment to operate provides for added employee protection. In this case, the barrier provides the safeguard needed to physically protect those on foot. Paint hand-rails yellow for greater visibility.

- Operator training can reduce or eliminate those injuries or close calls that are the result of someone physically walking into a parked lift truck. How could something that's parked harm someone? Imagine someone tripping over forks that are not flat on the floor. Forks have also been known to cause harm to someone's knees, groin, chest, face, and head as a result of the forks being left up at different heights, rather than lowered when parking or stopping. Forks must always be lowered when operating the lift truck. When parked, the forks are to be flat on the floor to provide the least amount of hazard as possible. True, someone can still trip over forks that are flat on the floor, but flat forks provide for added safety, not a zero hazard situation.

- Where possible, always park lift trucks and all powered equipment out of the way, especially forks or runners. It is a good practice to park a lift truck in such a way that provides for the forks to be hidden or covered up with product, instead of left in the open.

- Provide for separate pedestrian doors when entering other rooms or dock areas. When large doorways are used for equipment, those on foot tend to use the same doors. Separate doors that are marked and identified for employee safety should not only be provided, but used.

- Heat curtains should be clear, so that an operator could see someone on foot before driving through. These curtains are usually mounted on large doors which enter to the outdoors or dock area; they are intended to keep heat loss to a minimum. The plastic strips allow the piece of equipment and load to pass through without allowing a significant amount of heat loss. One of the problems with curtains such as these is that they tend to get dirty and scratched, thus the visibility becomes restricted. When employees are near these doorways, accident potential is high. Figure 8.1 provides a graphic illustration regarding safe entry for pedestrians.

Figure 8.1. Safe Entry/Exit Door for Employees

Provide for safe entry and exits for those on foot at large doors (Courtesy of Rite-Hite Corporation).

- Signs or labels on the inside doors that enter into an aisle that allows for lift truck traffic will provide greater awareness for employees and visitors. It's very easy to become distracted out in the areas of a facility; and every safeguard helps to reduce the margin of error by the powered equipment operators and pedestrians.

- Physical awareness as to the forklift itself can be accompanied by providing additional safeguards. Flashing lights mounted on the piece of power equipment provides for a visual warning at all times. The lights should never interfere with the operator. That is, the lights shouldn't be a distraction to vision when lifting or lowering a load, or while operating in general. The purpose of the light is to increase awareness for other pieces of power equipment and pedestrian. Each piece of equipment should be evaluated regarding the need for a flashing light.

- Lighting is an important factor in the safety of all personnel. Where power equipment is involved, the ability to see and be seen are critical. If the operating area is dark, the equipment must be equipped with operating lights. This provides for the margin of safety for pedestrians. The operator must also be able to see pedestrians throughout the facility. Lighting should be as bright as possible for everyone's benefit.

Figure 8.2. Look Before You Back

Backing accidents can injure employees and visitors. The operator must look before backing. A backup alarm can help prevent accidents (Courtesy of the National Safety Council).

- Backup alarms can provide for a special edge on counter balanced trucks in pedestrian's safety. Backup alarms should be installed in such a way that they are tamperproof. The decibel level of the alarm should be adjustable in the event it is too loud for the operation. Regulations do not require backup alarms on forklifts, but this added feature can help reduce incidents and injuries. The counterbalance lift truck is the piece of equipment that most commonly is equipped with an alarm. Figure 8.2 serves as a graphic reminder that a backing lift truck can be dangerous. It should be noted, however, that there are no definitive studies on the effectiveness of back-up alarms in relation to the prevention of injuries. Under certain circumstances, a back-up alarm can be very effective while in others it may not be as effective. OSHA regulations do not require back-up alarms.

Figure 8.3. Fork Lift Counterweight Hazards

The rear end swing of a lift truck can injure employees and damage equipment (Courtesy of the National Safety Council).

- Rearview mirrors mounted on the lift truck are another feature that can provide for improved pedestrian safety. Rearview mirrors should be the convex or panoramic type to provide for as broad an area of visibility as possible for the operator prior to backing up. With tight aisles, product being stacked around machines and aisles, blind corners, and hectic dock activity, looking in a mirror can be very helpful. Even though one of the standard operating rules is to look to both sides and behind before backing, many operators simply trust their luck and back up. To be struck by a lift truck or pinned by the rear-end swing of a lift truck is not unusual. In almost every case, the forklift operators will admit that they didn't see the victim. As with alarms, the use of mirrors should be evaluated prior to installation. They can be effective in some applications but not necessarily all applications. See Figure 8.3 which serves as a reminder of the movement of the lift truck.

- The rearview mirror becomes an asset because it is very easy to look into the mirror before backing. Even though the operator should physically look to both sides and behind before backing, the mirror provides an edge. This safeguard is inexpensive and easy to install. As with backup alarms and flashing lights, OSHA does not require mirrors. Any organization interested in accident prevention can safeguard their equipment with these devices.

- The horn on the lift truck serves as a ready-to-use alarm for warning pedestrians. Operators should ensure that it is functional prior to start up on each shift. They should use it only when necessary; the horn should not be abused. The use of a horn isn't meant to startle; it's meant to alert. Operators can overdo a good thing so employees don't pay as much attention to the sound of the horn. Where horns are used sparingly and at specific times, they tend to be most effective.

TRAINING SAFEGUARDS

The preceding section identified methods and features which could physically make the facility and pieces of powered equipment safer for pedestrians. It's important to provide these physical improvements for the welfare of everyone involved. However, if employees aren't properly trained in pedestrian safety, the potential for injury will continue to be an issue.

Operator/Employee Comments on Pedestrians

A survey was completed in a manufacturing plant of 350 people to determine if pedestrian safety was an issue. The results to management came as a surprise. In almost every case, when asking powered equipment operators for their opinion, they stated that the pedestrians were the problem in safety. Pedestrians step out in front of them, suddenly appear behind them when they are backing up; and don't obey signs, warnings, and lights. It is the pedestrian that acts unsafe, so the operators said.

Management now asked the plant employees if their personal safety was affected by the operators. Almost exclusively, the employees said that the operators drove too fast; did not slow down for them in different parts of the plant; failed to use their horns to alert the pedestrians; and sometimes parked their lift truck in front of doors, exits, passageways, or vending machines.

What's the moral to the above informal study? Possibly that both sides, pedestrians and operators, are both right, and, wrong. Much of this misunderstanding is the result of a lack of training, focus, and enforcement.

Where operators are trained on equipment safety, a session on pedestrian safety must be included. The typical hazards that are apparent between machine and pedestrian have to be discussed. Operators should be given a session with visual aids (video or slides), discussion, and a written exam on this special focus. Several of the major manufacturers of powered equipment are now including pedestrian safety in their training programs. Separate sessions, videos, and quizzes regarding safety are being used or are being developed to aid in pedestrian safety.

Pedestrian Safety

Factors for operators to be aware of should include:

- Use of warning devices to alert pedestrians including backup alarms.
- Proper inspection of horn and, where present, alarms and lights before the shift starts.
- Maintain safe operating speeds at all times.
- Allow for slow-down at intersections, aisle ways, blind corners, where doors open to an aisle, dock areas, and where lighting is impaired.
- Anticipate pedestrians stepping out in the path of travel, taking short cuts, or being oblivious to pieces of power equipment.
- Where possible, drive in the middle of an aisle; this would allow for reaction time should someone step out in front of the equipment.
- Be prepared for the movement of pedestrians when entering a building from outdoors. An adjustment to the light could take a few minutes on the part of the operator. Figure 8.4 illustrates the potential hazards when entering a building.

Figure 8.4. Fork Lift Entry From Out of Doors

When operators enter a building from out of doors they must be prepared to allow their eyes to adjust to the differences in light before proceeding (Courtesy of Clark Material Handling Company).

National Safety Council Guidelines

The National Safety Council offers the following tips for pedestrian safety:

- Never challenge a lift truck or piece of power equipment. Never challenge a loaded truck to attempt to swerve to avoid being struck. A load being carried could easily tip over or slide off of the lift truck and strike the pedestrian. Even though pedestrians have the right of way, they still have the duty to stay out of the way. Lift trucks cannot stop on a dime. By the time operators decide to apply the brake, they have already traveled a considerable distance.

- Stop and look when approaching a corner before proceeding—look both ways.

- Walk on the right side of aisles.

- Keep well away from a lift truck so you aren't pinned by the swing. Note the tight turning radius of the lift truck in Figure 8.5; this could easily pin someone.

Figure 8.5. Tight Turning Radius

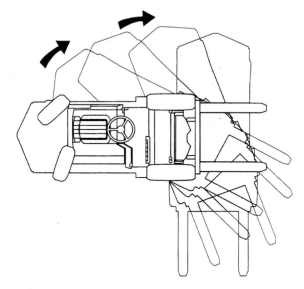

A lift truck performs tight turns. Operators must be aware of this hazard potential (Courtesy of Clark Material Handling Company).

- Unauthorized operators must not drive unless trained to do so.
- Never ride as a passenger on a piece of power equipment.

- Stay off of the tilt mechanism.
- Keep hands away from the mast and pulleys.
- Never walk or stand under the mast or load.
- Forks on a lift truck when parked pose a tripping hazard. Raised forks pose a strike against or spearing hazard.
- If on a ramp, be cautious of a lift truck; you could be pinned or struck by a load.
- Be aware of stacking procedures, you would be struck by falling loads—on both sides of the racking.
- A backing truck poses a hazard.
- Keep from cluttering aisles with hand trucks and carts. People, machinery, and tools take up space and the pedestrian could easily be struck.

SUMMARY

Safety for pedestrians is often overlooked in safety programs. Statistics show that many pedestrians are struck and killed by lift trucks each year. The number of close calls to pedestrians from lift trucks each day would be in the thousands.

A facility has to include pedestrian safety in their overall program. Proper aisle ways, markers, signs, lights, alarms, and training are needed to protect all pedestrians. This subject has to be a part of every lift truck training program.

REFERENCES

Augensten, K. "Industrial Lift Truck Accidents," *Modern Materials Handling Magazine*, January 1986, pp. 42-47.

Clark Equipment Manufacturing. *Employers Guide to Material Handling Safety*," Lexington, KY, 1990.

National Safety Council. *Rx for Plant Pedestrians, Industrial Supervisor*, January, 1981, p. 9.

National Safety Council. *Forklift Truck Operators Safety Training Program*, Itasca, IL, 1988.

Occupational Safety and Health Administration. OSHA, 1910.178, Powered Industrial Trucks.

9

PERSONAL PROTECTIVE EQUIPMENT (PPE)

Powered equipment operators are exposed to many hazards on their job. Where controls or safeguards can be provided to prevent injury, the operator is further protected. In some cases, however, the operator can't be protected from all hazards and therefore, must wear personal protective equipment (PPE).

THE OSHA PPE STANDARD

On July 5, 1994, OSHA developed a comprehensive PPE standard so that more employees were protected during the course of their job. These specific requirements can be found in 1910.132 to 1910.138 in the OSHA regulations.

OSHA developed this new set of regulations as a result of past accident statistics and employer surveys. Despite the original PPE requirements dating back to May 1970, the annual toll of injuries because of PPE issues in the workplace have steadily climbed. OSHA recognized this accident problem and developed a comprehensive survey which was sent to over 5,000 employers nationwide. The goal was to solicit information from each employer on their PPE program. The survey took approximately 45 to 60 minutes to complete. OSHA then followed up by calling those employers that had participated in the survey. Data was gathered on what PPE was being provided and worn at the employers facilities. All of this information helped OSHA to create the new PPE standard. This standard was not created in a vacuum.

Overall compliance with PPE throughout industry was not acceptable. What OSHA discovered was that the accident statistics were high as a result of:

- Many employees were not wearing PPE.
- Some employees were not wearing PPE properly.
- Some employees were wearing the incorrect PPE for the job.
- Management failed to enforce the wearing of PPE.
- Management lacked an understanding of what PPE was required.

- Employees were not trained properly in PPE.

The new standard, greatly expanded over that of the original, now requires employers to have an appropriate risk assessment program. Employers are required to:

- Carefully select PPE based on the job's application and requirements.

- Train employees in why PPE is necessary, how to properly wear required PPE, and how to care for PPE. A form is included in this chapter for this purpose; see Figure 9.1.

- The employer is to conduct a hazard assessment of each facility to document all PPE potential exposure areas. A survey form (see Figure 9.2) has been included in this chapter.

- Provide historical accident data from the facility to demonstrate why the PPE is necessary.

- Consider PPE based on the following:
    - Impact.
    - Heat.
    - Radiation light.
    - Pinching objects.
    - Penetration.
    - Falling objects.
    - Sources of high or low temperatures.
    - Chemical sources of motion.
    - Sharp objects.
    - Harmful dust.
    - Electrical hazards.
    - Rollover compression.
    - Plant layout.

Surprisingly, powered equipment operators are exposed to many, if not all, of the hazards listed above. Note the areas of focus that are a part of a PPE survey:

- **Chemical.** The operators can be exposed to the acid in the battery in their equipment, to products being stored or transported, or to chemicals in their immediate work environment.

- **Falling Objects.** Not every piece of powered equipment has an overhead guard. Employees could be in an area where products are stored on shelves or racks. The product could fall and strike the operator. The operator, even when protected by an overhead guard, must periodically get off the equipment for various tasks; at this point, employees are vulnerable to falling objects. In addition, an overhead guard has holes large enough to allow smaller items to fall through and strike the operator. This is why heavy-duty mesh is recommended for mounting on the top of the guard. Hard hats provide additional protection.

- **Roll-Over Compression.** It is not unusual for operators to be off of their piece of equipment during a work shift. It is also not unusual for any employees, including equipment operators, to have their feet near the tires of equipment. Imagine two pieces of powered equipment approaching each other, one of the operators gets off of the lift truck and approaches the other operator. They discuss something and the operator in the vehicle begins to pull away. The operator on foot happens to have a foot under the vehicle. The rear wheel runs over the foot, resulting in a serious injury. This is not an unusual occurrence. Steel-toe shoes could help prevent such an injury. In addition, a safer- thinking operator shouldn't stand so close.

- **Radiation Light.** In those areas where welding takes place, an unprotected operator can be the victim of welders flash. Weld screens would be the type of protection needed for anyone near a welding process. This would be a potential hazard on which management should focus during their documented hazard inspection tour.

- **Pinching/Sharp Objects.** The following exposures to employees can be reduced if the proper gloves are worn: when manually handling forks as they are being adjusted, getting off the vehicle to pick up product or move obstacles in their path, passing a chain around a mast to secure a safety cage, or picking up lumber with nails and splinters on it.

- **Sources of High or Low Temperatures.** In the course of performing their daily tasks, operators may have to drive through areas that are very hot or very cold. In this case, the operators need to have protective clothing that protect them from the extremes of weather or work processes. Though not entirely a part of the PPE standard, this subject should be addressed by management to assist in operator protection.

Those are just a few of the ways in which management must perform a comprehensive analysis of the workplace and exposures to each worker. PPE must be provided for worker protection when these hazards are present.

OSHA has updated the PPE standard, which clearly places the burden on the employer to ensure that all employees whose jobs require the use of PPE are, in fact, wearing it. This amendment, enacted on June 3, 1996, was necessary because the original rule, which was effective in July 1994, did not include the introductory phrase "the employer shall ensure"when stating that employees wear the proper PPE for eye, head, foot, and hand protection. The PPE rule is now more enforceable as a result of this amendment.

The PPE standard does not address exposure to noise and air contaminants. These were already covered by OSHA 1910.95, Occupational Noise Exposure and OSHA 1910.134 Respiratory Protection. Both of these standards have gone through several evaluations at OSHA and have been broadened.

130 / Forklift Safety

Figure 9.1. PPE Employee Training Form

Employers must document PPE training per the OSHA guidelines.

_____

The following employees have been properly trained on the proper use of the PPE requirement of care of PPE and disciplinary procedures for the PPE program.

Check training received:

Eye ☐	Hand ☐	Respiratory ☐
Face ☐	Ear (Hearing) ☐	Other ☐
Foot ☐	Head ☐	Other ☐

Operator Names/Signatures

Name (Print) **Signature**

1. _____   _____
2. _____   _____
3. _____   _____
4. _____   _____
5. _____   _____
6. _____   _____
7. _____   _____
8. _____   _____
9. _____   _____
10. _____   _____
11. _____   _____
12. _____   _____

Training Provided by:_____ Date: _____

Figure 9.2. PPE Location Survey for Power Equipment Operators

On (date)_____, (print name)_____

Conducted a facility survey to ensure:	Yes	No
■ PPE is being worn properly	☐	☐
■ PPE use is being enforced	☐	☐
■ PPE is clean/not damaged	☐	☐
■ PPE in use has been approved	☐	☐

Hazards Survey for Power Equipment Operators	Yes	No
■ Sources of motion—machinery or processes where an injury could result from movement of tools, machine elements or particles, or movement of personnel that could result in collisions, blows, or tripping around stationary objects.	☐	☐
■ Sources of high temperature that could result in burns, eye injury, or ignition of protective equipment.	☐	☐
■ Types of chemical exposures such as splash, vapor, spray, or immersion that could cause chronic illnesses or physical injury.	☐	☐
■ Sources of harmless dust that can accumulate or become airborne and cause a physical hazard to employees' eyes. If a material safety data sheet indicates that accumulated or airborne dusts of material used in the facility can be a pulmonary hazard, the employer is obligated to confirm the hazard through sampling under other regulation.	☐	☐
■ Sources of light radiation, such as welding, brazing, cutting, furnaces, heat treating, and high intensity lights.	☐	☐
■ Sources of falling objects or potential for dropping objects that could pose a compression or projectile hazard to employees' head, face, hands, or feet.	☐	☐
■ Sources of sharp objects which might pierce the feet or cut the hands.	☐	☐
■ Sources of rolling or pinching objects which could crush the feet.	☐	☐
■ Layout of workplace and location of coworkers.	☐	☐
■ Any electrical hazards.	☐	☐

In addition to the above-mentioned requirements for hearing protection and respirators, there are other pieces of personal protective equipment which operators may need. There may be specific exposures, which are part of the job or are present in the workplace, that pose a hazard. The following list of PPE is a broad representative sample, and may not be all inclusive of all workplace hazards.

- **Lasers.** A high risk, technical hazard that requires very special eye protection. There is not one specific type of glass or plastic that can protect from all laser wavelengths. Laser glasses or goggles alone can provide employees with a false sense of security; engineering controls are also very important. Powered equipment operators may be exposed to lasers. Management must ensure that not only are all employees protected from lasers by safeguards, but also that the required PPE is worn.

- **Protective Clothing.** It is possible that operators may be exposed to airborne hazards such as dust, mist, vapors, acids, and other potentially harmful particles. Impervious clothing would be necessary to protect the skin and all exposed areas of the body.

- **Fall Protection Device.** An operator may have to work in elevated situations. Any fall can be disastrous; therefore, the need for proper fall protection is imperative. Employees must be trained in the use of these devices. Follow manufacturer's guidelines. Full body harnesses are the safest form of fall protection. Of course, a proper anchorage point must have at least 5,000 pounds of static load strength; 6,000 pounds in Canada. Lanyards should be of sufficient length to minimize a fall, yet of sufficient length to allow the work to be performed from the elevated platform.

The following information is only a part of the new OSHA requirements for fall protection, as published in the MSA Safety Equipment Catalog for 1996.

A new OSHA standard entitled "29 CFR Parts 1910 and 1926: Safety Standards for Fall Protection in the Construction Industry," went into effect on February 6, 1995. OSHA believes that accidents and fatalities from falls will be reduced by establishing these guidelines.

In the February 6, 1995 standard:

- If a belt is used for fall arrest, OSHA limits the maximum arresting force to 900 pounds. Therefore, a decelerating device must be used. Where powered equipment operators are currently using safety belts, this practice should be substituted for fall protection harnesses

- Effective January 1, 1998, the use of a body belt for fall arrest will be prohibited.

- Only locking snap hooks are allowed as part of a fall arrest system.

Briefly, a few of the OSHA definitions are:

- All fall arrest systems are to be made up of four components:

    A = Anchorage
    B = Body Harness
    C = Connectors
    Plus at least one of the following:

    D = Deceleration device; or
        Lanyard; or
        Lifeline; or
        A suitable combination of the above items.

- **Anchorage.** A secure point of attachment for lifelines, lanyards, or deceleration devices. It has to be independent of the means of supporting or suspending the worker. A suitable anchorage is very important. A proper anchorage must be identified and evaluated by a qualified person at the job site (plant, etc.) before working with equipment. The anchorage must be capable of supporting at least 5,000 pounds for each attached worker.

- **Body Harness.** A harness offers better free-fall protection than a belt. A harness reduces the risk of injury to the worker, in much the same way as a seat belt with a shoulder strap secures a driver better than a seat belt alone. A body harness distributes the fall arrest forces over the thighs, pelvis, waist, chest, and shoulders. A body belt distributes the fall arrest forces only on the waist. It is recommended that a body harness be used on powered equipment operators rather than a safety belt.

- **Connectors, Lanyards and Lifelines.—**

    - D Rings and snap hooks must have a tensile strength of 5,000 pounds.
    - Lanyards must have a 5,000 pound minimum breaking strength.
    - Lifelines must have a 5,000 pound minimum breaking strength.
    - Deceleration Device—A deceleration device is any mechanism which serves to dissipate a substantial amount of energy during a fall arrest or otherwise limit energy impact on a fall during fall arrest. Examples of these devices are:

        —Rope grabs
        —Rip stitch lanyards
        —Specially woven lanyards, tearing or deforming lanyards
        —Automatic, self-retracting lifelines/lanyards

Self-retracting lifelines and lanyards are deceleration devices that contain a drum—wound line which can be slowly extracted from, or retracted onto, the drum under slight tension during normal employee movement; and which, after onset of a fall, automatically locks the drum and arrests the fall.

For further information on body harnesses, contact a reputable manufacturer of these products.

EQUIPMENT REQUIRED BY THE PPE STANDARD

An analysis of those pieces of equipment that are a part of the new OSHA PPE standard should be completed. Each piece of equipment should be known and understood by employees so that the proper wearing, care, and replacement can take place.

Cold Weather Clothing

Where powered industrial truck operators have to work out in the cold or in cold job locations, thermal clothing will help to protect from freezing and frostbite. If necessary, fire retardant clothing is also available.

Head Protection

Hard hats or helmets are designed to protect the head from impact of falling objects and collision with other objects. Helmets are approved by the American National Standards Institute (ANSI). The standard for helmets—"a device that is worn to provide protection for the head, or portion thereof, against impact, flying particles, electrical shock, or any other combination thereof and that includes a suitable harness (suspension)."

ANSI Z89.1 defines three classes of helmets.

- **Class A.** Helmets are intended to protect the head from the force of impact of falling objects and from electric shock during contact with exposed low-voltage conductors.

- **Class B.** Helmets are intended to protect the head from the force of impact of falling objects and from electric shock with exposure to high-voltage conductors.

- **Class C.** Helmets are intended to protect the head from the force of falling objects.

Helmets should be cleaned with soap and water. Solvent must never be used; this would weaken the shell. Painting a hard hat should not be allowed; paint would cover over any cracks or defects. The only holes that should be drilled in a helmet would be for the purpose of mounting brackets or attachments approved by the manufacturer.

The suspension in the helmet helps to absorb the blow or shock to the head. The suspension must not be altered. Regular inspections of the suspension should take place. Look for tears, frays, cracks, loose parts, or broken sewing lines. Wash the suspension regularly for sanitary reasons. Wear the suspension in the helmet as the manufacturer requires. Suspensions should be replaced at least on an annual basis.

Eye and Face Protection

The cost of an industrial eye injury can be very expensive. As an example, in the 1996 booklet from the U.S. Chamber of Commerce *1996 Analysis of Workers' Compensation Laws*, the ten states that have the highest costs for the loss of an eye are:

Pennsylvania	$139,975	Connecticut	$89,019
Maryland	$134,865	South Dakota	$81,450
Illinois	$121,681	Michigan	$80,838
Washington, DC	$112,245	Hawaii	$78,560
Iowa	$108,920	Delaware	$69,234

A print from the ANSI Z.87.1 standard illustrating the proper selection of eye and face safety equipment is included in this chapter, see Figure 9.3.

These costs do not represent any medical costs associated with the claim. The costs listed above are the maximum pay out to the injured employee depending on their earnings. In addition, indirect costs of the accident such as investigation time, claims processing, interruption of work in process, retraining of the injured employee upon their return, and costs associated with medical visits upon their return to work. There are other indirect costs that are a part of each workers compensation chain. Employers should recognize that for each dollar spent on direct costs, such as medical and lost wages, there are several times that amount paid out in indirect costs as just described. It pays in many ways to maintain a safe workplace. Management must ensure that enforcement of the rules takes place.

If employees, such as power equipment operators, are to wear eye protection on the job, then it must be comfortable. Manufacturers of protective eye wear have created products that are good-looking, lightweight and fully protective to the eyes. Each employee should have glasses that have been fitted to their particular size face, nose, and eye. Unfortunately, one size does not fit all, so comfort is obtained by using various sizes or models of eye wear. There may be a department or an entire facility where each employee is required to wear eye protection for an entire work shift. Power equipment operators would be included in this program.

ANSI Z.87.1 is the standard from which OSHA enforces the eye protection rule. Keep in mind that the Food and Drug Administration (FDA), is responsible for protective prescription eyewear for the public. Effective January 1, 1972, all prescription eye wear must be impact resistant. The ANSI standard, however, offers much more impact resistant requirements than the FDA. All eye wear worn in the industrial setting must be ANSI approved—not FDA approved.

Safety Shoes/Boots

ANSI Z41—1991, Personal Protection—Protective Footwear, is the standard that addresses safeguarding the feet and toes of workers. Steel-toes add little weight to the actual shoe or boot. Many of the current shoe and boot styles are fashionable enough to satisfy many employees. Selection is very broad, as well

Figure 9.3. ASSE/ANSI Eye and Face Protection Guidelines

AMERICAN NATIONAL STANDARD Z87.1-1989

SELECTION CHART — **PROTECTORS**

		ASSESSMENT SEE NOTE (1)	PROTECTOR TYPE	PROTECTORS	LIMITATIONS	NOT RECOMMENDED
IMPACT	Chipping, grinding, machining, masonry work, riveting, and sanding.	Flying fragments, objects, large chips, particles, sand, dirt, etc.	B,C,D, E,F,G, H,I,J, K,L,N	Spectacles, goggles faceshields SEE NOTES (1) (3) (5) (6) (10) For severe exposure add N	Protective devices do not provide unlimited protection. SEE NOTE (7)	Protectors that do not provide protection from side exposure. SEE NOTE (10) Filter or tinted lenses that restrict light transmittance, unless it is determined that a glare hazard exists. Refer to OPTICAL RADIATION.
HEAT	Furnace operations, pouring, casting, hot dipping, gas cutting, and welding.	Hot sparks	B,C,D, E,F,G, H,I,J, K,L,*N	Faceshields, goggles, spectacles *For severe exposure add N SEE NOTE (2) (3)	Spectacles, cup and cover type goggles do not provide unlimited facial protection. SEE NOTE (2)	Protectors that do not provide protection from side exposure.
		Splash from molten metals	*N	*Faceshields worn over goggles H,K SEE NOTE (2) (3)		
		High temperature exposure	N	Screen faceshields, Reflective faceshields. SEE NOTE (2) (3)	SEE NOTE (3)	
CHEMICAL	Acid and chemicals handling, degreasing, plating	Splash	G,H,K *N	Goggles, eyecup and cover types. *For severe exposure, add N	Ventilation should be adequate but well protected from splash entry	Spectacles, welding helmets, handshields
		Irritating mists	G	Special purpose goggles	SEE NOTE (3)	
DUST	Woodworking, buffing, general dusty conditions.	Nuisance dust	G,H,K	Goggles, eyecup and cover types	Atmospheric conditions and the restricted ventilation of the protector can cause lenses to fog. Frequent cleaning may be required.	
OPTICAL RADIATION				TYPICAL FILTER LENS SHADE — PROTECTORS SEE NOTE (9)		
	WELDING: Electric Arc		O,P,Q	10-14 — Welding Helmets or Welding Shields	Protection from optical radiation is directly related to filter lens density. SEE NOTE (4). Select the darkest shade that allows adequate task performance.	Protectors that do not provide protection from optical radiation. SEE NOTE (4)
	WELDING: Gas		J,K,L, M,N,O, P,Q	SEE NOTE (9) 4-8 — Welding Goggles or Welding Faceshield		
	CUTTING			3-6		
	TORCH BRAZING			3-4	SEE NOTE (3)	
	TORCH SOLDERING		B,C,D, E,F,N	1.5-3 — Spectacles or Welding Faceshield		
	GLARE		A,B	Spectacle SEE NOTE (9) (10)	Shaded or Special Purpose lenses, as suitable. SEE NOTE (8)	

Personal Protective Equipment / 137

Figure 9.3 (continued)

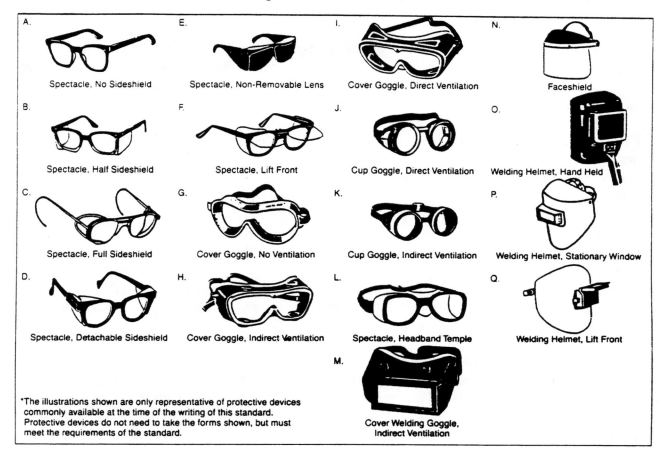

NOTES:
(1) Care shall be taken to recognize the possibility of multiple and simultaneous exposure to a variety of hazards. Adequate protection against the highest level of each of the hazards must be provided.
(2) Operations involving heat may also involve optical radiation. Protection from both hazards shall be provided.
(3) Faceshields shall only be worn over primary eye protection.
(4) Filter lenses shall meet the requirements for shade designations in Table 1.
(5) Persons whose vision requires the use of prescription (Rx) lenses shall wear either protective devices fitted with prescription (Rx) lenses or protective devices designed to be worn over regular prescription (Rx) eyewear.

(6) Wearers of contact lenses shall also be required to wear appropriate covering eye and face protection devices in a hazardous environment. It should be recognized that dusty and/or chemical environments may represent an additional hazard to contact lens wearers.
(7) Caution should be exercised in the use of metal frame protective devices in electrical hazard areas.
(8) Refer to Section 6.5, Special Purpose Lenses.
(9) Welding helmets or handshields shall be used only over primary eye protection.
(10) Non-sideshield spectacles are available for frontal protection only.

Eye and face protection is necessary for operators where the job requires it. (Reprinted from American National Standards Practice for Occupational and Educational Eye and Face Protection, ANSI Z87.1-1989, approved by the American National Standards Institute on February 2, 1989 and published by the American Society of Safety Engineers as the Secretariat of the standards project).

as price ranges. Powered industrial truck operators can be exposed to situations in which foot and toe protection is necessary.

There may be situations where metatarsal guards are required. A metatarsal guard protects the delicate bones of the instep. Any time these bones are bruised or broken, the employee would have difficulty standing or walking. The metatarsal guard, in addition to the steel toe guard, absorbs impact and distributes the blow.

Newer models of metatarsal guards are built into the shoe or boot and have a different look than older models with the same guard. Built-in guards are a plus to powered equipment operators in that the guard will not get caught on the pedals of the equipment or possibly trip the operator getting on or off of the vehicle.

Respiratory Protection

Though not a part of the new PPE standard, the need for respiratory protection for powered equipment operators may be needed. It is very difficult to perform an active day's work while wearing a respirator. Where operators may have to wear respiratory protection devices, management must ensure that they adhere to 1910.134 from the OSHA guidelines.

A complete respirator program must be developed, and employees must be properly trained for their use. Employees will have to have an exam by a physician to ensure they are capable to wearing a respirator. Respirators must be selected based on the hazard. Manufacture's and their representatives can provide specialized training in this area. NIOSH, the National Institute of Occupational Safety and Health, also produced specific guidelines on respiratory devices. Details can be found in *Respiratory Protection—An Employers Manual and NIOSH Guide to Industrial Respiratory Protection*.

There may be a need to wear a dust mask while performing certain jobs with a piece of powered equipment. As an example, operating a floor sweeper may create dust. Emptying the dust from the sweeper to a trash container might create unwanted or harmful dust. NIOSH- approved dust masks should be worn. As with respirators, the employees have to be trained in the use, fit, care, and disposal of the dust masks. A NIOSH dust mask has two bands, is designed for limits of how much of a contaminant is in the air, and has a TC (Technical Certification) number. Employers and employees are not to confuse NIOSH approved dust masks with nuisance dust masks, which offer less protection.

Hearing Protection

Just as respiratory protection is not a part of the PPE standard, employers should be thinking about making hearing protection a part of their overall program. Where noise levels exceed 85 decibels for an eight hour time-weigted period, audiometric testing and proper hearing protection should be required.

It is very likely that powered equipment operators would be required to operate their equipment in noisy areas. It is important to protect the hearing of employees, as well as comply with federal noise regulations.

Where possible, noise sources should be reduced through engineering improvements. If high noise levels, those over 85 decibels, are still present, choice of hearing protectors is important. When choosing, look for the NRR label on the container. NRR stands for noise reduction rating. Manufacturers place the NRR marking on containers to assist in choice of protector. The higher the number, the greater the rating.

As an example, a set of ear muffs is identified as having an NRR of 18. This means that there should be at least an 18 decibel reduction of the noise when the protection is worn properly per manufacturers guidelines. Always keep in mind that a perfect fit of hearing protectors may not be taking place. The NRR factor should be reduced by 10% to 20% in order to have a more realistic number for the noise reduction value of the protectors. For this example the ear muffs should be expected to reduce the noise by 14 or 15 decibels rather than 18.

Hand Protection

This equipment is a part of the new PPE standard. There may be work situations that require special hand protection for power equipment operators. As an example, rubber gloves should be worn when handling power equipment batteries. A cloth glove may be needed for handling chains or other devices. A heavier glove may be needed for picking up broken glass, splintered wood, or sharp parts that operators discovered in their route of travel; or perhaps they spilled these hazardous items by striking them with their powered equipment. Select gloves for the specific hazard; ensure they are worn as required. Consider glove selection based on exposure to:

- Severe abrasives (punctures and cuts).
- Absorption of harmful substances.
- Chemical burns.
- Thermal burns.
- Harmful temperature extremes.
- Severe cuts.

Of course, protection of the hands also involves the removal of all jewelry while at work. Rings and watches could pose a significant hazard if the jewelry is caught on something while the individual is in motion. It is not uncommon to have someone seriously injure a finger while jumping off a dock or machine; a ring can be caught on a nail or other item and result in serious injury. Necklaces and earrings can also be hazards; although this subject is not related to hand safety, employees should keep this in mind.

Hazard Assessment for PPE

There is a requirement in the OSHA PPE program that management is to conduct an extensive survey at least annually to determine if the proper PPE is being worn. Note the survey form on page 131; it is to be used throughout a facility to determine if employees are protected by PPE. In this case, of course, the focus is on powered equipment operators.

Employee Training

The PPE standard requires that employees be trained regarding the equipment they have to wear. A form is included with this chapter on page 130 to document powered equipment operator training.

SUMMARY

Powered equipment operators can be exposed to various workplace hazards. Where hazards cannot be corrected or eliminated by engineering controls, personal protective equipment (PPE) must be worn. Not only is this necessary because of the obvious reasons for injury reduction, OSHA requires employers to provide PPE where the job requires it.

There are many opportunities for injury where PPE can be used to minimize or reduce the risk. The information in this chapter can be beneficial for proper selection of the equipment. Employee training and a facility survey for PPE is necessary.

REFERENCES

American National Standards Institute. ANSI Z.87.1.—1989. Des Plaines, IL: ASSE.

Employers Are Responsible to Ensure That Their Workers Wear PPE, Kellers Industrial Safety Report. Neenah, WI: JJ Keller & Associates, July 1996, pp. 1-2.

Fall Protection: A Primer, MSA Safety Equipment Catalog, Pittsburgh, PA, 1996, pp. 218-220.

National Safety Council. *Personal Protective Equipment, Accident Prevention Manual*, Itasca, IL, 1992, 409-453.

National Safety Council. *The Eyes, Industrial Hygiene Manual*, Itasca, IL, 1996, pp. 112-116.

Occupational Safety and Health Administration. OSHA Standards, Subpart I–Personal Protective Equipment, 1910.132, 1910.133, 1910.134, 1910.135, 1910.136, 1910.138.

Occupational Safety and Health Administration. OSHA Standards, Personal Protective Equipment, OSHA #3077, 1994 (Revised).

U.S. Chamber of Commerce. *1996 Analysis of Workers' Compensation Laws*, Washington, DC, 1996.

10

PROTECTING THE HEALTH OF POWERED INDUSTRIAL TRUCK OPERATORS

With the intense focus on industrial safety, many firms, as well as employers, overlook the need for protective measures to protect the health of powered industrial truck operators. In the course of their daily work there are potential exposures to hazards that are present in their work environment. These hazards can harm the skin, eyes, lungs, other body organs and the musculo-skeleton system. There is no sure method of identifying each potential health risk within this publication, but some of the major issues will be explored in this chapter.

CARBON MONOXIDE

One of the most insidious health hazards in the workplace is carbon monoxide (CO). Carbon monoxide is easily produced by the internal combustion engine. Internal combustion engines are plentiful throughout industry. A report in *Michigan's Occupational Health* newsletter indicates that in 1993, 73% of all industry orders for sit-down lift trucks contained internal combustion engines. Of these, more than 80% were fueled by liquid propane gas (LPG), approximately 10% by diesel fuel and 6% by compressed natural gas (CNG).

When internal combustion engines are used indoors, driven into trailers, through tunnels or other enclosures, CO can easily exceed safe levels.

OSHA has established safe working levels of CO in the workplace. A time weighed average (TWA) of 50 parts per million, (PPM), is the current guideline. In Canada the established level is 35 PPM. The state of Michigan, for an example, has established a 35 PPM TWA. It is important to understand that 50 or 35 PPM includes an action level. This level is 50% of the permissible exposure limit (PEL) of CO established by OSHA. In other words where CO alarms are present, they should be set at or below the action level to ensure that enough warning is given to everyone to control CO before it reaches its PEL. The American Conference of Governmental Industrial Hygienists, (ACGIH), has established a TLV of 25 PPM. In addition, a level of 1200 PPM has been established by NIOSH to be Immediately Dangerous to Life and Health (IDLH).

Approximately 1,500 deaths from CO poisoning occur in the US each year. Some of these deaths are work related. Also, over 10,000 people each year are treated in clinics or hospitals for CO exposures. Most of the symptoms of CO poisoning go unnoticed because of the close resemblance to an illness like the flu.

Carbon monoxide is known as a chemical asphyxiant. This gas passes through the alveolar sacs (air sacs) in the lungs as it is breathed in and then passes directly into the bloodstream. During this passage through the lung, the gas does not harm the lung. The gas then ties up the hemoglobin in the blood so that it cannot accept oxygen. This causes oxygen starvation to the body tissues. Hemoglobin combines with CO much more readily than it does with oxygen by a ratio of approximately 300 to 1.

Carbon monoxide is a colorless, odorless, and tasteless gas. Early stages of CO poisoning are headaches, dizziness, and sleepiness. As more CO is absorbed into the body the symptoms can include nausea, vomiting, fluttering and throbbing of the heart, to name a few. Unconsciousness and possible death may occur. An exposure to high concentrations of CO may result in a rapid loss of consciousness or life without first producing any other significant symptoms. The heart and brain are especially vulnerable to oxygen depletion.

ACCIDENT FACT

A forklift operator in Colorado was transported to a local hospital because of chest pain. Propane lift trucks were being used to unload trailers containing palletized loads of beverages. The lift trucks continually entered and exited the trailers. The cause of his ailment was carbon monoxide poisoning. There was no evidence of tailpipe emission monitoring by the company on any of the lift trucks; CO levels were found to be above OSHA limits.

Individuals who are sensitive to CO exposures include:

- Pregnant women.
- People with heart, artery, or respiratory disease.
- People working at altitudes greater than 2,000 feet.
- Young adults and children.
- People who smoke.

It is apparent from the list above that some of the individuals listed could be powered equipment operators. Pack-a-day smokers often have 5-10% carboxyhemoglobin in their hemoglobin. A smokers carboxyhemoglobin is comparable to that produced by exposure to CO at the 50 PPM level.

According to the state of Michigan's newsletter mentioned earlier, there are recommended treatments for CO exposures:

- Remove the victim to fresh air immediately.
- Loosen any tight clothing.
- Administer artificial respiration, if necessary.
- Contact a physician.
- If necessary, administer oxygen as soon as possible.
- Prevent any cardiac or respiratory stress by requiring the victim to rest.

The carboxyhemoglobin concentration can be reduced by approximately 50% in 320 minutes when the person is moved to fresh air and away from any CO exposure. This half-life can be reduced to 80 minutes by administering oxygen. A hyperbaric chamber can reduce it to approximately 23 minutes.

Additional safeguards for CO would be to have heaters checked and approved at least once a year by licensed individuals. Also, educate all employees about the dangers and symptoms of CO poisoning.

OSHA has considered reducing CO levels to 35 or 25 PPM. The American Conference of Governmental Industrial Hygienists (ACGIH), has recommended that a threshold limit value (TLV) of 35 PPM for an eight hour TWA be adopted. A 200 PPM ceiling concentration must not be exceeded.

CO concentrations at the tailpipe of an LPG fueled truck at idle speed can produce 5,000 to 7,500 PPM. Under load and at full power, an LPG truck can produce levels of 2,000 to 5,000 PPM of CO. Fuel systems must be properly sized and tuned to attain proper readings. Where lift trucks have been poorly maintained, CO levels can reach 10,000 PPM at the tail pipe regardless of the fuel used.

Liquid petroleum gas (LPG), or compressed natural gas (CNG) is recommended over gasoline. Both of these fuels, LPG and CNG have clean-burning fuels with a high octane rating. Gasoline-fueled powered equipment requires a substantial amount of ventilation to maintain CO levels below the PEL. Diesel makes a poor fuel choice for powered trucks being used indoors, due to the soot and irritating gases it produces.

Control Methods for CO

There are methods which can be used to reduce or control CO exposures. Some of these are not directly controlled by the powered equipment operator himself, but should be considered by management for the overall health and quality of safety for employees. One sure-fire method to reduce CO levels is to provide adequate ventilation.

- CNG, compressed natural gas, could be a consideration for fueling. One organization reported a 20% reduction in CO levels while using CNG. The CO levels at the facility were below the OSHA PEL prior to going to CNG as a fuel.

- LPG, liquid propane gas, has shown to improve CO levels as compared to diesel or gasoline for powered industrial trucks.

- Batteries have zero CO emissions.

- Plant ventilation should be considered to reduce overall CO levels.

- Maintenance efforts such as properly sized carburetor, a serviced air cleaner, adjusted engine timing, proper diagnostic equipment, fuel system adjustment and maintenance, carbon monoxide emissions controllers, catalytic converters and the ensuring that replacement parts are from the original equipment manufacturer will reduce CO levels.

As with any program, device or machine, follow manufacturers guidelines. Ensure all employees are trained. Enforce these practices.

Testing for CO is highly recommended. Because CO is an insidious gas, it can be harmful in very small quantities. For this purpose, a carbon monoxide monitor can be used for testing the air. These devices have audible alarms and visible alarms. The alarms would alert the user that a certain level of CO had been reached. A passive CO badge can be worn by operators. The device reacts to the CO gas and changes color. The color is a function of the concentration of the gas as well as the exposure time.

Another device that could help warn operators of high CO levels is worn on or by the collar. It will sound an alarm when the CO level reaches 35 PPM. The alarm cannot be shut off until the operator leaves that area and the level of CO is reduced. A device like this can be very helpful for those operators that are entering trailers with equipment which use internal combustion engines.

Grab samples can be taken by using a hand-held suction pump with a special glass tube inserted into it. The ends of the tubes must be broken off to allow air to be drawn into the tube. The CO in the air would then stain the tube when the device is pumped by hand which allows for an instant reading of the ambient atmosphere.

In all cases employees should not be smoking when air monitoring is taking place because this would distort the numbers. In addition, all manufacturers testing guidelines have to be followed. Only qualified individuals should be authorized to conduct air sampling. Allow for the instruments numbers to be calculated in a range—not as a direct number. As an example, grab-sample testing can be a plus or minus 20% of the number indicated on the glass tube. More expensive and sophisticated instruments will provide for more accurate data.

NOISE

The chapter on personal protective equipment (PPE) discusses the devices needed for noise control. This section will discuss the subject of noise within the environment that may expose the operator to harmful exposures. As a part of their daily work, many powered equipment operators wear various forms of hearing protection. Loss of hearing is a health issue and must be addressed for employee protection.

When industrial hearing loss occurs, the employee may have been exposed to sound levels at ranges that caused permanent damage to the inner ear. The outer ear acts as a funnel to direct sound vibrations into the eardrum through the ear canal. The eardrum now vibrates as the sound waves strike it. This vibration process is transmitted to the three small bones in the middle ear. The inner ear now receives these vibrations and microscopic hair cells are stimulated. Nerve impulses are now sent to the brain for interpretation. The process just described is very technical and complex and has been simplified for this writing.

When noise levels exceed federal guidelines, hearing protectors must be worn to protect employees. Engineering practices should be employed to reduce noise levels as low as possible. Note the graph in Figure 10.1. The shaded area allows for noise levels to be observed with hearing protection being optional. Once noise levels are out of the shaded area (above the curved line) hearing protection is mandatory. However, noise levels of 90 decibels for eight hours are very loud and should be accompanied by hearing protection. The chart is only a guide. It is far better to wear hearing protection at 85 decibels and above.

Hearing protection must be worn to help eliminate occupational hearing loss. Protectors can consist of:

- Ear muffs.
- Ear plugs.
- Canal caps.

Employees exposed to noise levels of 85 decibels, TWA, or higher are to receive a base line audiogram. Annual audiograms are required for all employees exposed to 85 decibels or higher for an 8 hour time-weighted average. OSHA estimates that some 2.9 million employees in the workplace are exposed to noise levels above 90 decibels. An additional 2.3 million employees are exposed to noise levels above 85 decibels.

It is recommended that powered equipment operators be monitored for noise exposures. In addition, the OSHA standard, 1910.95, should be adhered to for full protection for all employees. The OSHA standard requires that the full text of 1910.95 be placed in a convenient location such as on a bulletin board for employees when sound levels exceed 85 decibels.

Back up alarms on counter balance trucks could help contribute to overall noise levels. This should be considered when deciding on the use of these devices.

Figure 10.1. Noise Exposure Graph

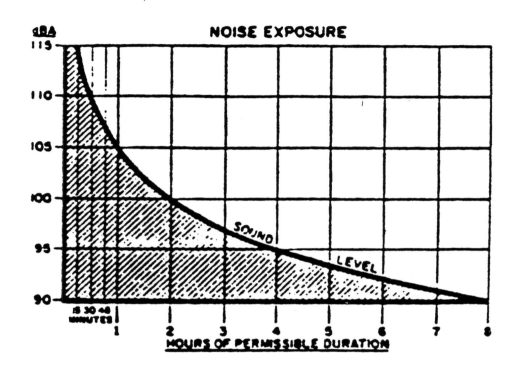

The graph illustrates that the shaded area corresponds with the OSHA guidelines on decibel exposure and exposure time.

CHEMICALS IN THE WORKPLACE

In the course of daily lift truck operation, operators can come in contact with various chemicals and environmental hazards such as vapors, gases, fumes and dusts. These airborne hazards can cause harm if not controlled properly. To protect employees, OSHA passed into law the Hazard Communication Standard, 1910.1200 in 1986. This standard gives employees the "right to know" regarding chemicals they are exposed to.

Responsibilities of this standard break down as follows:

- Manufacturers of the chemicals must determine what dangers their chemicals may have. Products must be labeled and a material safety data sheet (MSDS) must accompany each chemical.

- Employers must determine which chemicals in the workplace are hazardous to employees.

- Employees must be properly trained as well as provided with information regarding each chemical. Employees are taught to protect themselves from harm from the chemicals.

- Employers have an obligation to follow prescribed safety rules, guidelines and the federal standard.

Approximately 90% of all harmful substances enter through the lungs where they can either harm the lung or travel through the blood stream or to a target body organ such as the kidneys to cause further harm. Approximately 8% of harmful substances enter the body through the skin and may travel through the blood stream or to body organs and cause harm. Approximately 2% of harmful substances are actually ingested.

Chemicals can cause damage by chronic exposure, which takes a long time to cause harm (such as asbestos), or acute, which causes immediate harm (such as battery acid on the skin or eyes).

Chemicals can fall under two specific classifications:

Physical Hazards

- **Flammables.** Those with low flash points that can catch fire easily. Reactive or unstable chemicals that can explode or burn. Vapors or gases released can be harmful when exposed to air, heat or to specific other chemicals.

Health Hazards

- **Corrosive.** Can burn the eyes or skin.
- **Toxic.** Can cause serious illness or death.

It is apparent that powered equipment operators could be harmed from a variety of chemicals that are associated with their jobs, such as:

- Working on batteries.
- Product being carried spills or breaks.
- An overhead pipe or line is struck by a load.
- Unknown product arrives at a dock.
- A large puddle of an unknown substance is in the center of an aisle.

In all of the above cited incidents there is a potential for harm from the chemical in question. Hazards can be:

- Batteries are explosive, and batteries contain acid, a corrosive.

- A wide variety of products are handled and stored throughout industry. Products in containers can fall and break open. Bags, boxes or barrels are speared by forks from a lift truck. The chemical spills out or leaks out onto pallets, etc.

- Overhead pipes and lines can carry acids, corrosives, gases, steam, hot fluids or be covered with asbestos. If struck, the chemical can be released and the operator can be harmed.

- Unknown products arrive at a dock. No one knows what it is, or how harmful it is. There are no labels or MSDS's to accompany the load.

- An operator is driving down an aisle and spots a large puddle. Should he go around it, go through it, get a clean up kit, or tell his supervisor? It could be dangerous but no one knows. The operator should know how to safely clean up the spill.

The list could go on. There are numerous workplace hazards that are associated with chemicals. The powered equipment operator may easily become involved in the handling of dangerous chemicals. With this in mind, there are specific parts of the hazard communication law that operators must focus on. The standard requires that employees are trained for the following:

- Explanation of the haz-com program.

- A review of the chemical hazards where the employee works.

- Knowledge of how the employee should know of the presence of hazardous chemicals—how to recognize, how to detect.

- An understanding of how chemicals can cause harm to the human body.

- An explanation and understanding of the labeling systems.

- Education on how to read a material safety data sheet (MSDS) and where to obtain them for use.

- Knowledge of where a comprehensive chemical inventory is for all of the MSDS's.

- The wearing of PPE so that the person is protected from the hazard.

- An understanding of hazards associated with non-routine tasks and chemicals in pipes and lines.

- What procedures employees will have to use to protect themselves from the hazards.

Operators will recognize that as a group they may be exposed to more chemicals than the average worker in a facility. Because operators are moving around the facility and handle a variety of loads and products, potential for exposure or incident is rather high.

Regarding chemical labeling, there are two systems being used; a diamond format and a horizontal bar format. The labeling concept is similar for both labels. The diamond format, as illustrated in Figure 10.2, is the National Fire Protection Associations (NFPA) #704. The upper diamond is red, this identifies flammability. The blue diamond on the left represents health, the yellow diamond represents reactivity. The white diamond at the bottom is for special identification. As an example—a W with a line through it warns that water should not be used if a fire occurs with that chemical.

The second labeling system, as illustrated in Figure 10.3, is called the hazardous material identification system, (HMIS). The top horizontal bar on the label is blue, the second bar is red, the third bar is yellow, the fourth bar is white. As with the NFPA diamonds, numbers ranging from 0-4 go on the bars, similar to the numbers in the diamonds. The numbers represent the degree of risk with the chemical. As a guide the numerical rating is:

- 0 - no hazard
- 1 - slight hazard
- 2 - moderate hazard
- 3 - serious hazard
- 4 - severe hazard

Figure 10.2. NFPA 704 Label

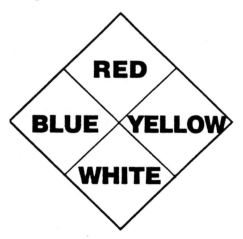

The NFPA 704 Label provides a diamond format for employee awareness of container contents. The diamond at the top identifies flammability, the diamond on the right identifies health hazards, the left diamond identifies reactivity, the lower diamond identifies special hazards.

Figure 10.3. HMIS Label

BLUE	
RED	
YELLOW	
WHITE	

The HMIS label. The top bar identifies health hazards, the second bar identifies flammability. The third bar denotes reactivity and the bottom bar the appropriate PPE.

Any employee, just by looking at a completed NFPA or HMIS label should have a grasp on how harmful the chemical is without handling it in any way. Guidelines on what PPE to wear is a part of the HMIS label on the fourth bar—white in color. PPE could be a respirator, goggles, face shield, glasses, rubber gloves, or rubber apron.

Manufacturers of the chemicals have an obligation to provide specific information on the container label.

- The identity of the chemical.

- Complete address and names of the manufacturers or importers.

- Identify the physical hazards of the chemical such as it's being explosive, flammable, etc.

- Identify the health hazards. Can it cause cancer? Can it affect the respiratory system, skin or eyes? Is it toxic?

- What PPE is required, what procedures are to be followed for a spill or splash. Basic first aid guidelines are also needed.

As an additional guide employees should be trained to identify chemical spills. Proper clean-up and disposal of all waste must take place. Proper clean-up kits, PPE and disposal containers are to be available. A video could be made which would identify the proper training, procedures and safeguards for clean-up. Be sure to follow EPA guidelines.

The proper eye wash unit with continuous running water for at least 15 minutes must be open and available where the emergency would exist. Full showers may be necessary as a result of exposure to other chemicals that would be required of a material handler with a piece of power equipment. Employees must be trained to use these devices.

RADIATION

There may be situations where industrial truck operators are exposed to radiation. The subject of radiation is rather complex and this book is not equipped to deal with the broad exposures and regulations governing radiation. It is important to say that a few basic reminders for radiation safety would be:

- Proper radiation signage.
- Monitoring equipment such as personal dosimeters, film badges and Geiger-Mueller counters.
- Employee training.
- Written procedures.
- Health monitoring.

The harmful types of ionizing radiation are:

- **Alpha particles** can be very harmful internally but externally are not as harmful. The particles can be blocked by a sheet of paper.
- **Beta particles** can be more penetrating and could cause skin burns. The use of light metals like aluminum for shielding is recommended.
- **X-radiation and gamma radiation** can be very harmful. This depends on time of exposure and power of the x-ray. Lead is an excellent shield.

Where operators must perform near any radioactive sources, management must ensure that all of the proper safeguards are taken to protect them.

ERGONOMICS

The subject of ergonomics is finally coming into its own when it comes to machine design for operators. It is not unusual to see a very tall person operating in a tight compartment of a lift truck. Just lifting a foot to brake suddenly is a chore. These tight quarters can have a big impact on safety of the vehicle as well as for the protection of others.

As another ergonomic example, when driving forward with a typical lift truck it is sometimes difficult to have a clear field of vision when peering through the mast. Operators sometimes have to lean to the sides to see where they are going.

Some of today's forklift manufacturers are keeping pace with the technology in automobiles. Trucks are better designed and feature improvements in quality, durability, maintenance and ergonomics. As an example, a new narrow aisle truck allows for the operator to sit sideways for access to fingertip controls and a greater view of the forks. Another allows a lift truck to be operated all day without using the brake. By using a variable displacement pump and hydraulic motor instead of a power shift transmission, the hydrostatic truck travels, stops and reverses without requiring an accelerator pedal. A single pedal that rocks back and forth operates the controls.

Manufacturers continue to apply ergonomic principles to lift trucks. Purchasers can consider the following improvements:

- **Vibration Characteristics.** Whole body vibration increases the risk of back injury and other musculo-skeletal injury. Where seats and cab compartments are designed properly, cushioning for the operator is used. At present, existing modifications can be made by using vibration-dampening floor mats and shoe inserts.

- **Seat Design and Structure.** Seat adjustments should include adjustments both vertically and horizontally. The back rest, seat pad and back rest are a part of this change.

- **Control Designs.** Where operators have to continually look at instruments, this condition has been improved. Locations of controls help to influence operator visibility, which helps the posture of the operator. The ergonomics positioning of controls helps minimize extended forward and side reaching, helps minimize forearm rotation and keeps the wrist straight.

- **Sitting/Standing.** A rule of thumb here is that if an operator must stand for more that four hours per shift, a seated model is preferred. If the operator must get off the truck more than once every five minutes or so, then a stand up model is recommended.

- **Physical Handling.** The handling of LPG trucks can be physically demanding for operators. Battery operated trucks require less manual or physical handling.

All operators should be trained and assisted in understanding how ergonomics can help them. Lighting can be poor in a location to where it affects safety as well as productivity. Operators have been known to raise the mast only to run into pipes, doorways or to knock down stored pallets because of the difficulty in seeing.

Forklift seats are notorious for taking abuse. Operators spend hours at a time in a seat. Many are not properly designed to ensure comfort and support for the operator. Proper padding as well as lumbar support would result in less fatigue for the operator.

Operator controls can be different from model to model. This is comparable to renting a different model car during different trips and finding that you have to search out how various controls function before you pull out. Fortunately, lift truck manufacturers are designing controls to be more user-friendly. Newly designed controls now fit the hand grip better. Operator compartments are now padded and include a forearm rest. Lettering and directional messages have finally been recognized as being important to the operator. Pedals are larger for the shoes and provide better surfaces. It is not unusual to see the two rubber pedal covers on a narrow-aisle truck to be bent and curled. A tripping hazard is easily created here. New rubber pads are needed from the manufacturer and the pads should be glued in place. It is apparent that the narrow-aisle operator must stand continuously to operate the machine, the surface he is standing or should be as comfortable as possible.

Ergonomics involves modifying the workplace to fit the demands of the worker. One size does not fit all. Where proper ergonomic practices are utilized, workers experience less fatigue, greater productivity, less absenteeism and fewer accidents.

It is recommended that when considering the purchase of a new piece of power equipment, first lease it. Allow employees to operate it and ask for feedback. Let the manufacturer or their representative know what design changes you would like to see. It is possible that the changes can be made to fit your specifications. The facility will see a greater return on investment by using this system for purchases.

Proper lifting is another area of concern for industry. Back injuries are very common throughout industry and tend to be the most expensive type of claim. The National Safety Council states that back injuries account for one-fifth to one-quarter of all workplace injuries. For workers under 45 years of age, back injuries top the list of workers compensation claims. Experts estimate that the annual costs of back injuries are $20 billion to $50 billion. For repeat claims, back injuries will occur in about 50% of all injured workers within two years. Back injuries can be reduced by following proper procedures. The concern might arise that questions the need for proper lifting by powered equipment operators. There are numerous times during the day that an operator must get off of his equipment to load a pallet, restack a pallet or boxes, move an object out of the path of the truck, work on a battery, etc. Opportunity does exist for the potential of back injuries.

The following recommendations should be followed to help in your manual material handling program:

- Employees are to be trained in the proper techniques for safe lifting and handling.

- Where possible, provide scissor-lift devices or other such mechanisms which reduce any bending or lifting by bringing the load off of the floor. The safest area for lifting is between the knees and the chest. Workplace design can be a positive benefit to the overall safety program.

- Keep loads close to the body and get a good grip. Get help if the load is too heavy. If handles or other devices can be placed on a load, use them. Figure 10.4 illustrates that the pallet has been raised to allow for ease of manually lifting.

- Lift mostly by straightening the legs with the back flat and feet close to the load. Do not jerk on the load to lift it.

- Size up a load before lifting it. Check for nails, sharp corners or any other physical hazard that can harm you. Use PPE for added personal safety.

- If the load is too heavy—stop. Do not attempt to lift it.

- Never twist or turn without first moving the feet along with the body and load.

- Do not lower the load awkwardly. Do not keep the arms extended when lowering or lifting.

- Exercise and keep in shape. By strengthening the abdominal muscles, the back muscles receive additional support.

Figure 10.4. Safe Lifting from a Pallet

Rather than manually lifting product from the floor, the operator should use a raised pallet to help prevent a back injury.

In 1981, NIOSH published a formula for manual lifting that industry has used as a guideline for workplace design. In 1994 NIOSH developed a new lifting equation and published it in a 120 page booklet titled *Applications Manual for the Revised NIOSH Lifting Equation*. For those that are interested, these guides can help in preventing back injury.

The subject of back belts or back supports is still being scrutinized for effectiveness. OSHA does not recognize back belts as PPE at this time.

Lights, Alarms, Mirrors

Being that the operator has a difficult time seeing, he must ensure he is being seen. Though not an OSHA regulation, flashing lights on a moving piece of power equipment will pay dividends for safety. The flashing light should never be objectionable to the operator or to others. As an example, when raising a load high to place on racking, the operator must now look through the overhead guard. If a light is mounted there and is bright enough, it could compromise worker safety. The important thing is that the light can be seen by others but should not create a hazard.

The operator's piece of equipment should also be heard. A horn is necessary to alert others. A back-up alarm is not required by OSHA, however, back-up alarms on counter balance trucks can do much to prevent accidents. The volume of the alarm can be adjusted in such a way that it is evident but not offensive to others. A reminder that battery-powered equipment is very quiet.

Another ergonomic issue to consider is the use of rear-view mirrors. Convex or panoramic-view mirrors can go a long way to improve operator safety. Before backing up, an operator must look to the left and to the right. A mirror allows for a third look before backing. Once operators become used to this device they usually will not part with them and consider mirrors essential. Without mirrors, accident potential increases.

ACCIDENT FACT

After stacking a pallet on a 25' rack, the operator proceeded to back up so he could lower his forks. He made a half-hearted attempt to look before backing. At that moment a warehouseman was walking down the same aisle to pick up a part. The employee on foot felt it was safe to walk behind the lift truck which was close to the rack. When the operator backed up he did not see the person on foot. There was no back up alarm to alert him. Also, there was not a convex type mirror for a better field of vision. The employee was pinned by the forklift to the racking and suffered a fractured pelvis.

SUMMARY

Powered equipment operators, like many other employees working at various jobs, are exposed to potentially harmful items that could harm their health. Many times little consideration is given to occupational health issues. As the chapter pointed out, there must be an on-going concern for carbon monoxide, noise, chemicals, radiation and ergonomics; all of which could harm operators.

Carbon monoxide, CO is the most prevalent atmospheric poison in this country—both at home and at work. Forklifts and other machines are fully capable of producing CO levels that exceed regulatory guidelines. The danger is unusual because it strikes silently. This gas is invisible, odorless, colorless and tasteless.

Noise is another very common work-place exposure. Employees never get used to high noise levels, as they claim they do, they simply lose their hearing in small stages over a period of time. Powered equipment operators can be exposed to harmful noise levels and must be educated and protected.

Over a thousand new chemicals are introduced to the workplace each year. This adds to the several hundred thousand already there. Lift truck operators lift, carry, stack and manually handle various chemicals in an assortment of containers. A plan has to be in place to fully educate operators to the hazards of chemicals. Procedures in the wearing of PPE and proper clean up, disposal and medical emergency response are vital to a safety program. The Occupational Safety and Health Administration has cited employers for violations of the Haz-Com Standard more times than any other standard.

Ergonomics definitely affects operators of powered equipment. Research has shown that work place re-design for ergonomics has improved productivity, morale, absenteeism and has reduced injuries. The comfort and work pace of operators has much to do with proper design of the lift truck. Many manufacturers are including ergonomics in the placement and design of controls, foot pedals, operating zone, seats, masts and maintenance processes to allow for greater ease in operating.

All of these elements; CO, noise, chemicals, and ergonomics should be a vital part of each powered equipment training program.

REFERENCES

McCammon, J.B., L.E. McKenzie and M. Heizman. "Carbon Monoxide Poisoning Related to the Indoor Use of Propane-Fueled Forklifts in Colorado Workplaces," *Applied Occupational Environmental Hygiene*, March 1996, pp. 192-198.

Materials Handling Engineering Magazine, "LP-Gas Requires Common-Sense Training," June 1996, p. 18.

Material Handling Engineering Magazine, "CO Controller for Lift Trucks Clears the Air at Casting Plant," January 1995, p. 130.

Michigan's Occupational Health, "Industrial Lift Trucks," Lansing, MI, Volume 29, No. 1, Spring 1996.

National Safety Council. *Fundamentals of Industrial Hygiene*, Fourth Edition, Itasca, IL, 1996.

National Safety Council, Safety and Health Agenda. "How to Prevent Back Injuries," *Safety and Health*, June 1996, p. 42.

Natural Gas Applications in Industry Magazine, "Smart Company Makes Smart Move to CNG," Fall 1995, pp. A-11,12.

Natural Gas Applications In Industry Magazine, "Natural Gas Forklifts Clean the Air," Fall 1994, pp. A-10,11.

Payne, M. "Carbon Monoxide: Silent Suffocation," *Ohio Monitor*, February 1988, pp. 5-9.

Schwind, G. "Ergonomics:Expanding Toward 2000," *Materials Handling Engineering*, October 1995, pp. 141-146.

Schwind, G. "New Lift Trucks," *Materials Handling Engineering*, June 1996, pp. 40-46.

Selan, J. "Ergonomic Design and Lift Trucks," *Modern Materials Handling*, October 1995, p. 17.

Upp, P. "Lifting the Illusions," *Warehousing Management Magazine*, Warehousing Management, May/June 1996, pp. 43-45.

11

USING JOB HAZARD ANALYSIS FOR POWERED EQUIPMENT

One of the most important programs in any overall safety effort is the job hazard analysis program (JHA). Some refer to it as the job safety analysis (JSA) or safe job procedures (SJP). Despite the name given to this program, it can do much to educate and safeguard employees in any operation. Discussions with seasoned safety directors have revealed that if they could select one program that would provide them with the greatest benefits to their overall safety program, it would be JHA.

Surprisingly, this program can enhance the safety of powered industrial truck operators and operations. Management must be able to recognize the benefits of the overall JHA program so an assessment can be made of all those jobs that operators perform in the course of their daily tasks. This process of management evaluation will require:

- Establishing a listing of all jobs performed by powered equipment.
- Utilization of the proper forms.
- Actual JHA development.
- Management review and publication of documents.

The JHA process must involve employees and cannot be performed solely by management. The reason for this requirement is that learning and education by employees is a vital part of the development of that final document—a JHA for each separate job.

This chapter will discuss the background and foundation of JHA, how management should get started on the process, proper forms to use, management training, employee awareness of the program, how to complete a JHA, the review process, and how to provide ongoing maintenance of the program.

WHAT IS JHA—AN OVERVIEW

To start, a review of what JHA is and what the program sets out to accomplish is important. JHA is a continuous management process. It is a program that allows for step-by-step procedures to determine the safest, as well as the most effective, way of performing a job. A job is a segmented set of steps to complete a task. As an example, a lift truck operator must check and top-off a large battery with water on an electric truck. This process contains a few dangerous steps. The battery is heavy, it contains acid, it may contain surface corrosion, and the hands or fingers could be pinched in the movement of the battery. It's important to assess exactly how to pull the battery out properly, check and fill it, and return it to its original position without harming the operator, or for that matter, without damaging the battery.

The above-mentioned battery filling is called a job. It is a task that the lift truck operator may perform. This is one task among many that would be required of power equipment operators. So when the term "job" is used in JHA, it is not referring to the operator's occupation; it is referring to tasks that an operator completes in the course of their work.

Management must then meet with at least one employee and ask for assistance in conducting a JHA. Management asks questions, the operator assists in relating to management personal knowledge of the task, and the job is detailed on the proper form. Once completed, someone that has been properly trained will then review this rough draft and dress up the copy. The final JSA will then be typed onto a form for filing, perhaps posting, and later for additional employee training and review.

It's important to note that JHA is best accomplished by the actual completion of a JHA—start to finish. Occasionally someone will ask, "why doesn't someone prepare the best JHAs and then give them (or sell them) to the various plants and facilities?" The answer is that the program can only provide benefits, as well as educate both management and labor, when the JHA is manually completed by both parties. Much more is learned by management and employees when they must complete a JHA by themselves. The learning takes place in the doing.

More than one employee can participate in the development of a JHA. For efficiency, it's best to select the most skilled and most cooperative employee to complete a JHA. If more than one employee is used, be sure to remind everyone that time is important and the supervisor cannot spend more time than necessary. As far as employee cooperation is important, this does not infer that employees are uncooperative; it simply means that for the sake of efficiency, quality, and time, an employee should be selected who will move along with succinct responses to questions, as well as who will lend overall expertise to the document.

Starting the Process

The starting point in the knowledge gathering process is to alert supervisors to the purpose of the program and overall benefits—mostly the reduction and elimination of accidents. Explanations of what a job (task) is must take place up front so everyone has the correct mental framework of the program. Supervisor's training is essential in the initiation of the program.

Next, a list of jobs should be developed. Included in this list can be jobs for any occupation in a facility, but for the sake of this book, the jobs should cover powered industrial equipment. Note the sample list of jobs below. You can add (to this list or delete from) as necessary. It's important to think through the process because some jobs that are performed infrequently could easily be left off of the list. Also, new jobs can be added to the list at any time. Keep in mind that JHA is a continuous program. Once started, the process should continue indefinitely. There will be more about this later.

This list of jobs is not in any specific order or priority:

1. Operating a counterbalance forklift.
2. Operating a narrow—aisle forklift.
3. Cleaning up a battery acid spill.
4. Recharging a battery.
5. Unloading a palletized trailer.
6. Operating a powered pallet truck.
7. Filling a battery.
8. Washing down a battery.
9. Using a fire extinguisher.
10. Using an eye wash unit.
11. Spotting a trailer at the dock.
12. Chocking the wheels of a trailer.
13. Using a truck restraint device.
14. Taking inventory from a scissors lift.
15. Operating a stock chaser.
16. Proper inspection of lift truck.
17. Stocking palatized material.
18. Etc.

The list, developed by supervisors and safety committee members, contains only seventeen JHAs and is intended to be an example of how to focus on a job or a task. It can easily be expanded to include many other jobs. Each one of the jobs has a beginning and an end. The JHA will detail how to safely perform each step so the complete job is completed without injury. The list is not in any form of priority or ranking, but is developed in a random manner. note: It's important to develop such a list before actually starting JHA. To aid in the development of a job list, a sample form has been placed in this chapter; see Figure 11.1 for Form #1 . This form will help track JHA activity.

162 / Forklift Safety

Three forms should be used on the JHA program:

- **Form #1.** JHA master list and tracking form.

- **Form #2.** Rough copy form that the supervisor can use on the shop floor to record a JHA that was completed with an employee.

- **Form #3.** Final copy form, one in which the JHA, after being approved, is typed onto the form for posting at key work stations, during training of new and current employees, and for filing.

Proper Forms

The first working form (Form #2) should be a copy that the supervisor can use on a clipboard to document a safety discussion with the employee. Figure 11.2 identifies a blank Form #2. This would be considered a rough draft copy, see a completed sample in Figure 11.3. Form #3 should be the formal, typed copy that will be placed in a binder, posted at a key location (such as the battery charging area), and used for safety meetings or for assisting in accident investigations. Figure 11.4 illustrates a blank Form #3. It can also be used for training new or transferred employees. This final form should only be completed after the review process has taken place to ensure it is accurate and complete.

Note the top of the JHA forms. It's important to enter specific information on both Form #2 and Form #3:

Place the JHA number on the form specific to the number on the JHA master list. This should include:

- The title of the JHA.
- The facility location.
- Who completed the JHA; who helped (employee).
- Date of completion.
- What PPE is required for the job.

This chapter contains a sample copy of all three forms; the forms can easily be duplicated. The rough draft form, #2, will contain handwritten details on JHA #17, "Stocking Palletized Material." Note that the number of the JHA from the master list serves as the identification for the JHA. JHA's would be selected from the master list on a priority basis.

Figure 11.1. JHA Master List

JHA Number	Title	Job Assigned to	Date Assigned	Completed Date	Review Date

A list of potential jobs will have to be created for the JHA program. This form provides guidance for this project.

Figure 11.2. JHA Form #2 - Blank

JHA # _____	Title of Job _____
Location (Plant) _____	
Supervisor _____	
Employee(s) _____	
Date Completed _____	Date Revised _____
PPE Required _____	

Basic Job Steps	Hazards	Safe Job Procedures

HAZARD SECTION: SB = Struck by, CW = Contact with, CBy = Contact by, CB = Caught between, O = Overexertion, E = Exposure, FS = Fall same level, FB = Fall to below, SAG = Struck against, CO = Caught on, CI = Caught in

Figure 11.2 *(continued)*

SIDE 2
FORM #2

A supervisor will need a form to complete a JHA with an employee at the worksite.

Figure 11.3 JHA Form #2 - Completed

JHA # 17	Title of Job STOCKING PALLETIZED LOADS
Location (Plant) WAREHOUSE #74	
Supervisor SAM BELANGER	
Employee(s) DOMENIC SERVIO	
Date Completed 11-20-96	Date Revised
PPE Required HARD HAT, STEEL TOE SHOES	

Basic Job Steps	Hazards	Safe Job Procedures
1. LOCATE PALLET	1. SBY - MOVING EQUIPMENT SAG - RACKING	1. DRIVE FORKLIFT TO DEPT. WHERE PALLET IS LOCATED. BE AWARE OF OTHER EMPLOYEES WHILE MOVING. FOLLOW PROPER DRIVING PROCEDURES.
2. LIFT PALLET	2. - - - -	2. VISUALLY INSPECT BOTH LOAD AND PALLET BEFORE LIFTING. PROPERLY PLACE FORKS IN PALLET AND PLACE THE LOAD AGAINST THE MAST. LIFT AND TILT THE LOAD TO SECURE IT. KEEP THE LOAD AS LOW AS POSSIBLE.
3. DRIVE TO DESIRED LOCATION	3. SBY - LOAD, MOVING EQUIPMENT, SAG - RACKING, PRODUCT	3. FOLLOW SAFE OPERATING RULES. BE OBSERVANT OF OTHER EMPLOYEES AS WELL AS RACKING AND PRODUCT. CHECK BIN OR RACKING AREA TO BE READY TO STORE THE PALLET.
4. RAISE PALLET	4. SBY - FALLING PARTS OR PRODUCT	4. PROPERLY POSITION THE LIFT TRUCK IN FRONT OF THE RACK/BIN. RAISE THE LOAD TO THE DESIRED HEIGHT. BE ALERT FOR OTHER EMPLOYEES IN THE AREA.
5. POSITION PALLET	5. SBY FALLING PARTS OR PRODUCT	5. BE SURE THE PALLET IS AT THE PROPER HEIGHT BEFORE MOVING FORWARD. CHECK ON BOTH SIDES OF THE LOAD BEFORE PULLING FORWARD. BE SURE THAT NO EMPLOYEES ARE NEAR THE LOAD ON THE OPPOSITE SIDE OF THE RACKING. TILT FORWARD AND LOWER THE

HAZARD SECTION: SB = Struck by, CW = Contact with, CBy = Contact by, CB = Caught between, O = Overexertion, E = Exposure, FS = Fall same level, FB = Fall to below, SAG = Struck against, CO = Caught on, CI = Caught in

Figure 11.3 *(continued)*

		PALLET ONTO THE RACK OR SHELF.
6 BACK OUT AND LOWER FORKS	6 SBY - FORKLIFT	6 ONCE LOAD IS SECURED, CHECK ALL SIDES OF THE LIFT TRUCK FOR EMPLOYEES AND BACK AWAY. LOWER THE FORKS WHILE THE LIFT TRUCK IS COMPLETELY STOPPED. LOWER FORKS AS CLOSE TO THE FLOOR AS POSSIBLE.
7 PROCEED TO OTHER AREAS	7 SBY - FORKLIFT SAG - PRODUCT	7 CAUTIOUSLY PROCEED TO OTHER STOCKING AREAS. BE ALERT FOR OTHER PIECES OF MOVING EQUIPMENT, PRODUCT AND INDIVIDUALS. USE THE PROPER PROCEDURE FOR CARRYING AND STOCKING OTHER LOADS.

A sample of a completed Form #2 on a job selected from the job list.

168 / Forklift Safety

Figure 11.4 JHA Form #3 - Blank

JHA # SUPERVISOR: LOCATION: PPE REQUIRED:	TITLE OF JOB: DATE COMPLETED:	EMPLOYEE: DATE REVISED:
Basic Job Steps	Hazards	Safe Job Procedures

HAZARD SELECTION: SB = Struck by, CW = Contact with, CBy = Contacted by, CB = Caught between, O = Overexertion, E = Exposure, FS = Fall same level, FB = Fall to below, SAG = Struck against, CO = Caught on, CI = Caught in

A final form on which to type the completed JHA is needed.

Priority Basis Rationale

When jobs are being selected they should be based on:

- Jobs that have produced serious injuries in the past.
- Jobs that have a high potential for injury.
- Jobs that are new or seldom performed in the facility.
- Jobs that do not meet the above criteria.

Starting a JHA

As an example of a job having a beginning and an end, note these examples. Let's say that you are going to trim the hedges at your house. You will need a set of trimmers, possibly an electric extension cord, gloves, and safety glasses.

Your first step would be to collect your appropriate tools. Then select a starting point for trimming. Once completed, return the tools and clean-up. This job would contain several steps. The point here is that each job or task has a beginning and an end.

If your Saturday afternoon chore is to do yard work, you wouldn't complete an entire JSA for "doing yard work"; this choice is much too broad. When you take the yard work a section at a time, you would end up with several JSA's, not one.

Each job has a starting point and an end. Most jobs have five to eight steps. Some have less, some have more. The number of steps will be determined by the complexity of the job. The example on the "rough draft" form, #2, will provide information on the style and the process.

A few key points to keep in mind before starting—before you put a pen to the paper:

- All JHA's must utilize a three column format, this can be noted by looking at forms #2 and #3.
- The left-hand column is for the basic steps—the very beginning of the JHA. In the basic step, there is no need to discuss the how, when, where, or why. This is a basic step—a few quick action words to describe what is about to happen. Keep this description to only a few words.
- The middle column identifies the hazards associated with the basic step. These are real hazards, not imagined or remotely possible. Note the code index which is used in the middle column to define the hazards. It's important to learn this code —not just for JHA, but for hazard awareness and accident investigation. The abbreviations are used to

shortcut the use of the form as well as to provide immediate recognition by the user of the JHA as to the hazards. As an example, SB or struck by, indicates that something is moving and has struck someone. If a step doesn't have any apparent hazards, a dotted line across the middle column after the number is sufficient.

There are only eleven injury types or exposures for any job. Note the hazard abbreviation and definition.

CBe   **Caught between**—usually a pinch point of some kind.

CO   **Caught on**—usually involves something moving and a part of the body is caught on something.

CW   **Contact with**—usually contact with a chemical, electricity, or a hot surface.

SB   **Struck by**—usually a struck by injury involves something in motion; an employee is struck.

SAG   **Struck against**—usually the employee strikes against something, such as tripping and falling against a machine—you are injured by striking against it.

E   **Exposure**— usually involves noise, carbon monoxide, chemicals, etc. Most times this is an unseen hazardous agent.

O   **Overexertion**— usually an injury from lifting or pulling, a strain.

CI   **Caught in**—usually the result of confinement of a body part, such as a foot being caught in a hole while walking.

FS   **Fall same level**—a slip on the floor, a trip, the employee is injured as a result of a fall.

FB   **Fall to below**—usually a more serious fall involving a fall off a dock, pallet, or landing.

CBy   **Contacted by**—something such as electricity, or a splash from a chemical.

Each of these capital letter hazard identifiers is to be used in the middle column of the JHA. Note the examples below:

FB   Employee could fall off the dock.

SBY   Pedestrians could be struck by the forklift.

SAG   Employee could strike his head on the bottom of the trailer while placing wheel chock.

The third column of the JHA requires detail and information that are a part of the first two columns—all three columns are tied together. Note the example from our list, JHA #4, Table 11.1.

Table 11.1. Step 1 in Recharging a Battery

Basic Step	Potential Hazards	Correct Procedures
1. Remove retainer clips.	1. CW—Potential battery corrosion on clip.	1. While wearing rubber gloves and goggles, reach down and pull straight up on the battery retainer clip on the open side of the lift truck. Set clip to the side.

Let's review the key points in this first step in JHA #4:

- Note that there is a number "1" for each of these columns; this links all three items together—sort of a marriage.

- There are only three words in the basic step.

- CW indicates that there is a hazard to the skin and eyes if the battery has allowed corrosion to form on the battery retainer clip.

- The correct procedure describes exactly how to remove the battery retainer clip without suffering an injury.

Continue with the rest of the steps in the JHA until the job is fully completed. Note the typed example of the entire JHA for job #4 or Form #3. Simply follow this format for each job. Note Figure 11.5, a completed JHA on Form #3.

Management Training

Who should be trained in JHA? How long does formal training take? Who should conduct JHA reviews?

First, those supervisors who are responsible for employees who operate powered industrial trucks should be trained in JHA. Because this program is so beneficial to a plant safety program, it should receive the blessing of top management. Top management should sit in on the training sessions so that they gain an understanding of how the program functions. It's an investment in time that will pay big dividends for the facility.

Because supervisors are closest to their employees, first-line supervisors should be conducting JHA's. Supervisors know the operation, they know their employees. When starting JHA, most supervisors know exactly who they will select to assist in various JHA's. They know who is the most knowledgeable, and has the best grasp of operating power equipment or completing any other job.

172 / Forklift Safety

Figure 11.5. JHA Form #3 - Completed

JHA #: 4	TITLE OF JOB: Recharging a Battery	
SUPERVISOR: John Belanger		EMPLOYEE: Mike Walker
LOCATION: Coil Mill Department	DATE COMPLETED: 6/15/96	DATE REVISED:
PPE REQUIRED: Hard Hat, Steel Toe Shoes, Rubber Gloves, Goggles		

Basic Job Steps	Hazards	Safe Job Procedures
1. Spot Lift Truck	1. SB - Moving lift truck SAG - Chargers	1. Cautiously drive lift truck to battery charging area. Be alert for other pieces of equipment and employees. Pull lift truck into recharge area directly in front of charger. Pull in slowly so as not to hit the charger or stand.
2. Prepare to Hook-up	2. FS - Trip over forks	2. Inspect battery charger plug after putting on gloves and goggles. Be alert as to where you walk near the lift truck. Be sure charger is "off".
3. Hook up to battery	3. CW - Battery corrosion FS - Trip over forks	3. Unhook battery cable from lift truck. Inspect battery plug for defects. Be alert to battery corrosion. Place male and female plugs together and lock in place. Be aware of where you are walking to keep from tripping on forks.
4. Turn on charger	4. FS - Trip over forks	4. Turn charger to "on" position. Set timer for proper charging time. Be alert for trip hazard. Remove gloves after washing them off in utility sink. Remove goggles and put away.

HAZARD SELECTION: SB = Struck by, CW = Contact with, CBy = Contacted by, CB = Caught between, O = Overexertion, E = Exposure, FS = Fall same level, FB = Fall to below, SAG = Struck against, CO = Caught on, CI = Caught in

A completed Form #3 on recharging a battery.

ACCIDENT FACT

Several electricians entered a large warehouse to work on overhead lights. They needed to be lifted up to the lights. A lift truck operator who was driving by was asked if he could lift them. He secured an appropriate lifting cage and lifted both of the visitors. While placing them as near to the lights as possible, the operator tilted the basket forward to make it level. He tilted it too far. The cage almost slid off of the forks because no one bothered chaining and latching the cage to the mast. Both men could have fallen approximately thirty feet to the warehouse floor. Luckily, the cage was stopped by nearby racking. The job did not have a JHA review for many years and the lift truck operator was new and was never briefed on the full details of the job.

The basic learning part of JHA should only take a few hours of classroom time. The real learning takes place on the factory floor with the employees. Once supervisors have completed the classroom phase of JHA, they should be ready to go out and complete their first JHA. History has shown that this first attempt will be confusing and the initial draft copy on Form #2 will undoubtedly be rough.

Before the supervisor continues, someone who has a firm grasp of the JHA program should review all work. A critique of the form should take place and feedback should be given. JHA is a learning process and it's not expected that the first try will be a polished JHA. Feedback should be given as soon as possible.

A typical JHA should take 20 to 30 minutes once the hurdle of learning the correct method has taken place. At first, a supervisor may take twice this long and not yet arrive with an acceptable copy. Learning takes place by doing. The more JHA's completed, the more knowledge is gained.

The review process is very important for many reasons. The completed JHA must be as accurate as possible. After all, it's serving to define the exact procedure in which to safely perform a job. An approved JHA should be able to stand the test of time.

Additional Comments on JHA

OSHA is considering legislation which will require formal safety and health programs throughout the industry. Safe job processes will be a part of this program. When California passed its new safety and health legislation, Senate Bill 198, the program guidelines required safety procedures for jobs. The format used is very similar to JHA. The Bureau of Mines, in an effort to have zero fatalities by the year 2000, has embarked on an extensive JHA (JSA) program.

For years, it has been recognized that JHA is one of the most significant parts of any safety endeavor and deserves full attention—especially where powered equipment operators are involved.

SUMMARY

Many safety professionals agree that the job hazard analysis program is probably the most important program in safety. Organizations many times fail to include JHA in the programs. Much of this lack of attention is due to not fully understanding the benefits of JHA.

The program does involve time, which is needed to train supervisors and employees. Forms have to be created and used properly. Each analysis is important to not only safety, but to productivity.

This chapter contains all of the information and forms to develop and implement a JHA program; thus it can enhance powered equipment safety. A note of caution: JHA is not an easy program. In many cases, the program has a slow start because of the learning curve. Once the program begins and the JHAs are reviewed and corrected on a regular basis, the program then becomes easier. This difficulty, however, should not be reason to avoid this important program.

REFERENCES

Armco Steel Corporation. *Accident Prevention Fundamentals*, Job Safety Analysis, Chapter 5, Middletown, OH, 1975.

Business & Legal Reports, Inc. "Job Hazard Analysis," Madison, CT, 1996, pp. 145-1; 145-15

DeReamer, R. "Job Hazard Analysis," *Modern Safety & Health Technology*, John Wiley & Sons, NY, 1980, pp. 160-173.

U.S. Department of Labor. *Job Hazard Analysis*, OSHA, OSHA #3071, 1988.

National Safety Council. "Job Safety Analysis," *Instructor Manual*, Itasca, IL, 1995.

National Safety Council. " Safeguarding—Job Safety Analysis," *Accident Prevention Manual*, Itasca, IL, pp. 377-382, 1992.

National Safety Council. *Job Safety Analysis—Identifying & Controlling Hazards*, Itasca, IL, 1994.

12

INVESTIGATING ACCIDENTS AND INCIDENTS

It's inevitable that where there are forklifts, there will be accidents or incidents. Forklifts can be destructive to property, product, machinery, and people. A typical forklift weighs three to four times more than an automobile. It's no wonder that it's so easy for the operator to bump, dent, or damage things. Also, when making contact with a person, the forklift always wins the fight.

Of greatest concern is the safety of individuals. As identified in an earlier chapter, the statistics developed by OSHA illustrate that there are many individuals who are struck by various pieces of power equipment. A forklift can be very unforgiving when it comes to striking an individual. Keep in mind that the only good thing that can come out of an accident is the knowledge gained in preventing the next accident.

When an accident occurs, an immediate investigation is essential. Evidence must be gathered and documented while the facts are still fresh. Where there is damage to the building, premises, product, or machinery, an investigation of the incident can yield surprising results. An accident where bodily injury is involved usually gets the attention of many people. Daily incidents that involve damage to the plant or product sometimes fail to raise any concern. Both incidents and accidents are a problem within industry that cannot be overlooked or avoided.

For accidents that involve bodily harm, the accident should be investigated immediately. Obviously proper medical care should be given to the injured employee. Witnesses should be interviewed and facts should be gathered. The ultimate concern is to prevent the next accident from occurring.

It's a commonly accepted principle in industry that there will be hundreds of incidents in relation to each accident that occurs. Note the illustration in Figure 12.1; to maximize the safety efforts in a plant, management should focus on the investigation and correction of the day-to-day incidents.

A pioneer in the safety profession, William Heinrich, researched many thousands of workplace accidents and concluded, statistically, that for each very serious injury, there were 29 less serious injuries and 300 minor injuries. Later, the concept of 6,000 incidents or opportunities would also be a contributing factor. The idea is to focus on the day-to-day incidents; this will help prevent all injuries. This concept is illustrated in Figure 12.1.

Figure 12.1. Accident Pyramid

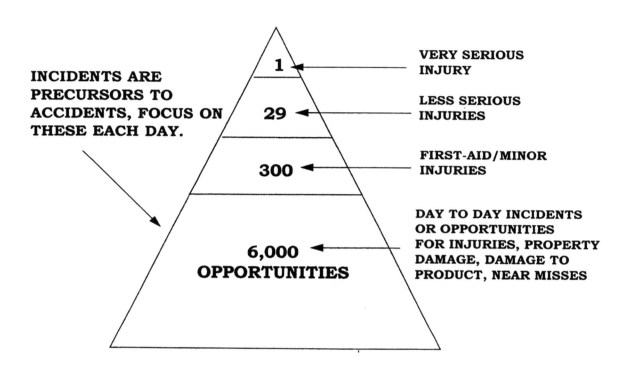

For accident reduction, focus on the prevention of day to day incidents.

William Heinrich researched several thousand accident claims during this study in the 1930's. His conclusion was that most accidents are the result of unsafe behavior. This study was developed in further years to focus on the investigation and prevention of daily incidents to aid in accident prevention.

To do an effective job on the investigation of incidents and accidents involving powered industrial equipment, management should focus on the following:

INCIDENT INVESTIGATION

One look at a typical forklift truck will reveal scratches on the paint and other tell-tale signs that the vehicle has made contact with various items. Each dent or scratch on the lift truck is an indicator that something else beside the paint was damaged in the process.

The dollar losses involved in daily incidents in a plant can be significant. Many of these costs go unnoticed by management. Management would most likely display an awareness over a dropped pallet full of television picture tubes or damage to bags of food products. These incidents would result in high costs because of the nature of the item being transported. Management will get excited, the operator may get disciplined, and in a few weeks, the episode will be forgotten.

A close look at the entire details of these incidents may reveal that the forklift had faulty brakes and has damaged other product and equipment recently. The operator knew he had bad brakes and reported the condition to his supervisor. His supervisor was too busy and failed to take the forklift out of service to have the brakes repaired. Figure 12.2 illustrates damage to a building as a result of faulty brakes.

Figure 12.2. Damage to Wall From Forklift

Forklifts can cause damage to buildings that can cost an organization more than that of accidents.

A look at the contributing causes to the incident would most likely be that:

- There were no formal requirements for a daily operator check for inspections of power equipment.
- The operator continued to operate the lift truck despite the recent incidents.

- He has had close calls while driving, almost hitting other employees because he couldn't stop fast enough.
- The employee was not disciplined for his unsafe behavior.
- The supervisor failed to take the lift truck out of service.
- The plant has no awareness in formally reacting to daily incidents so that it can control its losses.

The damage to the pallet full of product costs several thousand dollars. It was noticed by many people because the incident was so costly. Despite the commotion at the time, the cycle will repeat itself because the company did not take any action to prevent reoccurrence. Most of all, the episode will be forgotten because no one was injured.

Incidents are very common. Incidents are a daily part of the everyday industrial setting. When management fails to act upon incidents, these events will eventually turn into accidents with injury. Note the damage to the battery chargers in Figure 12.3.

Figure 12.3. Battery Charger Incident

A forklift operator drove his lift truck into a battery charging station and caused severe damage; fortunately no one was injured.

To help resolve the issue of ongoing incidents, a special form should be developed to document the incident and recommend methods of prevention. Note the sample form in this chapter. Of course, it's important to investigate daily incidents of any kind in the workplace. Effort must be put forward to correct incidents so that the accidents are prevented. Since this book deals with powered industrial equipment, it would be very important to focus on incidents involving power equipment. Appropriate forms have been included in this chapter. Figure 12.4 illustrates a blank incident reporting form. Figure 12.5 cites a real incident on the form.

The form allows for the recording of the events involved in the incident. It allows for a listing of the damage incurred, the down time involved, and the corrective action. The sample form has been filled out with information as a result of a real incident from a factory. Note the details involved; problem-solving is evident. With management taking corrective action on the incident, it's less likely that it will occur again. The form should be completed by the supervisor who manages that particular department in which the incident took place. The form would only take a few minutes to complete because of the simplicity in its design. Supervisors could use the forms during employee safety meetings to remind and reinforce the subject matter. Completed forms can also be placed on a safety bulletin board as a reminder to all employees.

Employee involvement is key in an effective safety and health program. Employees can assist in the investigation process and completion of incident reports; this will allow employees to think more seriously about the daily incidents they encounter. Forklift operators can focus on the inspection and correction of forklift incidents. The ultimate goal is to have fewer and fewer incidents. Knowledge is gained through the utilization of the completed forms. Accident prevention is gained by correcting those elements that contribute to incidents.

ACCIDENT INVESTIGATION

An employee slips and falls on some hydraulic oil that was left on the floor. The hydraulic oil has been leaking from a forklift. Many people knew that the forklift was defective. The employee that slips and falls gets up, shaken but not injured.

In another part of the plant, supervisor slips and falls as a result of the same oil being dropped by the same forklift. In this case, the supervisor breaks his wrist as a result of the fall.

Two incidents; two separate outcomes. Why? Because not all incidents or exposures result in the same consequence. The first employee who fell could have broken his wrist. The supervisor could have fractured his skull in the fall. Many times, chance plays a part in the outcome of an incident. Either it will result in a close call, as described earlier, or result in a personal injury to someone.

Once the injury occurs, there will be an investigation by management. A plant with an effective safety program would do more than fill out the required state forms for the employer's first report of injury; these forms are for reporting the accident for the purpose of workers compensation processing and filing, not necessarily for investigating.

Figure 12.4. Blank Incident Reporting Form

**NON-INJURY INCIDENT REPORT
PROPERTY DAMAGE REPORT
AND/OR
DISCOVERY OF POTENTIAL HAZARD**

(CHECK ALL THOSE THAT APPLY)

☐ Near Miss ☐ Non-Injury ☐ Property Damage ☐ Incident/Event ☐ Fire Loss ☐ Unsafe Condition

Facility/Location: _____ Department: _____ Date Occurred: _____

1) Describe the incident or what occurred: _____
   _____
   _____
   _____
   _____

2) Machine or equipment involved: (number/name) _____
3) Extent of damage/cost of repair, replacement, etc. (describe): _____
   _____
   _____

4) How much lost time or down time was involved: (Describe) _____
5) How many employees were involved in the incident: _____
6) Was anyone injured: Yes ☐   No ☐
7) Was first aid administered: Yes ☐   No ☐
8) Was there a fire: Yes ☐   No ☐
9) Was employee wearing required personal protective equipment: Yes ☐   No ☐
10) What steps were taken by you to prevent recurrence: _____
    _____
    _____

11) What steps were taken by others (maintenance, management, outside sources, etc.) to prevent recurrence:
    _____
    _____

12) Was a health hazard or exposure involved: Yes ☐   No ☐
13) Can this incident or event take place again: Yes ☐   No ☐   Explain on back side.
14) Identify additional factors or comments on back of this report.
15) Have you drawn a sketch of the details on the back of this report or attached a photo: Yes ☐   No ☐

Signature: _____   Date: _____

Forward A Copy to: Corporate Safety

3400-64A   Revised 8/92

Each incident should be properly investigated and corrected. This form provides the proper information and format.

Investigating Accidents and Incidents / 181

Figure 12.5 Completed Incident Reporting Form

NON–INJURY INCIDENT REPORT
PROPERTY DAMAGE REPORT
AND/OR
DISCOVERY OF POTENTIAL HAZARD

(CHECK ALL THOSE THAT APPLY)

☒ Near Miss ☐ Non–Injury ☐ Property Damage ☒ Incident/Event ☐ Fire Loss ☐ Unsafe Condition

Facility/Location: WAREHOUSE Department: SHIPPING Date Occurred: 9-27-96

1) Describe the incident or what occurred: EMPLOYEE WAS ROUNDING A CORNER IN "G" AISLE AND DID NOT SLOW DOWN. PLANT MGR WAS ESCORTING VISITORS THROUGH WAREHOUSE AT THE TIME. PLANT MGR NARROWLY ESCAPED INJURY.

2) Machine or equipment involved: (number/name) FORKLIFT # 202

3) Extent of damage/cost of repair, replacement, etc. (describe): NO DAMAGE. COULD HAVE BEEN SERIOUS IF SOMEONE WAS STRUCK BY THE LIFT TRUCK

4) How much lost time or down time was involved: (Describe) 5 MIN OF LECTURE

5) How many employees were involved in the incident: ONE

6) Was anyone injured: Yes ☐ No ☒
7) Was first aid administered: Yes ☐ No ☒
8) Was there a fire: Yes ☐ No ☒
9) Was employee wearing required personal protective equipment: Yes ☒ No ☐

10) What steps were taken by you to prevent recurrence: IMMEDIATELY MET WITH OPERATOR AND HAD A SERIOUS DISCUSSION. WRITTEN FIRST WARNING WAS ISSUED

11) What steps were taken by others (maintenance, management, outside sources, etc.) to prevent recurrence: ASKED MAINTENANCE TO EVALUATE BLIND CORNER FOR A CONVEX MIRROR. BRAKES WERE CHECKED ALSO.

12) Was a health hazard or exposure involved: Yes ☐ No ☒
13) Can this incident or event take place again: Yes ☒ No ☐ Explain on back side.
14) Identify additional factors or comments on back of this report.
15) Have you drawn a sketch of the details on the back of this report or attached a photo: Yes ☒ No ☐

Signature: Janny Arnold Date: 9-27-96

Forward A Copy to: Corporate Safety

3400–64A Revised 8/92

A sample of an incident report which has been completed by a supervisor.

A plant with a good safety program would have taken the forklift out of service after a daily inspection revealed a hydraulic leak. Imagine how many other employees were exposed to those hydraulic leaks on the floor all over the plant. As fate would have it, no one else slipped and fell. Perhaps the employees became attuned to this dangerous condition and adjusted for it by stepping over or around the puddles. They should all realize that accidents are prevented when the contributing conditions are corrected.

Incidents are precursors to accidents. An accident is a sign that something has broken down in the overall management of the plant. In a sense, management allowed the hydraulic oil to be on the floor and did nothing to correct it. It's expected that forklifts will leak hydraulic oil. A preventive maintenance program, administered by staff or outside services, could have prevented the broken wrist suffered by the supervisor.

Was it "an accident"? According to Webster's dictionary, an accident is an unforeseen event. Was the leak unforeseen? Were the puddles unforeseen? How about the lack of an inspection program or more employee awareness to report and correct hazards—were these also unseen? Perhaps the word accident is inappropriate as it is used in industry. An event yes. An incident with serious consequences yes. An accident—perhaps not.

When investigating an accident, it's important to identify all contributing causes. All accidents have a combination of unsafe conditions and unsafe behavior. It's important to understand each. Management must focus on these two contributors to the accident *before* they take place, so corrective action can be taken.

Basic types of unsafe conditions:

- Defective equipment, tools, machines, working surfaces.
- Improper machine guarding.
- Hazardous arrangement or storage.
- Improper procedures and methods.
- Improper lighting and ventilation.
- Improper protective equipment or apparel.

Basic types of unsafe behavior:

- Working on moving, unguarded, or dangerous equipment.
- Operating equipment without authority.
- Failure to warn others or give proper signals.
- Making safety devices inoperative.
- Taking unsafe postures or positions.

Investigating Accidents and Incidents / 183

- Operating or working at unsafe speeds.
- Destructive horseplay.
- Failure to properly handle objects or product.
- Unsafe loading, carrying, lifting, or walking.
- Failure to use safe attire or personal protective equipment.
- Using unsafe equipment, or, using equipment in an unsafe manner.

Figure 12.6 illustrates an unsafe practice; a foot is dangerously out of the rear of a narrow aisle lift truck.

Figure 12.6. Hazards of Feet Extending from a Lift Truck

A very common accident potential situation, an employee's foot is out of the lift truck while backing up in a narrow aisle lift truck. Foot injuries are very common in this case.

Management must dedicate their efforts to the prevention of all types of accidents. Accidents are costly. Studies have shown that for every insured dollar spent on medical and workers compensation costs, an additional four to five dollars could be tacked on to the accident in indirect or uninsured costs. Depending on the industry involved, say a high-tech employer who employs highly skilled employees, the incident costs could easily reach 10 to 15 times that of direct costs. With this in mind, an employer should do all that's possible to eliminate accidents.

Figure 12.7. Accident Investigation Form

REPORT OF ACCIDENT INVESTIGATION				
Date of Report:		Employee Name:	Employee Title:	
Date of Injury:	Time Shift Started:	Shop Location:	Time of Injury:	Length of Employment:
Body part injured:		Was first-aid administered? Yes ☐ No ☐	Social Security Number - -	
Description of injury:				
Description of how injury occurred:				
Shop Managers Estimation of Accident: ☐ Work Related ☐ Non-Occupational ☐ Undetermined ☐ Disabling	Employee able to return to work? ☐ Yes ☐ No	Doctor or hospital treating employee:	How long will employee be disabled?	
Was Employee performing regularly assigned job? ☐ Yes ☐ No Was Employee instructed in this particular job? ☐ Yes ☐ No	Machine, tool or device involved:	How much production time was lost?		
Injury was caused by: ☐ Inattention of Duty Being Performed ☐ Inadequate Instruction ☐ Failure to Follow Procedures ☐ Lack of Knowledge & Skill ☐ Inadequate Enforcement of Rules ☐ Unsafe Act or Behavior ☐ Improper Job Instruction ☐ Unsafe Mechanical Condition ☐ Other - Describe: _____		Protective Equipment: Required Being Worn Hard Hat ☐ ☐ Glasses ☐ ☐ Goggles ☐ ☐ Safety Shoes ☐ ☐ Gloves ☐ ☐ Long Sleeves ☐ ☐ Face Shield ☐ ☐ Ear Plugs/Muffs ☐ ☐ Other? ☐ ☐>		
Give details on who, what, where, why and how, of accident:				
What will you do to keep this type of accident from reoccurring?				
What do you feel were the main causes of the accident?				
List any additional details or comments here:				
Shop Manager Signature: _____ Date: _____		List any costs associated with the accident; medical, workers comp payouts, damaged equipment, etc.:		

Each accident should be investigated and the appropriate form, such as this one, is to be used.

To assist you in proper accident investigation, a sample form has been provided (see Figure 12.7). This will provide you with the necessary information which you may need to help you with your own investigation.

A few key points to keep in mind when conducting a proper investigation are identified below:

- Ensure that the injured party receives proper medical care.
- Contain the area where the accident occurred—don't allow the scene to be altered.
- Take photos and possible measurements of skid marks, etc.
- Talk to witnesses, don't lead those that are willing to make a statement, don't interrupt.
- Take extensive notes, record time of day, weather conditions, and any additional details.
- Speak to the injured employee as soon as possible to get first-hand information.
- Compile the report based upon all of the facts that have been gathered.
- Complete the appropriate state workers compensation form in a timely fashion.
- Log the claim on the proper OSHA form.
- Proceed to correct all of those elements that contributed to the accident.
- Share this information with all employees.

Effective safety programs result in fewer accidents, fewer incidents, lower costs, better working conditions, and improvements in morale. Forklift accidents can be deadly. Everyone in a facility who uses any form of power equipment must dedicate themselves to the prevention of forklift accidents.

SUMMARY

The proper investigation of accidents and incidents is an important part of any accident prevention program. Once the causes of the accident or incident are discovered, appropriate intervention by management must be taken to prevent future accidents of the same type.

Powered equipment can be destructive to property and product. Employees can be easily injured as a result of something as minor as being bumped by a forklift. Accidents are costly. Incidents are equally expensive.

This book discusses numerous methods for accident prevention. When used properly, these guidelines will help reduce the needless waste that is the result of an accident or incident. However, when an accident or incident does occur, first provide medical care; then a professional approach should be used to investigate and correct those associated problems. If corrective action isn't taken, rest assured these same conditions or behavior will produce future accidents.

REFERENCES

Heinrich, H.W. *Industrial Accident Prevention*, 5th Edition, New York: McGraw Hill Book Company, 1980.

National Safety Council. "Accident Investigation: Analysis and Costs," *Accident Prevention Manual*, Itasca, IL, 1992, pp. 283-302.

Smith, S.L. "Near Misses: Safety In the Shadows," *Occupational Hazards*, September 1994, pp. 33-36.

Swartz, G. "Incident Reporting: A Vital Part of Quality Safety Programs," *Professional Safety*, December 1993, pp. 32-34.

13

TIPS FOR TRAINING OPERATORS

Within the various chapters of this publication are many sources of information, suggestions, guidelines, and tips to safeguard employees and visitors. The information is broad and requires special skills for each equipment operator to master. Not only is compliance necessary and important to an operation, but the prevention of needless accidents and loss of product and resources are of paramount importance.

Having a book full of safety rules and regulations is a good start. However, each employee must be properly taught what these rules consist of, why they exist and how to properly implement each rule.

Each member of management must visibly support operator training. Time must be allocated to provide the training. Productivity is interrupted when operators must leave their jobs and spend time in the classroom. Is this really lost production or an investment in the future?

If management completes a comprehensive evaluation of past accidents and their associated costs, as well as the lost revenues from product damage, damage to the building and equipment and OSHA related citations, it should be apparent that training is essential. Operator training *is* an investment in the future. The intent, as with any business venture, is to maximize profits and to reduce losses. What would the costs be if no operator training took place?

An evaluation of past records regarding losses from powered equipment accidents obviously looks to the past. Efforts for employee training are intended to prevent any similar accidents from occurring. William Heinrich's statement made in the 1930's is just as appropriate today in that "past events forecast the possibility of future and potentially more severe incidents."

Training of operators will help ensure that employees know how to operate a powered industrial truck by using the proper methods. Quality training can change the undesired behavior of operators and help to reinforce the desired safe behavior. It must be remembered, however, that training can be used only to correct skill deficiencies or to improve operator knowledge. If poor engineering practices or poor operator attitudes exist, training will most likely not be successful.

UNDERSTANDING TRAINING

The training of operators involves the two senses that are used most when one is learning—sight and hearing. Humans obtain some 83% of what they learn through sight and 11% through hearing. Only 1% is learned through taste, 1.5% through touch and 3.5% through smell.

Additional research by psychologists points out that humans retain 20% of what they hear, 40% of what they see and hear and 70% of what they see, hear and do. This is why it is so important to provide the following overall training to powered equipment operators:

- Classroom lectures.
- Visual aids such as charts, slides, graphs, flip charts and films or videos.
- Show-and-tell demonstrations on how power equipment functions.
- Skills driving evaluations to determine levels of operating knowledge and ability.
- Follow up quizzes.
- Job safety observation.
- Job hazard analysis.

In addition, ensure that the classroom has adequate space, is well ventilated, has comfortable seats and good lighting.

When a new employee is hired or transferred, this is an important time to brief him on the need for safe vehicle operation. This induction of the new operator into the company or department is actually a part of training. Training is necessary to make sure that employees do their jobs safely and efficiently.

Managers and supervisors play a key role in accident prevention. According to research in studies of accident case histories, the underlying causes of most accidents have been the result of management being:

- Lax in the enforcement of safety rules.
- Remiss in requiring inspections to correct hazardous situations.
- Remiss in properly informing and training employees.
- Indifferent in promoting employee participation in the safety program.

What about the operators themselves? The above focuses on management's failures in the prevention of accidents. The employees play an equal role in self preservation. If we analyze the above-mentioned failures, how can these situations be placed in the employee training program to obtain the desired results?

- Employees have to be taught the various safety rules associated with their jobs; in this case, operating the pieces of powered equipment. Supervisors must keep in mind that required safety rules will not be followed if they are not understood.

- Are the forklifts safe? Are operating areas safe? The employee plays a key role here because they are being asked to:
  - Properly inspect their equipment at the start of each shift *and* document the findings.
  - Report to their supervisor if the equipment is defective or if their work areas contain any hazards. Being that supervisors cannot be everywhere, one must rely on the operators to assist in accident prevention by stepping forward to make the problems known to the right person.

- Operators must be properly trained. The amount of information that one must have to safely operate a lift truck is significant. Supervisors must recognize that safe work habits, safe operating skills and an understanding of the need for overall safety must be taught. The employee should not learn this by trial and error—it is far too dangerous.

Training Tips

Thinking on the part of the employees has to be accompanied by learning. When presenting materials in a classroom, allow for employee participation, allow everyone to contribute. At no time should an employee be criticized or ridiculed. This type of behavior on the part of the instructor can have a significant negative impact on the entire training class. Instead of "putting an employee in his place" the instructor has alienated the employee and caused the rest of the class to turn off to the program.

Always compliment, even when the employee makes unfavorable comments on the current safety program. Be a good listener and really listen to what is being said; it may have a positive impact on the program. The more the employees participate, the more knowledge will be gained. Compliments, as long as they aren't overdone or are insincere, will encourage others to contribute.

When a hands-on demonstration is necessary, the instructor must follow all of the required safety rules. When it comes to demonstrating how to operate a piece of power equipment, the instructor should be properly skilled. The knowledge of the machine's functions will help convince employees that what they have heard in the classroom is realistic. When the instructor or supervisor cannot operate the equipment because of a lack of skills, a qualified employee can assist. It would be advantageous to brief the employee ahead of time on the need for his assistance and obtain his consent. Peer acceptance is very important here.

The instructor cannot use too many technical terms without an explanation. The message may be too difficult for operators to understand. As an example, the materials and information which identify the point of stability and the stability triangle on a lift truck are difficult concepts for some operators. This subject of tip-over deals with real circumstances that are life-threatening but the concept of the stability triangle is difficult to comprehend because the triangle is invisible.

Records of training would have to be maintained during the training process for proper documentation. The proposed OSHA training standard for powered equipment operators requires detailed training records. The rear of this chapter contains a form which can be used to record the various scores of written tests, driving skill evaluations and overall scoring for each operator. The form can be used to document the results of operator training.

Use Your Listening Skills

Effective listening is a much-ignored skill. Most people underestimate the amount of time they spend listening. Studies indicate that we spend 80% of our waking hours communicating. At least 80% of that time is spent listening. Listening takes up much more time in everyday communications and as a result, many people believe that they are good listeners. This is not true.

Immediately after listening to a 10 minute presentation, the average listener has heard, correctly understood and retained only about 30% of what was said. This is obviously very important when it comes to the training of powered equipment operators. Within 48 hours, retention of oral communication drops to about 20%.

Anyone can become a better listener if he or she has the proper attitude toward listening skills. Anyone that is instructing employees on fork lift skills has to be a good presenter as well as listener. If one has a negative attitude toward listening, he will have a difficult time becoming a good listener.

One bad habit to avoid is to assume you know what the person you are listening to is going to say. Important information can be lost this way.

Another bad habit is to interrupt others while they are speaking. It keeps one from hearing what the other person was going to say. It is also rude to interrupt. Often people interrupt because they are concentrating on what they are going to say rather than listening to the message.

Powered equipment trainers should follow these rules for effective listening:

- Give your full attention to the person speaking. Avoid distractions.
- Focus on the speakers message by looking for the central concept. Try to get the gist of what someone is saying, rather than trying to remember every fact that was mentioned.
- Indicate your interest by leaning forward toward the speaker, nod or encourage the employee speaking by saying "yes" or "I see" in a quiet voice.
- If necessary, recap the message by repeating it.
- If the employee is not being clear enough, ask questions but do not interrupt.
- If the employee pauses do not jump in with a totally unrelated comment or a joke.
- Be sure the subject is being presented in such a way that it is interesting. Employees will be listening at different levels of interest.

- When asked a question try to ask the employee how he felt or reacted in a specific situation.

Training of Supervisors

Training of supervisors must be included in the powered equipment operator program. It is not unusual for an organization to provide considerable training efforts for the first-line supervisor. After all, the supervisor is the main link between upper management and the work force. With the proper training, supervisors can do much to assist an organization in identifying the needs and concerns of employees. They can also help to resolve problems and minimize grievances. The supervisor also has the responsibility to always set the proper example.

Needless to say, when it comes to knowledge of lift truck rules and guidelines, the supervisor should know as much as the operator. The supervisor may not have those same operating skills that most operators have, but, knowing the operating rules and safeguards is a necessity. When operators are given classroom training, supervisors should be included. They can assist in answering questions, cite first-hand information on accidents and incidents, and demonstrate to operators that they do in fact know the safety rules.

Unless it is necessary, supervisors should not be a part of any testing of operating skills as those required of operators. The proper operation of a piece of powered equipment, however, should be known by supervisors. They should know how a piece of equipment handles so they are able to assist operators in trouble-shooting or accident prevention. The daily operation of powered equipment should be left to those that are best trained for the job.

Developing a Training Plan

With the proposed OSHA training guidelines, the employer is responsible for providing the proper training on a timely basis which includes all of the segments required by OSHA. The first step would be to develop a set of program objectives. If an organization is planning to train operators by using an in-house trainer, a set of objectives will provide guidelines on what the company is planning to achieve. If outside training sources are used, such as those provided by manufacturer's representatives, that program plan should have flexibility. In addition, the plan should not only provide for the necessary training and information required by regulations, the plan should also be able to fit the needs of an organization.

A word of caution regarding trainers; not everyone knows how to properly teach someone how to perform a job just because they have experience in performing the job. Teaching is not only a skill that requires training, it also requires performance and practice evaluations. Some organizations are fortunate to have training departments that focus on various forms of employee safety training. Some firms have sent employees to local colleges to learn training techniques before the actual teaching takes place for powered equipment operators. Some organizations use a key supervisor who knows the job well and has the ability to relate to operators and provide the proper program instruction.

The lesson plan represents the organizations blueprint for training. The purpose of a lesson plan is to help the instructor:

- Present lift truck material in the proper order.
- Emphasize training materials and their importance in the program.
- Keep from omitting any necessary materials.
- Programs and classes should be scheduled properly.
- Allow for operator participation and feedback.
- Allow for supervisory participation and feedback.
- Provide for the appropriate visual aids.
- Allow for proper testing procedures—both written and operating skills.
- Allow new or transferred employees to become certified before starting to operate a piece of powered equipment.

Those that are providing the training and developing preparations for operator training may wish to complete the self-check list below (Figure 13.1). This exercise could prove beneficial to the outcome of the training program.

Group Methods of Training

To be successful in a training program, it is important to include those elements that have been successful through the years.

- **Group Discussions.** It is important to allow for group discussions in training sessions. Open, free-flowing dialog is important. All employees should be encouraged to participate. As mentioned earlier, supervisors should be a part of classroom training. Upper-level management should be available in the classroom to show support.

- **Lectures.** A qualified and reliable instructor will not only provide the necessary information on powered equipment training, he/she should be able to fire up the group and make the training exciting.

- **Group Interaction.** Allow for questions and answers during training sessions. Certain technical points are sure to arise and operators must be provided with exact information. Instructors should be prepared to respond to questions of all sorts from the group. Any questions that cannot be answered should be written down on a flip chart or board and resolved during a break or for the next training session. Always allow time for employees to reflect on the information. Operators may need time to prepare their thoughts before they can ask questions.

Figure 13.1. Instructors' Self Check List

A good instructor should be able to answer "Yes" to at least 20 of these questions. Under 15 would be below average.

		Yes	No
1.	Do you check your classroom before the session for proper ventilation, lighting and seating arrangement?	☐	☐
2.	Do you preview all films or slides before showing them?	☐	☐
3.	Do you prepare and test equipment before class begins?	☐	☐
4.	Do you keep the classroom clean and orderly?	☐	☐
5.	Do you introduce yourself to each class?	☐	☐
6.	Do you attempt to know your students and learn their names at the beginning of each course?	☐	☐
7.	Do you state and clarify the course objectives in the first session?	☐	☐
8.	Do you introduce each subject and explain its importance to course objectives?	☐	☐
9.	Do you vary your teaching methods according to the material you are presenting and the students taking the course?	☐	☐
10.	Do you speak directly to the class and avoid distracting mannerisms such as chewing gum, pacing, and juggling change or chalk?	☐	☐
11.	Do you use words easily understood by the students and explain all technical terms?	☐	☐
12.	Do you refrain from using sarcasm or vulgarity in class?	☐	☐
13.	Do you stay on the subject during each session?	☐	☐
14.	Do you cover all the material in the lesson for each class period?	☐	☐
15.	Do you allow time for questions and for clarification of the material?	☐	☐
16.	Do you encourage student participation or use group discussion methods, if appropriate?	☐	☐
17.	Do you use audiovisual aids whenever possible?	☐	☐
18.	Do you demonstrate techniques or work procedures clearly and ask students to repeat the steps as you showed them?	☐	☐
19.	Do you summarize each lesson at the end of the class time?	☐	☐
20.	Do you give regular assignments with clear instructions?	☐	☐
21.	Do you keep to the class schedule throughout the course?	☐	☐
22.	Do you test students to identify your weaknesses as an instructor, and do you use the results to improve your skills?	☐	☐

A self-check test can help a supervisor rate teaching methods.

To ensure that an instructor is well prepared for training they may wish to use a check list such as this (Courtesy of the National Safety Council).

194 / Forklift Safety

Figure 13.2. Log for Lift Truck Quizzes

D-2 FORKLIFT TRAINING -- EMPLOYEES AND SUPERVISORS

EMPLOYEE NAME	T-SHIRT SIZE	SCORE: QUIZ 1	SCORE: QUIZ 2	SCORE: QUIZ 3	SCORE: QUIZ 4	SCORE: QUIZ 5	TOTAL 5 QUIZZES	SCORE: DRIVING SKILLS	FINAL TOTALS/ SCORE
1									
2									
3									
4									
5									
6									
7									
8									
9									
10									
11									
12									
13									
14									
15									
16									
17									
18									
19									
20									

Dates of quizzes:

To provide for a convenient means of logging operator training for quizzes and skills driving, use the appropriate form.

Hands-on Training

A must in all operator training is the inclusion of hands-on training. Each piece of equipment that will be used must be included in the training. This is especially true for new or transferred employees. Various models of powered equipment can have significant differences and require new techniques or coaching. The instructor may have to demonstrate how to operate any piece of the equipment. The representative from the equipment manufacturers can be of assistance for this training.

Quizzes

To ensure students have understood, and can apply the information they have received during the training programs, quizzes should be administered. Any test must be fair and understandable. Tests that are poorly designed can undermine the objectives of the program and cause employees to become frustrated. The quizzes are to be reviewed, scored and the incorrect responses are to be discussed with the students at the next session. A form to record the quiz process is illustrated in Figure 13.2. Retain all copies of any tests and allow for any recognition or awards to be distributed before each session begins. (See Appendix B for miscellaneous quizzes).

Audiovisuals

When developing an operator training program, the instructor must visualize the intended audience and how they will react and accept the information. Some of the students may have knowledge of the subject, others may be just beginning to recognize and understand powered industrial trucks. There may also be language difficulties which the instructor must take into consideration.

A few tips on some common audiovisual aids:

- **Video.** Video that is being used can be purchased from many of the equipment manufacturers. In-house videos can be created but a word of caution: the program may be of such poor quality, that the message can be lost. This is not to say that the development of in-house programs should not be attempted. Through patience, proper camera work and editing, good programs can be created. The benefit of video is that it can be used over and over again as well as the tape being stopped to make a point about a specific issue. Duplicates can be made for other facilities; however, do not violate copyright laws. Videos can be used to gather information about accidents. The camera is portable and with the proper lighting, can create an effective visual aid.

- **Slides.** A 35 mm camera can be used to shoot slides of actual workplace situations. This is a very inexpensive type of program. A script could be created and the appropriate information developed to match the slides which will be used. The photographer can snap the slides at his leisure and include additional shots which can be used even though they may not be a part of the script. The slides should be reviewed and placed properly in a tray for the presentation. Those slides that are not acceptable can be re-taken. A slide that is depicting a safe operation should not be used if the slide also displays an unsafe condition or unsafe behavior. This would apply to video, film or any other media. As an example, many forklift safety films demonstrate proper lifting techniques which illustrate how to lift a container and properly stack it. In the video, the operator is usually moving forward and lifting the load at the same time. This is a potentially dangerous situation because the human mind, to be effective, can only focus on one subject at a time. The operator should first be concentrating on positioning the lift truck prior to lifting the load into position. Yes, this movement does save time, completing the two movements at once, but accidents can easily be the result of these types of operating errors.

Keep in mind that each slide has to be narrated and it is possible to give different groups a variety of comments on each slide. This inconsistency could be a disadvantage.

Slides can be placed on video with a voice-over treatment. To create some action with the still slides, the camera can move in and out or place a halo around a specific part of the slide so as to provide a greater variety of visuals to the trainee.

- **Overhead Projectors.** This form of media is very popular and the least expensive. The use of an overhead can be limited, however. It would be difficult to provide all of the necessary regulations and visuals needed by the student with this method. For the most part, words, graphs and charts can be effectively displayed with an overhead projector. An overhead transparency cannot offer the same depth and color that 35mm slides or video can offer.

Be sure the transparency slide is straight and balanced on the screen. Do not make the mistake of putting an image on the screen and forgetting to look at it. This can be very disruptive to the audience if the slide is upside down or crooked. Once the slide is set in place, do not move it. Also, do not block the screen with your body. Use a pointer to focus on key topics, do not use your finger as a pointer. Be sure the slide is properly focused and the projector is properly positioned in front of the screen.

The trainer can also write on a transparency if a specific point has to be made. Color transparencies can add a professional touch to the presentation.

- **Flip Charts.** The best flip charts are those that exhibit quality images and are prepared in advance of the presentation. When time can be taken to neatly prepare charts which are a part of the training program, the program will look more professional. The prepared flip charts can be used over and over for different groups. An individual that has a good artistic touch should be the creator. There should always be a separate flip chart in the classroom to record questions or comments from the group. The training should not be completed until questions are properly answered.

- **Computer Based Interactive Video.** This may be the most expensive audio visual program discussed so far, but there are definite advantages. Students come away with a higher retention of the materials presented. For powered equipment operators this is very important. Training can be easily documented with this system for record keeping purposes. The message is repeated in the same way for each student and the instructor does not have to be present. Figure 13.3 illustrates the use of interactive video.

This method trains students one at a time, thus it would take longer to train all of the operators when only a single unit is available. Also, the student has to be briefed in the operation of the rather simple controls and keys; this could cause some apprehension on the part of some operators. A major video training company has developed a new CD ROM program that has enhanced operator training. The program requires no special computer skills and is structured for a 6th to 8th grade reading level. Employees can review the material at their own pace and are able to view the results of their answer decision. Employees pay more attention and cannot proceed until they have mastered the last session.

The CD ROM disk allows the purchasers to add their own policies and procedures to the disk. A company could custom design its own program. Photos can be taken and added to the training program, other material can be deleted, if necessary. The program is cost effective and will help satisfy OSHA requirements and properly document the training.

Figure 13.3. Interactive Video in Use

Interactive video has resulted in a high retention rate among students (Courtesy of Material Handling Engineering).

SUMMARY

For a training program to be successful, it should correct deficiencies in the workers performance. Good training calls for behavioral objectives. Specific changes in the workers behavior is expected. Pre and post tests are one way of testing.

Any individual that is conducting operator training has to be familiar with the myriad of regulations and the most current teaching techniques. Students have to come away from the training program with a greater comprehension level of the safe operation of powered industrial trucks.

Various methods of multi-media programs can be used with great effectiveness if used properly. Video, 35mm slides, overhead transparencies, flip charts and computer-based interactive programs provide for excellent communication. The choice of which system to use can many times be influenced by budget considerations.

Supervisors as well as operators should participate in the training. To effectively prevent accidents and incidents associated with powered equipment, supervisors must have a working knowledge of the program. If supervisors are expected to enforce safety rules associated with powered industrial trucks, it makes sense that they should first know and understand these same rules.

REFERENCES

Breecher, M. "Now Hear This: Effective Listening Takes Skill," *Chicago Tribune*, August 28, 1983.

Coastal Video, Clarity Multimedia, Virginia Beach, VA, 1996.

DeReamer, R. *Modern Safety and Health Technology*, New York: John Wiley & Sons, 1980, pp. 151-175.

Grimaldi, J. and R. Simonds, *Safety Management*, Homewood, IL: Richard D. Irwin, Inc., 1975, pp. 457-488.

National Safety Council. "Audiovisual Media," *Accident Prevention Manual*, Itasca, IL, 1992, pp. 383-411.

National Safety Council. "Safety Training," *Accident Prevention Manual*, Itasca, IL, 1992, pp. 365-382.

Swartz, G. "Forklift Safety Training," *Professional Safety*, Des Plaines, IL, January 1993, pp. 16-21.

14

FORKLIFT RODEOS AND FORKLIFT RALLIES

What better way to generate enthusiasm and personal pride in the skills of forklift operators than to fully test their skills in a competitive event? An ideal method to use for skills evaluation is a forklift rodeo or rally. Whatever you call the competition is not as important as the event itself. The objective is to establish specific operating guidelines, select who will participate, and hold a competition. This chapter will outline the details involved in forklift rodeo competition.

WHY HOLD A FORKLIFT COMPETITION?

An organization should consider competition between operators for specific and beneficial reasons:

- There are operators who have excellent skills in handling and maneuvering a forklift; they have a reputation of being skilled. Most plant employees usually know who these special operators are. These special operators would most likely be able to demonstrate these skills to fellow operators, competitors, and spectators.

- Management knows that forklifts can do a great deal of harm to employees and to company property. Operators can demonstrate that they can perform with specific skills which can then easily serve as an indicator that safe driving can save lives, prevent injury and prevent a loss of property. Management's interest should be serving both needs: safety and loss of property.

- Operators can show off their skills to outsiders and demonstrate to these "challengers" that the company for which they work has an excellent safety program. It's unlikely that individuals who qualify to participate in competition would go out onto the factory floor and operate in a reckless manner. Rodeos do make for safer operators.

- Employees will discuss the program with their peers. The higher visibility of the program and a heightened awareness of plant safety will be helped. For those employees who travel to a site with the company picking up the costs, there will probably be some envy on the part of other employees, as well as the desire to be the lucky one selected for the next program.

ELEMENTS OF CONSIDERATION FOR COMPETITION

For an organization to properly conduct a forklift rodeo, specific details must be outlined and carried out to ensure success. Consider the following details:

- Management commitment to the program.
- Site location.
- Determination of skills testing and judging.
- Selection of participants.
- Set up of test site(s).
- Judging of the event.
- Determination of winners.
- Awards and recognition.

Management Commitment

For an operation such as a forklift rodeo or rally to be successful, senior management must be firmly behind the endeavor. Time will have to be allocated for planning and completing the program. Time is money. Management must weigh the return on this investment to the benefits the program will provide.

A program such as this generates considerable hype among the employees. a sense of anticipation is generated. Who will compete, how well will the operators do, will they be successful and win a trophy or other award?

Employees will talk among themselves regarding the competition. There will also be a heightened awareness before and after the program for forklift safety. A program such as this generates a tremendous amount of good will.

Additional management considerations regarding costs will involve wages and possible fringe benefits for the employees during the competition. If the program takes place during work hours, then the employees won't be performing their regular jobs; therefore, they won't be productive during this time. If the program takes place after work or on weekends, overtime costs may be involved.

Consider travel expenses for employees who may have to be brought to the site of the competition. There may be additional costs for hotels, parking, meals, rental cars, and perhaps other items. This should be a part of the overall planning process. When presenting program costs to management, be sure to consider all related costs; be as accurate as possible. Money can be saved by having two employees share a hotel room, rather than one person. Weekend travel usually results in lower air fares. Advance purchase of tickets also helps reduce costs. These savings can be significant.

A serious consideration should be made to videotape as much of the competition as possible. The final tape can be edited and an abbreviated copy can be given to each participant.

Site Location

A site will have to be selected for the actual operator testing. Consideration should be given to where the competition would serve the needs of the employees and forklifts. Will the obstacle course be established outdoors? Will weather be a factor? Can the competition be held indoors? Is there adequate space for the power equipment, spectators, and overall obstacle course?

A site layout plan should be developed to ensure the safety of everyone present, as well as to ensure the functionality of the operating course. Where weather is a factor, the facility may have to set the program up indoors. Of course, adequate operating space for the powered equipment must be available or the program could be ineffective.

For outdoor programs, the driving surfaces should be flat and free of even the smallest hazards. After all, the rodeo is intended to allow the most skilled to demonstrate their highest abilities; therefore the conditions should be as good as possible.

Tents should be set up outdoors to provide shade for spectators and others that may be attending. Cold drinks should also be provided for everyone. Judges and those individuals who are part of the program's administrative efforts should have tables and shade where necessary.

Demonstration of Skills Testing and Judging

Many drivers have exceptional skills on different pieces of equipment. It must be determined what equipment will be selected and used. It makes sense to select forklifts and other pieces of power equipment that are the most commonly used. Also select those pieces of equipment that are the most reliable and free of defects. Evaluate operators for performance prior to the rodeo as in Figure 14.1.

What challenges shall be identified for the operators? What standard should operators be held to? If an operator is competing, that person should be tested on the full knowledge of the job, not just operating skills.

Consideration should be given to written and oral testing. Written questions which contain true/false and multiple choice questions are commonly used to evaluate employees' knowledge of forklift operating rules. A set of 100 multiple choice forklift questions and 175 true or false questions have been placed in Appendix B. They can be used selectively for testing purposes.

202 / Forklift Safety

Figure 14.1. Evaluate Operators in Advance

The supervisor is using a checklist to evaluate this operator for his driving skills.

An oral examination is another good method to test the overall job knowledge of the operator. The oral examination can use either one or both of the following examples. Figure 14.2 illustrates the oral testing requirements.

Have the operator give the judge(s) a step-by-step walkthrough of exactly how to check out their power equipment. The judges could then check off their proper responses on a specific form. Be sure that other operators are not near this testing area, so they don't pick up any "tips" on their own. Note the form on lift truck pre-inspections.

Next, the operators should be asked key questions regarding safety in battery re-charging, changing propane tanks, safety at the dock, fire safety, etc. This will also test the overall job safety knowledge of the operator. Management can decide which questions best suit their operations when developing these tests. Questions should be uniform and written so each contestant is given the same material.

Forms and specific guidelines should be developed and reviewed in advance to ensure that the aspect of subjectivity is reduced or eliminated so that the judges can rate everyone as fairly and honestly as possible. Timing of events is important to separate winners, a timing element must not be the sole determination of the winners. Operators must always exhibit the proper use of safety rules when competing—there should be no compromise here.

Figure 14.2. Show and Tell Test Form

Truck Type:		Powered By:	
Sit Down	☐	Electric	☐
Stand Up	☐	Propane	☐
Rider Picker	☐	Gasoline	☐
Handjack	☐		

EMPLOYEE NAME: _____

FACILITY/LOCATION: _____

TESTING DATE: _____

OBJECTIVE OF PAGE The objective of this side of the rating sheet is to ensure that the employee has an understanding of the mechanics of the lift truck as well as all of those items that involve standard checking prior to driving the lift truck. The operator should also have an understanding of various safety features within the facility that relate to lift truck operations.

PHYSICAL EXAMINATION OF LIFT TRUCK (TOUCH AND TELL)

The operator should be familiar with the features of the lift truck. The operator must demonstrate and describe.

	CORRECT	INCORRECT	DNA
1. Proper use of tilt.	☐	☐	☐
2. Proper use of raise/lower.	☐	☐	☐
3. Sounded the horn.	☐	☐	☐
4. Checked for oil leaks.	☐	☐	☐
5. Checked mast chains.	☐	☐	☐
6. Checked the brakes.	☐	☐	☐
7. Checked the tires/wheels.	☐	☐	☐
8. Checked the hour meter.	☐	☐	☐
9. Checked the scissors reach.	☐	☐	☐
10. Checked the warning light.	☐	☐	☐
11. Checked the rear view mirror.	☐	☐	☐
12. Checked battery retainer.	☐	☐	☐
13. Checked the discharge indicator.	☐	☐	☐
14. Checked the back up alarm.	☐	☐	☐
15. Checked hoses/hose reel.	☐	☐	☐
16. Checked the overhead guard's light.	☐	☐	☐
17. Knows the capacity of the lift truck.	☐	☐	☐

TOTAL POINTS THIS SECTION: _____

When testing operators there should be a means of testing their knowledge of the equipment they operate. Note the form.

For judges, it might serve the program well to have at least one member of their local management team accompany each operator to the test site. Of course, this may not be practical in every case, but this consideration does provide for more uniform judging. When multiple judges are being used to judge specific phases of the program, it's not likely that any of them would be biased or negatively influence the scoring.

Judges should be completely briefed on the entire program, including the wearing of PPE. Have handouts available for everyone. Allow for questions, especially if the competition involves employees from organizations other than the sponsor. a rodeo should include multiple events, therefore; the judges should be aware of all of these details before starting. Also, some of the judges may favor a particular phase of the testing and may wish to volunteer for specific assignments. As an example, a judge may have a great deal of experience with a narrow-aisle truck and be more comfortable judging this event.

Once judges have scored their forms on an individual basis during the competition, they should meet as a group after each contestant has completed their test. This ensures that each knows what the others saw. This serves several purposes:

- It will help others in looking closer for specific violations that operators may be committing.
- The brief meeting of the judges between the performance of individual operators will help ensure that none of the judges change any of their scores because of what someone else has pointed out to them. Scoring for that operator should be final at that point and not changed. The completed forms should then be given to a head judge. This ensures that the score that was initially given to an operator—right or wrong—stays that way.

It should be noted that a master form could be created which would identify each operator's name on the far left horizontal column. The individual events could be listed in vertical form. As each operator attains a certain score or incurs penalty points, the score could be placed on this master form. This method has potential drawbacks. Operators may constantly come over to look or inquire as to what score they achieved. It's much better for each judge to complete a rating form for each operator after each event and then immediately turn it into the head judge.

Skills testing should consider the following:

- A written examination with 50-125 questions.
- An oral examination regarding forklift daily checks.
- An oral examination regarding selected plant safety issues.
- The possible use of various pieces of power equipment for skills testing.
- A skills driving for trailer loading-unloading.
- A skills test for stacking multiple containers or pallets.

Figure 14.3. Speed Stacking Illustration

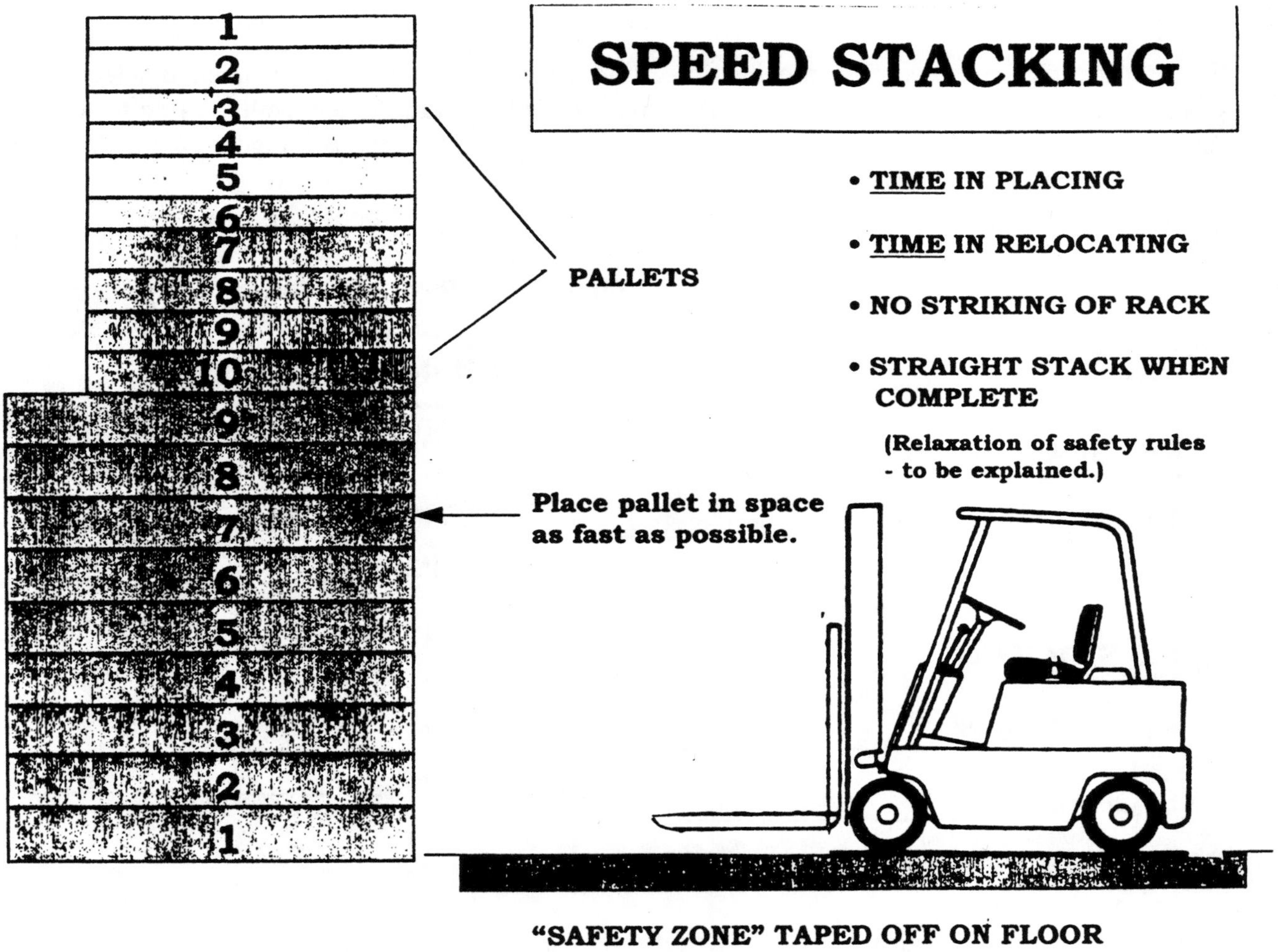
SPEED STACKING
1
2
3
4
5
6
7
8
9
10
9
8
7
6
5
4
3
2
1
PALLETS
• TIME IN PLACING
• TIME IN RELOCATING
• NO STRIKING OF RACK
• STRAIGHT STACK WHEN COMPLETE
(Relaxation of safety rules - to be explained.)
Place pallet in space as fast as possible.
"SAFETY ZONE" TAPED OFF ON FLOOR

Speed stacking can test the operators skill while handling empty pallets in racking.

206 / Forklift Safety

- A skills drive—forward and reverse—with and without a load, between obstacles such as pallets, barrels or cones.

- A skills test regarding speed stacking and unstacking of pallets in racking.

In regards to the speed stacking, time and precision are very important. Some standard safety rules should be waived for this exercise. This information would be a part of the overall briefing for judges and contestants. Refer to Figure 14.3, which identifies this particular exercise. Figure 14.4 is the form the judges would use for this event.

Figure 14.4. Judges Scoring Sheet - Speed Stacking

SPEED STACKING/UNSTACKING

REMOVAL

START TIME _____

STOP TIME _____   TOTAL TIME: _____

RE-STACKING

START TIME #4 _____

STOP TIME #4 _____   TOTAL TIME: _____

YES		NO
☐	Did operator call "GO" when 1st starting?	☐
☐	Did operator call "GO" on 2nd starting?	☐
☐	Did operator strike the rack at any time while stacking or removing pallets? (Place marks here [Example: 𝖧𝖧𝖧 | |] _____)	☐
☐	Was the completed stack of pallets straight and secure?	☐

Judges would need a scoring sheet for judging the speed stacking event.

When safety rules are waived in a particular exercise, this does not mean that safety is being ignored. For speed stacking, yellow tape is placed on the floor to indicate the boundaries of the movement of the lift truck. No one but the operator should step inside this barrier. Because there is a standard rule that requires an operator to look behind when backing up, this particular rule is waived. Operators must focus on removing a pallet from the top of the stack and continuing until all of the pallets are spotted on the other racks. After a brief wait, the process is repeated for the judges. Speed and precision are the main considerations here—because of this, several safety rules have to be waived in this controlled environment. This is one event where timing is important—judges should be made aware of this.

Selection of Participants

Surprisingly, many skilled employees are reluctant to display their forklift driving skills in front of an audience. These levels of skills being displayed can make an operator nervous because of being watched. To determine who should compete in the program, management can use several guidelines.

The first method in the selection process could be the result of each plant performing their own testing of their operators. During the past year, if the plant conducted forklift training, those individuals who scored highest on written tests and had the most points for their driving skills testing should be considered as front-runners for the competition. Annual training should always be accompanied by scoring sheets. Figure 14.5 has been included in this chapter for assisting in the safety skills testing. Operators who have had accidents in the past year should not be allowed to participate.

It's possible that potential contestants can be selected by their peers or to personally volunteer for the program because they feel they have the skills. All contestants must also be willing to wear all required PPE.

If employees know that their classroom skills and driving skills will land them a trip to another city, or offer a chance for an award of some kind, they may try even harder to excel. This means they must pay more attention to their day-to-day driving in the plant and pay greater attention at forklift training sessions. Where travel and monetary awards or trophies are concerned, employees may seize this opportunity to really demonstrate their know-how.

It's important to realize that there are operators who excel at taking written exams. There are others who will excel at driving skills testing. If an organization looks closely at what criteria to use for participant selection in the rodeo, a combination of both skills can be used. The driving portion could be a percentage of test-taking versus more or less of a percentage of points, or whatever fair and equitable means the organization uses to make their selection of contestants.

Note—let the employees know this process before starting any testing procedures, so everyone starts with a clean slate.

Figure 14.5. Skills Evaluation Forms

(VALUE: THIRTY POINTS)

(PLEASE PRINT)

Truck Type:	Powered by:
Sit Down ☐	Electric ☐
Stand Up ☐	Propane ☐
	Gasoline ☐

EMPLOYEE NAME: _____
YRS. OF EXPERIENCE: _____
FACILITY/LOCATION: _____
TESTING DATE: _____

OBJECT: RATE EACH LIFT TRUCK OPERATOR BY HOW THEY PERFORM BASIC FORK LIFT DRIVING SKILLS. EACH OPERATOR IS TO BE GIVEN THE SAME DRIVING ASSIGNMENT AND INSTRUCTIONS. ANY INFRACTION OF THESE GUIDELINES DURING THE SKILLS SESSION SHOULD BE COUNTED AS A VIOLATION AND DEDUCTED FROM THE POINT TOTAL.

	Pass	Fail
1-8 PHYSICAL EXAMINATION OF LIFT TRUCK: (Operator must demonstrate and describe inspection of each of these items) — Tilt, Raise/Lower, Horn, Tires, Oil Leaks, Mast chains, Brakes, Hour Meter	☐☐☐☐☐☐☐☐	☐☐☐☐☐☐☐☐
9. Did the operator pull forward toward designated section of racking without striking anything?	☐	☐
10. Did the operator place the forks under the pallet properly?	☐	☐
11. Did he raise or tilt the load properly.	☐	☐
12. Did any of the posts strike any section of racking while removing the pallet?	☐	☐
13. Did he lower the pallet before moving/backing out? (Don't drive + lower together)	☐	☐
14. Did he drive at a safe rate of speed?	☐	☐
15. Did he slow down or stop at cross aisles?	☐	☐
16. Did he sound his horn?	☐	☐
17. Did he pull into the area of racking properly to place the pallet back in the racking?	☐	☐
18. Did he strike any racking on the way up or going into the rack?	☐	☐
19. Did he back out and lower his forks before moving?	☐	☐
20. Did he _always_ look behind him before backing up?	☐	☐
21. Was he wearing his protective equipment?	☐	☐
22. Did he drive around the block of wood or get off the lift and remove it?	☐	☐
23. Did he set the load flat on the floor before getting off?	☐	☐
24. Did he put on his hardhat before getting off?	☐	☐
25. Did the operator perform any moves that were potentially dangerous?	☐	☐
26.(a) Ask the operator for five safety rules to follow at a loading or receiving dock. They should include: — Chocking wheels, Wearing equipment, Keep clear of others, Hold onto handrailing, Operater safety, Proper lighting, Warn others where necessary, (Other)	☐☐☐☐	☐☐☐☐
— or —		
26.(b) Ask the operator for five safety rules to follow at the battery charging stations. They should include: — Proper equipment, No smoking, Clean up procedures, Eye wash stations, Proper plug/unplug procedures, MSDS, use of, JSA, use of, (Other)	☐☐☐☐	☐☐☐☐

TOTAL POINTS: _____

NOTE: ON NO. 26(A) OR (B) THE OPERTOR MUST FULLY DESCRIBE HOW TO PERFORM THESE FUNCTIONS. IF THE EMPLOYEE FAILS TO NAME FIVE ITEMS, DO NOT GIVE THEM FULL CREDIT.

SUPERVISOR: _____   DATE: _____

TOTAL POINTS (SCORE): _____

This form can be used for evaluating skills of the operator. A means of numerical scoring can be used.

Set Up of the Test Site

In order to ensure maximum demonstrations of operating skills, the obstacle course and test site must be properly developed. Space is a key consideration. If a facility has little space and tight quarters, the competition will be of little value for a major program.

Figure 14.6. Maneuvering Skills for a Rodeo

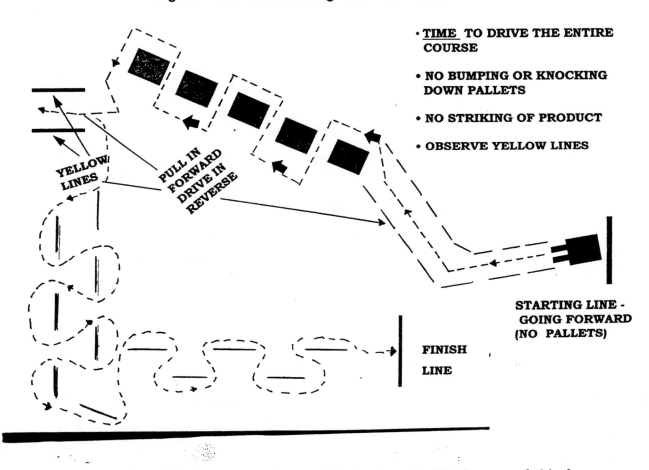

Testing of operating skills during a rodeo can be developed with the use of this form.

Refer to Figure 14.6, which identifies a maneuvering course layout for a large test site. The obstacle course is set up to test the operating skills of each contestant while maneuvering through pallets which have been placed on edge. The test area will need a substantial amount of square feet to effectively serve its purpose. Operators must first drive forward without a load. When they get to the end of the course, they double stack two pallets of stable product (you may wish to shrink-wrap the load to keep it from spilling or shifting). The operator then proceeds to pick up both pallets and drive in reverse through the maze without bumping or knocking down any pallets. Figure 14.7 represents the judges' scoring sheet for this event.

Figure 14.7. Judges Scoring Sheet - Maneuvering Skills

| OBSTACLE COURSE – MANEUVERING |

START TIME _____

STOP TIME _____   TOTAL TIME: _____

YES		NO
☐	Did operator call "GO" at start line?	☐
☐	Did operator properly steer through "S"?	☐
☐	Did operator bump any pallets?	☐
☐	Did operator bump any stacks?	☐
☐	Did operator knock down any pallets?	☐
☐	Did operator complete the entire required course?	☐
☐	Did operator pull into the square turnaround area without touching the lines?	☐

(How many pallets were bumped _____)

(How many loads were bumped _____)

(How many pallets were knocked down _____)

| ☐ | Did you observe any violations of any prescribed safety rules? | ☐ |

List: _____

This form can be used by judges to score the maneuvering during testing.

Another test for operators should be the loading and unloading of trailers at docks— this is quite common in daily operations in any facility. Data published by OSHA indicates that dock areas contribute significantly to injuries and fatalities in the workplace. It stands to reason that contestants should demonstrate a high level of awareness and ability when working at a dock.

In the planning of the program, be sure to allow for food and beverages. Coffee, rolls, fruit, soft drinks, and perhaps lunch or dinner should not be omitted from the planning.

A partially loaded trailer could be used while spotting an empty trailer next to it. Note Figure 14.8 on "basket stacking;" this drawing identifies most of the details needed for a dock safety driving challenge. Operators actually enter or simulate going into a partially loaded trailer, picking up two pallets of product, and backing out. They then proceed to the empty trailer (real or simulated) and safely transfer all of the remaining pallets. As with all of the events, judges could rate this part of the competition from the appropriate judges' sheets. Figure 14.9 illustrates the judges' rating form for these events. Judges would look for smoothness of operation and any violation of safety rules.

Figure 14.8. Basket Stacking Illustration

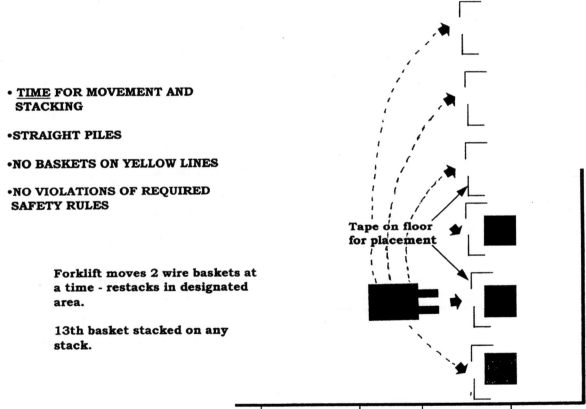

- **TIME FOR MOVEMENT AND STACKING**
- **STRAIGHT PILES**
- **NO BASKETS ON YELLOW LINES**
- **NO VIOLATIONS OF REQUIRED SAFETY RULES**

Forklift moves 2 wire baskets at a time - restacks in designated area.

13th basket stacked on any stack.

Stacking of pallets or baskets will provide for skills evaluation of operators.

212 / Forklift Safety

Figure 14.9. Judge's Scoring Sheet - Basket Stacking

BASKET STACKING

START TIME _____

FINISH TIME _____   TOTAL TIME: _____

YES   **NO**

☐   Did operator call "GO" when starting?   ☐

☐   Did operator raise forks properly under each basket?   ☐

☐   Did operator look before backing up?   ☐

☐   Did operator lower before moving in forward or reverse?   ☐

☐   Did operator stack in the prescribed lines on the floor?   ☐

☐   Did operator stack straight and secure?   ☐

☐   Did operator place #13 basket properly?   ☐

☐   Did operator complete the entire stacking program as required?   ☐

☐   Did you observe any violations of prescribed safety rules?   ☐

List: _____

The pallet or wire basket stacking exercise would require a sheet for scoring for the judges.

There may be a need to utilize powered hand trucks or powered hand jacks in the competition. Key elements to consider are smoothness of operating, ability to stack, ability to safely transport a load, and overall skill of operating. Keep in mind that the operator will most likely be walking while operating the hand truck. Safety is important while competing; operators could be vulnerable to injury from the very obstacles they are attempting to navigate around. There is little protection for a person on foot.

Testing of power hand truck operators could be a part of the overall competition but shouldn't be a testing process unto itself. There just aren't enough challenges involved with a power hand truck versus all of the challenges involved with a forklift. It's obviously important to train operators in all of the pieces of equipment they will have to operate. a rodeo, on the other hand, has to offer a broad enough challenge for full testing of all of the operator skills. Refer to Chapter 4, Powered Hand Trucks, and Appendix B, Test Questions for Powered Hand Truck Operators.

Judging of the Event

Prior to the actual event, there should be a complete set of guidelines drawn up to provide for effective judging. At the end of this chapter are forms for the events identified in this section for proper judging. The forms are a result of the coordination of what events are being offered, as well as the layout of the course and the contest that will be judged. Figure 14.10 provides the written guidelines for each event.

The forms being illustrated represent the course challenges as outlined in this chapter. If other course challenges are developed, the appropriate forms should parallel the responsibilities of the judges. Judging forms should be simple, yet effective, for proper evaluation.

Properly brief all of the judges on what to look for from each participant. Stress fairness and strict judging. Have them score the operators as they perform. Finalize the sheets between events through a head judge. Timing of the event allows for a method in separating winners should there be a tie. Time should not be a sole factor in any of the programs unless there is no other method to determine a winner. Keep in mind that the purpose of a rodeo or rally is to determine who the safest operators are—not who the fastest operators are.

Each contestant should know how the judging will take place and who the judges are. a decision should be made as to whether or not the scoring of each event should be announced to each contestant at that time.

Determination of Winners

The objective of the event is to score as many points as possible in all of the events. Each error committed by the contestants takes away from the total. Depending on the particular event, points should be deducted for:

- Bumping a pallet, cone, or obstacle.
- Knocking over a pallet, cone, or obstacle.
- Striking racking or other stationary objects.
- Failing to properly tilt a load, elevate a load, or travel with the load at the correct height.
- Failing to fully stop when changing direction, raising, or lowering a load.
- Failing to look behind before backing up.

Figure 14.10. Guidelines for Judges

SPEED STACKING/UNSTACKING

- Lift pulls forward and removes top pallet in a stack of 10 pallets.
- Lift backs up and lowers pallet to allow placement into the 9 or 10 inch opening.
- Lift backs up and raises forks to remove another pallet.
- This procedure continues until all pallets are inserted in the rack.
  Time is called. Driver is allowed to stop and get ready.
- Lift then continues to reverse the procedure and restack all the pallets on top.
  Time is called.
  Time—While pallets are inserted.
  Time—While pallets are re-stacked. No striking or racking—finished load of pallets should be as straight as possible.
  (No safety rules are followed here—a blocked off area on the floor is the safety zone.)

SHOW AND TELL LIFT TRUCK

- Operator will complete a pre-start inspection on a sample lift truck (sit down/stand-up) at their leisure.

  No time involved—accuracy and completeness are important.

WIRE BASKET STACKING

- Lift moves three stacks of four wire baskets to designated area next to original stacks. Check for tape on the floor for placement.
- Stacks are to be straight and even.
- The last basket (#13) is to be placed on top of any of the three completed stacks.

  Time—All baskets lifted, carried, and placed—must be done by the book. Check for crooked stacks. Be sure fifth basket is safely placed.

FREE MANEUVERING COURSE

- Lift starts at start line, steers through an "S" shaped path to double stacks of pallets.
- Lift steers through loads and at the end pulls forward into a square for proper turn around.
- Lift continues to back through the zig-zag course and then crosses over finish line while in reverse.

  Time—Must not touch any pallet or load while driving. Must pull squarely into box on floor to back up. Time starts when first line is crossed while going forward. Time stops when lift crosses other line while in reverse.

Specific guidelines for the judges should be spelled out for both operators and judges in advance for greater understanding.

- Failing to use the horn or other signals.
- Failing to wear required PPE.
- Incorrectly responding on the written exam.
- Incorrectly responding on the oral exam.
- Starting an event before a judge authorizes the start.
- Violating any workplace safety rule during the operation of the power equipment.

Particular points could be established by the judges prior to starting. For instance, one point should be deducted for bumping a pallet, two points for knocking over a pallet. Additional points could be determined depending on the event and the particular violation.

As mentioned earlier, a stop watch should be used to time a particular event to break ties or to score a particular event. The entire program should not be judged by the stop watch or the aspect of safety will be compromised.

Depending on the physical position of the judges in the various obstacle courses, they may or may not see each violation. This is the reason that more judges scattered throughout the course will arrive at a more complete determination of each operator's skill level. Each judge would then deduct the appropriate points. Station judges at strategic spots along the courses for maximum rating and point assessment. Figure 14.11 identifies the penalty points involved in the judging.

Awards and Recognition

When the competition has been completed, a time for celebrating and recognizing the operators with the fewest penalty points and/or most overall points. Consideration should be given to the following:

- Trophies for various placing among contestants.
- Cash awards.
- T-shirts, sweatshirts, and baseball caps.
- Savings bonds.
- Other incentives.

The awards ceremony can be held at the test site or at a location where an awards banquet can be held. As suggested earlier, if the scoring has been kept a secret from the contestants, the ceremony would have the impact of surprise on everyone. a moment of suspense as to who the final winner is could add a great deal to the competition.

Figure 14.11. Penalty Point Code for Judges

PENALTY CODE

- **1 point** — Bumped pallet (each one)

- **2 points** — Knocked down pallets – chain reaction counts as only 2 points. Knock downs in separate areas – 2 points each time.

- **1 point** — Violations of prescribed safety rules.
    - No hard hat
    - Moving while lowering or raising
    - Not looking while backing
    - Poor stacking
    - Damage to product
    - Not stopping before changing directions
    - Speed shifting
    - Any part of the body unnecessarily out of the lift
    - Not wearing steel toe shoes
    - Not parking or turning around in designated areas
    - Dangerous/excessive speed
    - Reckless operation of lift truck
    - Proper placement of baskets on top of each other
    - Not stacking a load in the designated area(s)

- **1/2 point** —
    - Not calling out "GO" when starting a testing session

- **1 point** —
    - Not properly identifying pre–start inspection items on the lift truck.

- **1 point** —
    - Each incorrect response on the 150 question quiz.

This sheet allows for the identification of penalty points for the competition.

Photos of the ceremony, as well as speeches by contestants and others would round out a very special event. Needless to say, employees would convey the entire program to their peers when returning to the facilities or departments. The long-term benefits of the program will continue when employees are reminded in some way of their participation in the program.

Rodeos or rallies can pay big dividends in many ways for any organization willing to sponsor such an event. The goal is to develop more safety awareness among employees, reduce injuries, and reduce costs associated with product and facility damage.

SUMMARY

Those organizations that sponsor forklift competitions in the form of a rally or rodeo speak highly of this exercise. There is probably no better way to showcase the skills and knowledge of operators than to provide for competition.

A rodeo can be a real boost to overall employee awareness regarding their powered equipment. Planning is necessary if the competition is to be held in your own facility. This chapter outlines those specific items that would be a part of preparation all the way through the awarding of prizes to the winners.

If an organization wishes to participate in competition held by an association, this material can be a real asset for the prospective competitor(s). The questions in Appendix B and illustrations in this chapter should assist in operator performance.

REFERENCES

"IMMS Fork Truck Rally Boosts Safety and Morale," *Modern Materials Handling Magazine*, August 1986, pp. 70-71.

"Lift Truck Rallies Promote Safety," *Modern Materials Handling Magazine*, February 1996, p. 24.

National Safety Council. "Forklift Truck Operators Safety Training Program," Itasca, IL.

"Operator Rodeos: Big Investment, Even Bigger Payoff," *Modern Materials Handling Magazine*, January 1995, pp. 12-13.

"Professional Activities, IMMS's 11th Annual Fork Truck Safety Rally," *Modern Materials Handling Magazine*, 1987, p. 47.

"Professional Activities, 1990 IMMS Lift Truck Rally: It's Safety First," *Modern Materials Handling Magazine*, February 1990, p. 43.

"Trained Operators: A Basic First Step," *Modern Materials Handling Magazine*, March 1988, p. 85.

15

MISCELLANEOUS ISSUES

There are additional concerns regarding safety in a lift truck program which are not included with the prior chapters. To be fully effective, management and operators must have an understanding of these additional issues. This chapter provides support material for the following:

- Reducing forklift damage.
- Lift truck forks.
- Pallet safety.
- Use of lift cages/working platforms.
- Environmental considerations.
- Fire fighting.
- Voluntary protection program (VPP).
- Job safety observation (JSO).

REDUCING FORKLIFT DAMAGE

Lift trucks can be destructive to the interior or exterior of a building. Much can be done to protect the building from damage. Despite the depth of training which can be provided to operators, there will still be bent racking, cracks in walls, bent posts, dented electrical boxes, etc. Many times the counter balanced truck or narrow aisle trucks are the units that cause the most damage. Power hand jacks are also notorious for causing significant structural and product damage.

To help prevent damage to a building, safeguards can be installed within the building.

- Lights and horizontal piping systems can be tucked into steel beams or joists. Where vertical beams are in place, conduit, wires and fire extinguishers can be placed within the beam itself.

- Braces and safeguards made of angle iron can be placed around disconnect boxes.

- Sprinkler pipes can be highlighted with color coding or with flags or streamers. Where protective bars or angle iron can be used, this added protection will help ensure the system doesn't get struck by a load or mast.

- Sprinkler heads can be guarded with open metal cages to give added protection.

- Large blocks made of wood or concrete can be mounted along walls, driveways and office areas to protect the building from the impact of forks or loads.

- Posts made of heavy duty steel can be placed in the floor on both sides of large roll up doors or other locations that need protection. Ends of racking are also worth protecting because of the possibility of collapse and being struck by product. The posts can be filled with concrete and painted yellow. Figure 15.1 is an example of a protective post for racking.

- Raise lights, where possible, to keep them clear of impact or from being pulled down.

- Bumpers or highway type barriers can be placed near walkways, offices or machines. Paint yellow and pad any sharp ends to protect pedestrians.

- Where an operator can strike an overhead obstacle, a barrier on the floor limiting the position of the lift truck can pay dividends.

- Keep areas well lighted so all obstacles, product, etc. are seen by operators.

- Place gongs, bells or other alarms on certain overhead items so that an operator will know he has struck it.

LIFT TRUCK FORKS

Neglected forks on a forklift truck can be potentially dangerous. Forks can wear a long time if they are not abused by operators. It is estimated that 46 percent of all forks do not meet safe guidelines. However, worn forks can appear to look like new forks if a trained technician does not know how to inspect them.

Forks may be damaged by:

- **Overloading.** Picking a load up that is too far out on the ends of the forks, or, by lifting loads heavier that the rated capacity of the truck.

- **Modifications.** Individuals may have performed welding on forks, drilled or used a cutting torch to put a hole in them, or, maintenance tried to bend them back into shape.

Figure 15.1. Protective Post for Racking

It is important to protect racking and other structures from impact by lift trucks. Heavy duty posts, anchored into the concrete, can help preserve property.

- **Attachments.** Forks can be stressed by adding drum clamps and booms which are supported by the forks. What a drum contains and what is connected to the hook of the boom are critical safety factors.

- **Damage and Stress to Forks.** Damage and stress to forks can be created when one fork lift tries to lift another fork lift. Railroad cars are opened by the forks at times, thus the abuse.

- **Hidden Damages.** Hidden damage may occur when excessive heat is applied to any part of the fork. The heat may be applied during repair.

Forks are designed to take a beating. The organizations that manufacture forks are careful and precise by providing checks and inspections. The factors that are critical for manufacturing forks are:

- The steel.
- The bend and thickness of the heel.
- The welding of the brackets that hold the forks to the carriage.
- The heat treatment of the finished product.

A basic rule of lift truck operation is to keep the forks low when traveling. This could wear forks if an operator allows them to drag on the floor. This practice can wear the heel and bottom of the fork which reduces fork capacity.

The Industrial Truck Association recommends that forks be withdrawn from service when fork blade thickness has been reduced by 10 percent. When a blade thickness is reduced by 10 per cent, this results in a loss of 20 percent of capacity. The safety factor of a pair of forks has also now been reduced by 20 percent.

Figure 15.2 illustrates fork wear versus capacity.

There are ASME/ANSI specifications for User Fork Wear Standards. They are a part of the B56 standards by which lift trucks are manufactured and tested.

Fork lift operators, maintenance staff and others may not know of the standards or understand that the forks must be inspected. Measuring forks should be a part of regular maintenance inspections of the lift truck.

How Forks Are Inspected

Forks should be inspected at least once a year and more frequently in severe applications or where multi-shift operations are taking place. To check for wear and distortion a fork caliper should be used. This device is a type of adjustable go/no-go gauge.

The fork itself consists of two sections:

- **The Shank.** The vertical part attached to the carriage.
- **The Blade.** The portion that picks up the load.

Figure 15.2. Fork Wear Versus Capacity

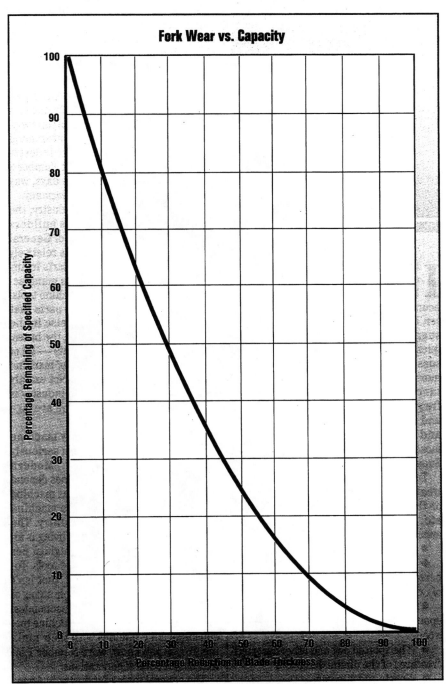

This chart identifies the amount of wear and capacity of forks (Courtesy of Material Handling Engineering).

224 / Forklift Safety

The caliper is placed on the shank and set properly. This part of the fork has little or no wear. Once set, that dimension on the caliper is used to check the shank back, near the heel of the blade. The four contact points of the special fork caliper automatically measure the wear placed on the blade.

Figure 15.3. Measuring Fork Wear with Calipers

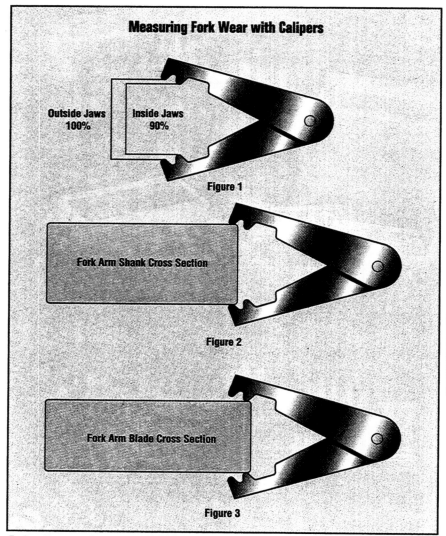

Fork calipers perform two tasks at once. They measure the thickness of the fork arm shank (A) and then automatically indicate what a 10% wear factor would be when the calipers are applied to the blade cross section (B).

A caliper can be used to measure forks in the manner illustrated (Courtesy of Material Handling Engineering).

The wear is checked by first measuring thickness of the vertical shank portion of the fork because this part of the fork experiences little wear. The interior part of the caliper has two additional points that automatically show a 10% reduction of the thickness of the shank.

These points set on the caliper are slid over the blade of the fork. If the caliper slides down the blade to the heel of the fork, the fork has been worn to the point of being unsafe and unuseful. Figure 15.3 illustrates the proper use of a caliper. See Appendix C for information on fork arm caliper purchases.

PALLET SAFETY

Operators must be aware of the condition of pallets that they handle on a regular basis. When a load arrives at a receiving dock, the condition of the pallet may be such that it becomes a hazard for all employees. Pallets are used as the holding base for storing and stacking material. A defective pallet could easily give way and spill a load. The load could be high and come crashing down on those below.

Pallets can be inspected by first looking at their condition when picking up the load. If the load is raised above eye level of the operator a quick visual check of the lower pallet should be made. If the pallet is sound, continue to stack it. Product must be safely stacked, note the stacking patterns in Figure 15.4. If the pallet is defective in some way, lower it and place it in a special area so the product can be transferred to a good pallet. Tags can be used to alert employees that a particular pallet is defective. The tag can be of a certain color or design to make it more visible and different from other tags.

The National Wooden Pallet and Container Association recommends that employees be trained to look for the following problems:

- Protruding, broken or missing stringers.
- Sagging deck boards and bowed stringer boards.
- Defects such as notched areas, including splits, decay and excess knots.
- Splits in the wood, especially around fasteners and stringer feet.
- Broken boards or missing wood exceeding allowable limits.
- Damaged wood or poorly repaired damage to wood and plastic pallet components.
- Decayed wood or unsound knots, especially around fasteners.
- Excessive use of companion stringer repair boards.
- Improper component replacement.
- Repairs above a notched area on stringer boards.
- Replaced pallet components that have notably different dimensions from companion pieces on the same pallet.

226 / Forklift Safety

- Pallets with non-conforming dimensions, including squareness or flatness.

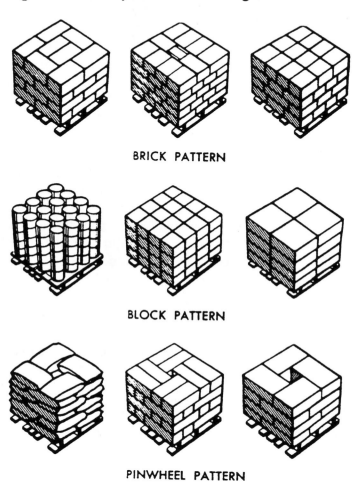

Figure 15.4. Proper Pallet Stacking Patterns

BRICK PATTERN

BLOCK PATTERN

PINWHEEL PATTERN

It is important to safely stack product on pallets.

The U.S. Postal Service has developed an in-depth checklist for the millions of reusable press wood and plastic pallets it has in use. If an inspector spots a pallet with any of the following defects, the pallet is taken out of service for corrective actions.

- Any corner leg broken such that the leg cannot rest on a level surface.
- Any corner leg that is missing.
- Any center leg that is broken or missing such that the legs cannot rest on a level surface.

- Any two support legs that are broken such that the legs cannot rest on a level surface.
- Any two support legs that are missing.
- If 25% or more of the pallet deck is deformed such that the deformation is greater than one-half inch below the normal surface of the pallet.
- Any crack 10 inches or longer which extends through the center of the pallet (this applies only to structural foam plastic and twin-sheet thermoform units).
- Any crack 6 inches or longer which extends through the center of the pallet (this only applies to press wood units).
- A hole in the pallet deck greater that the surface areas of the opening in the pallet deck for a leg.
- Pallet deck deformation which includes cracks and holes along the pallet edges of 3 inches or more into a leg assembly or into a pallet surface.

Figure 15.5. High Stack of Empty Pallets

Keep empty pallets stacked to a height of six feet for fire safety codes. The stack in this photo is too high.

Pallets should always be neatly stacked to prevent them from falling. If the stack is too high, employees may not be able to lift individual pallets from the stack. Wear gloves when handling pallets to protect the hands from nails and splinters.

Another consideration regarding pallet safety is fire safety. A fire in a stack of pallets can be very hot and easily spread within a building. Fire safety guidelines ask that pallets be kept to a height of six feet when stacked.

Note the high stack of pallets in Figure 15.5.

USE OF LIFT CAGES/WORKING PLATFORMS

There are times when someone has to be lifted to an upper level to perform maintenance. If someone is lifted in an unsafe manner or in/on an unsafe device, disaster could occur. OSHA statistics show that many employees fall from heights while performing forklift duties. It is not unusual to hear of an employee falling from the forks of a lift truck. Riding the forks is not that uncommon.

To safeguard anyone who must be lifted by the machine, some basic principles must be followed.

- Only use an approved lifting cage. Many are sold commercially and offer protection as required. Be sure the power equipment manufacturer approves of this process. Check their operating guidelines or personally contact them.
- Proper hand railing on two sides and the front.
- Mid rail and toe board.
- Moveable gate that locks into place.
- Non-skid walking surface.
- Slots to allow maximum width for the forks.
- A high back that includes small enough metal grid mesh that prevents a finger from being injured if placed in the operating mechanism.
- An effective means of securing the platform to the mast of the lift truck.
- A means shall be provided to allow personnel on the platform to shut off the power to the truck.

Note the safe use of an approved lifting cage in Figure 15.6.

Figure 15.6. Safe Use of a Lifting Cage

A safe lifting platform or cage is important for operator safety. Follow manufacturers guidelines.

These work platforms are usually stored up and out of the way until they are needed. A lift truck operator must retrieve the platform and can take it to the work site. Workers should not be carried to the work site while on the platform. The platform should be inspected before being used. A critical safety requirement is to properly anchor the platform to the mast. A word of caution, some manufacturers only allow for a chain hook to be passed over a link to lock the platform to the mast. This hook can easily become undone which could allow the platform to slide off the forks. A high tensile bolt and nut or an approved clevis should be used for maximum safety when fastening the chain. A chain manufacturer can assist in this situation.

> **ACCIDENT FACT**
>
> An employee needed to replace some burned-out bulbs approximately 15 feet high. He flagged down a passing lift truck, put a foot on each fork and held onto the mast of the truck. While he was being raised his little finger was caught between the roller and mast and was amputated. He knew he was doing the wrong thing in not using a lift cage but said he was in a hurry. The lift truck operator stated that it wasn't unusual to lift the mechanic this way, he did it all the time.

When the operator and vehicle is directly below the area of work, the worker should get into the platform and properly lock the gate. Once elevated the communication between the worker and the operator should be such that both have an understanding of what is being said. Hand-held talk devices are a plus here. At no time should the operator desert the lift truck. At no time should the operator drive to another location with the work platform elevated. An added safety rule should allow for the worker to anchor a support line to the mast and platform while wearing a safety harness. Check with the safety harness manufacturer and lift truck manufacturer to ensure that this procedure is safe.

Where the operator has to look up at the work platform, or at a load, small wire mesh placed over the overhead guard adds to operator safety. Most overhead guards have large openings in them and could allow small pieces of product or tools to come through the overhead guard. It is not unusual to hear of an operator being struck by a small object that has gone through the overhead guard. A hard hat becomes an added safety factor in this case to protect operators.

When complete, return the platform to its storage area. If it is in need of repair, properly tag the platform and leave it down so it can be properly repaired. If put back after use, even with a repair tag on it, the likelihood of repair is remote. When the platform is needed for the next project, employees may not take time out to repair the platform and an accident could be the result. Figure 15.7 illustrates how to properly store the work platform.

Where construction crews or outsiders are in a facility to make repairs, management should consider allowing the use of working platforms in the hands of non-employees. Only trained and authorized employees can operate power equipment. Will the non-employee have the same degree of skill? Will they use a lift platform safely? These issues should be resolved prior to any work being done. Both employees and visitors are being endangered in this case. A written program in contractor safety will help safeguard employees and visitors. Contractors, as well as employees, have been known to climb upon handrailing to obtain a higher elevation from a lifting platform. This practice must be prohibited.

Figure 15.7. Storing a Lift Cage

When a lifting platform is not in service, safely store it out of the way. This also protects it from damage.

ENVIRONMENTAL CONSIDERATIONS*

*The information detailed in this section is drawn from the Hyster Company's publication entitled *Environmental Guidelines for Clean Forklift Truck Operations*.

In addition to safety and health issues in the workplace, environmental issues are just as important. As an example, choosing the type of fuel or power for a lift truck can have a bearing on the environment. Lift truck selection, when done properly, can help reduce pollutants. Choice of fuel source involves many considerations:

- Emission levels.
- Noise levels.
- Ability to operate in hot and cold conditions.
- Fuel and maintenance costs.

Table 15.1. Advantages and Disadvantages of Lift Truck Fuel Options

Advantages **Disadvantages**

Gasoline

* Fast lift and travel speeds
* Easy maintenance and repair
* Short refueling time

* Exhaust contains higher levels of CO, NO and HC than liquid propane gas and compressed natural gas exhaust
* Storage costs dramatically increased by environmental regulations
* Trucks are relatively expensive to operate $9-$10 per eight hour shift.

Diesel

* Fast lift and travel speeds
* Short refueling times
* Diesel fuel usually costs less than gasoline or LPG

* Emissions contain possible carcinogens
* Difficult to start in cold operating conditions
* Noise levels are somewhat higher than other fuel options

Liquid Propane Gas

* Carbon monoxide emissions are half that of gasoline exhaust
* Easy refueling
* Good combustion under hot and cold operating conditions

* Tank and engine modifications make for a higher initial cost
* Relatively high fuel costs
* Specialized procedures needed to maintain the fuel system properly

Compressed Natural Gas

* Low emissions
* Fuel costs less to operate than LPG or gasoline

* Tank and engine modifications make for a higher initial cost
* High cost for refueling station
* Lengthy refueling time
* Specialized maintenance required

Electricity

* No exhaust emissions
* Relatively low fuel cost
* Low noise levels

* Slower lift and travel speeds than internal combustion engines
* Added expense for battery charging area
* Batteries require special disposal procedures

Maintaining Water Quality

If a plant is using solvents to clean lift trucks, it is possible that they are contaminating the water. Pollutants could consist of:

- Antifreeze.
- Used oil.
- Hydraulic fluids.
- Other chemicals.

All of the above require proper disposal according the Federal, State, and local EPA guidelines. Recycling or substitution is the best method of control.

When managing hazardous waste storage and removal, it is important to recognize the importance of control and monitoring. The Resource Conservation and Recovery Act (RCRA) regulates hazardous waste. The lift truck operator could easily become involved in the storage or removal of hazardous waste.

Storing Hazardous Waste

- Clearly mark each storage container with the words "Hazardous Waste" and the collection start date.
- Keep containers closed, except when filling or being emptied.
- Keep waste containers in good condition. Handle all containers with care and replace them as needed.
- Use tanks designed to hold specific waste. Check the condition of the tank before using it. Be sure the tank will not leak in any way. Consider tank corrosion or tank rupture. Mark all tanks with the appropriate signage. Maintain MSDS's as needed.
- Keep tanks covered or provide at least two feet of space at the top of the tank (referred to as free boarding).
- Inspect monitoring systems for corrosion or other signs of decay on a weekly basis.
- Be sure the appropriate fire fighting equipment is available if needed.
- Be sure the proper spill clean-up materials are available. Note the spill clean up kit in Figure 15.8.

Figure 15.8. Spill Clean Up Kit

Where chemicals and hazardous waste are stored or handled ensure a spill clean up kit is readily available.

Preparing Waste for Shipment

- First, consult with your hauler to determine a suitable transport container. Ensure that the hauler is reputable.
- Label each container carefully, stating the composition of the waste and the quantity.
- Document your shipment with a hazardous waste manifest.
- The multicopy shipping document helps track hazardous waste shipments from the generator, to the hauler, to final storage.
- Select an EPA approved hauler and a Subtitle D-designated facility.

Lift Truck Maintenance/Environmental Controls

Those who own lift trucks must be aware of the need to provide on-going maintenance. When strict procedures are controlled, costly repairs are prevented and environmental issues are also controlled. Operator safety is also enhanced.

Organizations should consider the following recommendations for environmental control:

- **Recycle Oil.** One gallon of improperly disposed used oil can contaminate one million gallons of water. One gallon of recycled oil provides 2.5 quarts of lubricating oil, the same amount produced by 42 gallons of crude oil. This program can lead to reduced material costs for a facility.

- **Recycle Oil Filters.** All oil filters should be hot drained for at least 12 hours. The oil should be placed in a used oil container. A puncture hole on the top or crown of the filter helps the draining process. Where possible, crush the filter to claim additional used oil prior to recycling or disposal. Before disposing of any filters, check with the local authorities for proper procedures.

- **Recycle Fluids.** In addition to the recycling of used motor oil, other fluids such as hydraulic oils, brake fluid, transmission fluids and antifreeze require special disposal. Do not mix any fluids unless the recycler has given authorization to do so. Be sure the waste handler for the recycled fluids is reputable.

- **Recycle Batteries.** Because of the lead acid in lift truck batteries, many states ban them from landfills. Where batteries are stored outside, contaminants from the battery can be washed down into drains and contaminate a stream or the soil.

    Most battery manufacturers are able to recycle 100% of a used battery.

    - The steel tray is separated and sent to a steel recycler
    - Battery acid is reclaimed and reused in new batteries
    - Plastic and lead are separated and reused for new batteries or other products.

Lift truck batteries are usually replaced every 5 to 7 years depending on the application. The life of a lift truck may require several replacements. Battery suppliers should be contacted for the return of used batteries.

- **Recycle Tires.** During the life of a lift truck, the tires will most likely have to be replaced at least once. Most landfills ban whole tires or charge a high fee for acceptance. For tires to be placed in a landfill, they may have to be shredded first.

    A few alternatives to landfilling tires are:

    - Properly equipped power plants and pulp and paper mills can sometimes burn shredded tires as a power source.
    - Shredded tires can be reprocessed and used to make rubber asphalt additives for street pavement.
    - Whole tires can be used for soil erosion control or as man made reefs and breakwaters. Ensure that your local, state or federal EPA office is contacted prior to any of the above recycling programs.

FIRE FIGHTING

Every employee in a facility should have some working knowledge of basic fire fighting. Not only does this knowledge help at work, it can be very beneficial at home, on vacation or anywhere a fire can occur.

Powered equipment operators should know and understand the fire extinguishers that they would be exposed to on the job. Some basic information regarding fire extinguishers can be helpful.

Types of fires and fire extinguishers:

- **Class A.** Fires need a water type extinguisher that can fight fires in cloth, paper, wood, coal or anything that can have hot embers. The water will absorb the heat and help cool the fire. Sometimes the ashes or items that have been burning may need to be moved around so a thorough soaking can take place. Water must not be used on chemical fires, gasoline fires or electricity.

- **Class B.** Fires involve a dry chemical or powder extinguisher that discharges fine powder granules to smother the fire. Use this extinguisher for chemical or gasoline fires, the powder provides a cover or blanket so re-ignition will not occur. This type of extinguisher will most likely not be able to put out a Class A fire. The fire may be knocked down by the powder, but the deep-seated embers will not be extinguished. B extinguishers, when used, will require extensive cleanup because of the powder.

- **Class C.** Fires involve extinguishers which are for fires in energized electrical equipment. The extinguisher shell, most times red in color, has no gauge. Inside the shell is carbon dioxide (CO_2) liquid. When the pin is pulled and the handle is squeezed carbon dioxide gas extinguishes the fire by removing the oxygen. This gas is also very cold; -80 to -100 degrees Fahrenheit. Unlike Class A and B extinguishers which have a gauge to alert anyone of a full or empty charge, the CO_2 extinguisher has to be weighed to determine capacity. CO_2 will knock down a fire in Class A materials but will only remove the oxygen. The fire may reignite. CO_2 can put out a flammable liquid fire but the oxygen is only temporarily removed. A spark or open flame can reignite the vapors.

- **Type ABC** is a multi-purpose unit which will extinguish all three types of fires; A, B and C. It should be noted that powder type extinguishers create a lot of powder and there may be extensive cleanup. The powder can also have a negative effect on motors or any other mechanical item.

For those pieces of power equipment that operate in remote areas, a fully charged fire extinguisher should be on the unit. Operators should know how to use the extinguisher to fight a fire on their trucks or one that may occur at any part of the building. Fire hoses should not be touched unless the user has had specific training and authorization to operate a fire hose. Note the fire extinguisher classification in Figure 15.9. Figure 15.10 illustrates a fully charged and available extinguisher.

Figure 15.9. Fire Extinguisher Classification

Letter-shaped symbol markings are also used to indicate extinguisher suitability according to class of fire.

Extinguishers suitable for Class A fires should be identified by a triangle containing the letter "A." If colored, the triangle should be green.

Extinguishers suitable for Class B fires should be identified by a square containing the letter "B." If colored, the square shall be colored red.

Extinguishers suitable for Class C fires should be identified by a circle containing the letter "C." If colored, the circle should be colored blue.

Extinguishers suitable for fires involving metals should be identified by a five-pointed star containing the letter "D." If colored, the star shall be colored yellow.

Extinguishers suitable for more than one class of fire should be identified by multiple symbols placed in a horizontal sequence.

Class A and Class B extinguishers carry a numerical rating to indicate how large a fire an experienced person can put out with the extinguisher. The ratings are based on reproducible physical tests conducted by Underwriters' Laboratories, Inc. Class C extinguishers have only a letter rating because there is no readily measurable quantity for Class C fires which are essentially Class A or B fires involving energized electrical equipment. Class D extinguishers likewise do not

Operators should know how to choose the proper fire extinguisher for fighting fires.

238 / Forklift Safety

Figure 15.10. Fire Extinguiser Location

Fire extinguishers are to be fully charged, readily available, and properly inspected.

VOLUNTARY PROTECTION PROGRAMS (VPP)

OSHA set out in 1983 to encourage organizations to excel at safety while recognizing their programs and statistics. A facility would be honored by OSHA by receiving their highest honors, the STAR or MERIT award. The MERIT award requires additional corrective work on the company's part; but the award has significance. STAR, OSHA's most prestigious certification lasts for three years. MERIT certification lasts for one year. Re-application and an OSHA team visit would re-start the program.

The awards, however, are not just handed to a facility. An extensive application has to be prepared by the facility and forwarded to the OSHA regional office. OSHA established specific program elements that the facility would have to be using in their safety and health programs. In addition, the incidence rate for the facility would have to be below the national average for their type of business.

A key ingredient in the OSHA requirements is employee training. Included in the required training program is powered industrial truck training. To receive credit for this particular program a facility would have to maintain an on-going program for operators. Each different piece of powered equipment would have an operator training program to match it. Annual training would be evaluated to meet the VPP criteria providing it included all operators and contained lectures, visual aids, testing, and drivers skills evaluation.

The quest for STAR or MERIT status is not easy. It takes dedication and hard work to qualify for VPP. Once attained, OSHA expects a facility to continue to make progress and maintain lower than average safety statistics. Of the 6.2 million facilities for which OSHA has responsibility, there have only been 300 sites selected as STAR or MERIT since 1983. Most of these sites maintain safety statistics much lower than the national average. It is not unusual for a VPP site to maintain an incident rate 50% lower than that of comparable industry.

For those employers that are small and wish to obtain recognition for their safety program they may wish to participate in state sponsored SHARP programs. Contact your state safety regulators for assistance.

To achieve VPP status, an organization's safety program must be comprehensive. A method of auditing should be in place to continually monitor all facets of the program.

Listed below are two segments from an audit of a major corporation. Note Figure 15.11, which is from the training section that identifies the various components of their forklift training program. It would be beneficial if readers developed this same type of program to assist in their overall program efforts.

The individual components of this audit section outline the requirements of the forklift program. Each facility that is audited in the corporation each year has to maintain these high standards. The outline provides guidelines to follow for full compliance. The completion of the program results in full points for lift truck training.

The second part of this audit illustration (Figure 15.12) outlines the scoring sheet for the entire training section. Section D, one of six, is worth 20 points in a 100 point audit. Note that forklift training has the most points attached to it, or 5 full points, from the 20 that are available in this section.

Figure 15.11. Audit Questionnaire - Forklift Training

D-2 Forklift training for employees and supervisors.
Note: Use Appendix B for necessary identification.

	YES	NO	N/A
Are all <u>full-time</u> and <u>part-time</u> operators trained or about to be trained through the corporate program? (10.0)			
Was each video segment unit and quiz given one at a time? (5.0)			
Are you allowing several weeks (at least) between each quiz?			
Has the driving skills portion or rodeo been completed? (5.0)			
Is a record of all scores available (who scored the most points)?			
Has each supervisor been shown the video segments, <u>and</u>, been given each written exam? (5.0)			

(Note: the corporate video/quiz/and driving skills program must be used to obtain full credit in this category.)

If program has not been completed, does management have a schedule?

N	F	G	E
0	1 TO 8	9 TO 18	19 TO 25

If an auditing program is being used, a list of questions should be used in evaluating the forklift program.

JOB SAFETY OBSERVATION, (JSO)

An organization has just provided a comprehensive training program to all of its operators. The training was costly to complete because it took time away from production. The firm wants to reduce accidents, prevent costly incidents and comply with regulations. How will they know that their program has been successful?

An ideal program to evaluate employees as they are safely working at their jobs is the job safety observation, JSO, program. The program works in this manner:

- A supervisor observes an employee performing his work for a short period of time. The observation time could be for 5 to 10 minutes.

Figure 15.12. Audit Scoring Form

GROUP "D"
SAFETY TRAINING

	NONE	FAIR	GOOD	EXCELLENT
1. First aid and CPR training.	0	1 / 4	5 / 10	11 / 15
2. Forklift training for employees and supervisors.	0	1 / 8	9 / 18	19 / 25
3. Supervisors safety training.	0	1	2 / 4	5
4. Job safety analysis (JSA).	0	1 / 4	5 / 10	11 / 15
5. Employee safety meetings.	0	1 / 3	4 / 8	9 / 10
6. Job safety observation program.	0	1	2 / 4	5
7. Use of films, slides, visual aids.	0	1 / 3	4 / 8	9 / 10
8. Human Relations materials for supervisors.	0	1	2 / 4	5
9. Fire committee or emergency team.	0	1	2 / 4	5
10. Accident repeaters program.	0	1	2 / 4	5
	+	+	+	=

Total point value of circled numbers   ____ x .20 = ____

This form provides information on how the forklift training segment is rated in the training section of the audit.

- The observation is recorded on a form; the employee is fully aware that he/she is being observed.

- Any unsafe behavior or unsafe conditions are to be noted by the supervisor on the form, and corrected.

- Each employee in a department should be observed at least once a month.

JSO's are not safety meetings, they are snapshots of what the employee is doing in his normal daily work. For those jobs that take a long time to complete such as unloading a trailer, hand picking individual parts, changing a battery or loading a trailer, the supervisor should observe that part of the job that is the most hazardous. If unsafe behavior is detected, another JSO should be completed within a week to ensure the employee knows and understands how to safely do the job. Note: the word "job" does not refer to an occupation, but involves a segment of a task. This same principle is used in the Job Hazard Analysis chapter.

The JSO program is an aid to the overall safety program. Does experience help prevent accidents? Consider a few of the following comments regarding worker behavior and erroneous beliefs:

- Experienced employees will take shortcuts.
- Accidents always happen to "someone else."
- Seniority will keep a worker immune from injury.
- Some workers actually deny hazards and rules—they feel they are above them.

Therefore, experience does not always guarantee safe behavior. Much is learned by observing.

What Jobs to Select

A good starting point for a supervisor in this program is to observe those jobs that have a history of:

- Producing serious injuries.
- Have caused the supervisor to fill out accident reports.
- Involve new or transferred employees.
- Have produced close calls.
- Have been a cause of concern in the past.
- Have involved product or building damage.
- Involve employees that bear watching as a result of returning from an illness, or are accident repeaters.

Informing Employees of JSO

This program is not intended to be undercover or mysterious. Employees are to be trained in what JSO is all about. The bottom line is to ensure that the job is being performed safely. The supervisor should be observing employees each day as a normal part of his job to confirm that they are doing the job safely and also correctly for production purposes.

The employee is to be told in advance that he is about to be observed. It would be unusual for an employee to work in an unsafe manner in front of a supervisor. Most individuals will do the best they can when they know they are being formally observed. When performing safely, the employee demonstrates that he can do the job safely.

An employee can always be observed informally; that is, by observing his work and not using the form in Figure 15.13 or discussing the process with him. If an employee has had an accident and has returned to the job, casual observations are needed to confirm that he is not once again endangering himself or others.

Scheduling JSO's

JSO's should be completed throughout the month. It is wrong to attempt to observe an entire department during a single day; much is lost in this process.

Select the jobs, have the form ready and proceed, look for any indications of unsafe behavior. In addition, if there are any unsafe conditions observed, the supervisor is required to correct them. Never hesitate to stop an employee from doing something that can harm him, a fellow employee or cause property damage.

Once completed, review the results of the observation with the employee. Never comment on the individual, only on the observed behavior. Mark the form properly.

A few tips for review are:

- Explain what was observed.
- Use "show and tell" techniques, if necessary.
- Be sensitive to employee comments about the job.
- Explain a safer alternative, if necessary.
- Try to persuade or convince the employee to follow safe procedures.

Figure 15.13. Job Safety Observation Form

JOB SAFETY OBSERVATION MONTHLY ACTIVITY REPORT
MONTH OF ____________, 19___

SUPERVISOR: ____________ LOCATION: ____________

MAKE ADDITIONAL COMMENTS ON THE BACK OF THIS FORM

EMPLOYEE NAME	JOB BEING PERFORMED	WAS JOB PERFORMED SAFELY?	IDENTIFY UNSAFE CONDITIONS	WAS EMPLOYEE RE-INSTRUCTED?	WAS A VERBAL WARNING ISSUED?	WERE UNSAFE CONDITIONS CORRECTED?	DATE OF OBSERVATION	COMMENTS	EMPLOYEE INITIALS*

* If employee does not wish to initial form, write D.N.I. (Did Not Initial)

Much can be accomplished by observing employees perform their jobs. The job safety observation form provides for proper documentation.

SUMMARY

The purpose of this chapter was to identify many of those potential hazards or programs that could impact powered equipment operators. The individual items were too brief to be a single chapter but were never the less important.

The section on reducing the damage which are caused by forklifts is essential to any operator. Millions of dollars are lost each year throughout industry as a result of product and building damage.

Proper care of the forks on lift trucks is often overlooked. The section on care and inspection of forks is important to any program. This information may come as a surprise to many readers regarding this subject. For years there has been a lack of understanding and attention toward forks.

Pallet safety is also very important to the safety of all employees in a facility. There are many guidelines and tips to help regarding pallet safety. A facility should develop a special program to impact and correct any deficiencies in their pallet program.

Many falls take place each year in the workplace. It is not unusual for workers to stand on bare forks or an empty pallet to be lifted. Falls can be deadly. There are OSHA statistics to validate this. The proper lifting cage and work procedures can safeguard workers. Involve the maufacturer for approval in using the equipment.

We are living in an environmentally concerned world. The issues associated with a forklift bear consideration because of the oil, filters, tires, fuels and antifreeze that it uses. It behooves an organization to be environmentally conscious and follow proper EPA procedures for the use and disposal of chemicals.

Do all operators know how to fight fires in their machines? Can this fire fighting be done in such a way to safely put out a fire without causing any personal harm? Management must make certain that the proper amount and type of fire fighting equipment is available and that employees have been properly trained.

Looking for a super-safe workplace? Would you care to invite OSHA in to evaluate your program and inspect the workplace? Would your employees like to announce that they work in a STAR or MERIT facility? Forklifts and other powered equipment are a part of this evaluation. Audit programs that parallel this OSHA program can be an asset to any safety program.

How can you determine if employees are working safely? The job safety observation program can pay big dividends. Supervisors are the key element in this program. Management must insist that the program is followed properly to be effective. Safe behavior of all powered equipment operators is the final goal.

REFERENCES

Auguston, K. "What You Can Do to Improve Pallet Quality," *Modern Materials Handling*, August 1996, pp. 34-37.

Hyster Company. *Environmental Guidelines for Clean Forklift Truck Operations*, Danville, IL, 1996.

"Take Heed of Warning Signs for Pallet Safety," *Material Handling Engineering*, October 1994, p. 74.

Schwind, G. "Lift Truck Forks: Maintenance and Inspection," *Material Handling Engineering*, March 1995, pp. 28-29.

Swartz, G. "OSHA's VPP Program; Are You Ready For It," *Professional Safety*, Des Plaines, IL, June 1995, pp. 21-23.

Taub, R.J. "Protecting Structures and Equipment from Lift Truck Damage," *Plant Engineering*, October 13, 1988, pp. 112-114.

APPENDIX A

OSHA Proposed Rule for Training Powered Industrial Truck

Operators in General and Maritime Industries

60 FR 13782

March 14, 1995

DEPARTMENT OF LABOR

Occupational Safety and Health Administration

29 CFR Parts 1910, 1915, 1917, and 1918

[Docket No. S-008]

Powered Industrial Truck Operator Training

AGENCY: Occupational Safety and Health Administration, Labor.

ACTION: Proposed rule.

SUMMARY: The Occupational Safety and Health Administration (OSHA) is proposing to revise the general industry safety standard for training powered industrial truck operators and to add equivalent training requirements for the maritime industries. The existing standard in part 1910 requires that only trained operators who are authorized to do so can operate powered industrial trucks and that methods of training be devised. The proposed training requirements would mandate the development of a training program that would base the amount, type, degree, and sufficiency of training on the knowledge of the trainee and the ability of the vehicle operator to acquire, retain, and use the knowledge and the skills and abilities that are necessary to safely operate the truck. A periodic evaluation of each operator's performance would be required. Refresher or remedial training also would be required, based primarily on unsafe operation, an accident or near miss, or deficiencies found in a periodic evaluation of the operator.

DATES: Written comments and requests for a hearing on this proposed rule must be postmarked by July 12, 1995.

ADDRESSES: Comments, information, and hearing requests should be sent in quadruplicate to: Docket Office, Docket No. S-008; Room N2624; U.S. Department of Labor, Occupational Safety and Health Administration; 200 Constitution Avenue NW; Washington, DC 20210 (202-219-7894).

FOR FURTHER INFORMATION CONTACT: Mr. Richard P. Liblong, Office of Information and Consumer Affairs, U.S. Department of Labor, Occupational Safety and Health Administration, Room N3641; 200 Constitution Avenue NW; Washington, DC 20210 (202-219-8148).

SUPPLEMENTARY INFORMATION:

I. Background

a. The General Industry Standard

On May 29, 1971 (36 FR 10466), OSHA adopted some of the existing Federal standards and national consensus standards as OSHA standards under the procedures described in section 6(a) of the Occupational Safety and Health Act (OSH Act) (29 U.S.C. 655, et.al.). Section 6(a) permitted OSHA to adopt, without rulemaking, within 2 years of the effective date of the Act, any established Federal standard or national consensus standard.

One of the consensus standards that was adopted under the 6(a) procedure was the American National Standards Institute (ANSI) B56.1-1969 Safety Standard for Powered Industrial Trucks. Among the provisions adopted from that standard was the operator training requirement codified at 29 CFR 1910.178(l), which states:

Only trained and authorized operators shall be permitted to operate a powered industrial truck. Methods of training shall be devised to train operators in the safe operation of powered industrial trucks.

In that consensus standard, a powered industrial truck is defined as a mobile, power-driven vehicle used to carry, push, pull, lift, stack, or tier material. One truck may be known by several different names. Included are vehicles that are commonly referred to as high lift trucks, counterbalanced trucks, cantilever trucks, rider trucks, forklift trucks; high lift trucks, high lift platform trucks; low lift trucks, low lift platform trucks; motorized hand trucks, pallet trucks; narrow aisle rider trucks, straddle trucks; reach rider trucks; single side loader rider trucks; high lift order picker rider trucks; motorized hand/rider trucks; or counterbalanced front/side loader lift trucks. Excluded from the scope of the OSHA standard are vehicles used for earth moving or over-the-road haulage.

b. The Maritime Safety Standards

In 1958, Congress amended the Longshoremen's and Harbor Workers' Compensation Act (LHWCA) (44 Stat. 1424; 33 U.S.C. 901 et seq.) to provide maritime employees with a safe work environment. The amendments (Pub. L. 85-742, 72 Stat. 835) required employers covered by the LHWCA to "furnish, maintain and use" equipment and to establish safe working conditions in accordance with regulations promulgated by the Secretary of Labor. Two years later, the Labor Standards Bureau (LSB) issued the first set of safety and health regulations for longshoring activities as 29 CFR part 9 (25 FR 1565, February 20, 1960). These regulations only covered longshoring activities taking place aboard vessels.

Passage of the OSH Act (84 Stat. 1590; 29 U.S.C. 650 et seq.) authorized the Secretary of Labor to adopt established Federal standards issued under other statutes, including the LHWCA, as occupational safety and health standards under the OSH Act. Accordingly, the Secretary adopted the existing shipyard employment and longshoring regulations and recodified these rules as 29 CFR parts 1915 and 1918 (39 FR 22074; June 19, 1974). Since the OSH Act comprehensively covered all private employment, the longshoring standards also applied to shoreside cargo-handling operations. (See 29 CFR 1910.16.) The requirements for the use of mechanically powered vehicles used aboard vessels were codified at § 1918.73. These provisions did not include a requirement for the training of vehicle operators.

In addition, in accordance with established policy codified at 29 CFR 1910.5(c)(2), OSHA has applied its general industry regulations to shoreside activities not covered by its older longshoring rules. Citations also have been issued under section 5(a)(1) (the General Duty Clause) of the OSH Act (84 Stat. 1593; 29 U.S.C. 654), since some serious hazards are not addressed by the requirements of part 1910, 1915, or 1918.

On July 5, 1983 (48 FR 30886), OSHA published its final standard for Marine Terminals. These rules were intended to address the shoreside segment of marine cargo handling. Section 1917.27 Personnel required that:

(a) *Qualifications of machinery operators.*
(1) Only those employees determined by the employer to be competent by reason of training or experience, and who understand the signs, notices and operating instructions and are familiar with the signal code in use shall be permitted to operate a crane, winch or other power operated cargo handling apparatus, or any power operated vehicle, or give signals to the operator of any hoisting apparatus.

Exception: Employees being trained and supervised by a designated person may operate such machinery and give signals to operators during training.

(2) No employee known to have defective uncorrected eyesight or hearing, or to be suffering from heart disease, epilepsy, or other ailments which may suddenly incapacitate him shall be permitted to operate a crane, winch or other power-operated cargo handling apparatus or a power-operated vehicle.

The Marine Terminal Standards also had requirements for powered industrial trucks at § 1917.43 Powered industrial trucks. However, these requirements were for the operation, maintenance and outfitting of those vehicles and did not expand upon the training requirements found at § 1917.27.

On June 2, 1994, OSHA published in the Federal Register (59 FR 28594) a Notice of Proposed Rulemaking (NPRM) for the revision of the longshoring and marine terminals standards.

That NPRM did not propose to amend significantly the aforementioned training requirements of § 1917.27 or to incorporate a training requirement for longshoring (on-board vessel) operations.

c. Updated Consensus Standard

Since promulgation of the OSHA standards, the consensus standard (ANSI B56.1) has undergone four complete revisions (dated 1975, 1983, 1988 and 1993). The current consensus standard (Ex. 3–1) states:

4.18 Operator qualifications.

Only trained and authorized persons shall be permitted to operate a powered industrial truck. Operators of powered industrial trucks shall be qualified as to visual, auditory, physical, and mental ability to operate the equipment safely according to 4.19 and all other applicable parts of Section 4.

4.19 Operator training.

4.19.1 Personnel who have not been trained to operate powered industrial trucks may operate a truck for the purposes of training only, and only under the direct supervision of the trainer. This training should be conducted in an area away from other trucks, obstacles, and pedestrians.

4.19.2 The operator training program should include the user's policies for the site where the trainee will operate the truck, the operating conditions for that location, and the specific truck the trainee will operate. The training program shall be presented to all new operators regardless of previous experience.

4.19.3 The training program shall inform the trainee that:

(a) The primary responsibility of the operator is to use the powered industrial truck safely following the instructions given in the training program.

(b) Unsafe or improper operation of a powered industrial truck can result in: death or serious injury to the operator or others; damage to the powered industrial truck or other property.

4.19.4 The training program shall emphasize safe and proper operation to avoid injury to the operator and others and prevent property damage, and shall cover the following areas:

(a) Fundamentals of the powered industrial truck(s) the trainee will operate, including:

(1) characteristics of the powered industrial truck(s), including variations between trucks in the workplace;

(2) similarities to and differences from automobiles;

(3) significance of nameplate data, including rated capacity, warnings, and instructions affixed to the truck;

(4) operating instructions and warnings in the operating manual for the truck, and instructions for inspection and maintenance to be performed by the operator;

(5) type of motive power and its characteristics;

(6) method of steering;

(7) braking method and characteristics, with and without load;

(8) visibility, with and without load, forward and reverse;

(9) load handling capacity, weight and load center.

(10) stability characteristics with and without load, with and without attachments;

(11) controls-location, function, method of operation, identification of symbols;

(12) load handling capabilities: forks, attachments;

(13) fueling and battery charging;

(14) guards and protective devices for the specific type of truck;

(15) other characteristics of the specific industrial truck.

(b) Operating environment and its effect on truck operation, including:

(1) floor or ground conditions including temporary conditions;

(2) ramps and inclines, with and without load;

(3) trailers, railcars, and dockboards (including the use of wheel chocks, jacks, and other securing devices;

(4) fueling and battery charging facilities;

(5) the use of "classified" trucks in areas classified as hazardous due to risk of fire or explosion, as defined in ANSI/NFPA 505;

(6) narrow aisles, doorways, overhead wires and piping, and other areas of limited clearance;

(7) areas where the truck may be operated near other powered industrial trucks, other vehicles, or pedestrians;

(8) use and capacity of elevators;

(9) operation near edge of dock or edge of improved surface;

(10) other special operating conditions and hazards which may be encountered.

(c) Operation of the powered industrial truck, including:

(1) proper preshift inspection and approved method for removing from service a truck which is in need of repair;

(2) load handling techniques, lifting, lowering, picking up, placing, tilting;

(3) traveling, with and without loads; turning corners;

(4) parking and shutdown procedures;

(5) other special operating conditions for the specific application.

(d) Operating safety rules and practices, including:

(1) provisions of this Standard in Sections 5.1 to 5.4 address operating safety rules and practices;

(2) provisions of this Standard in Section 5.5 address care of the truck;

(3) other rules, regulations, or practices specified by the employer at the location where the powered industrial truck will be used.

(e) Operational training practice, including:

(1) if feasible, practice in the operation of powered industrial trucks shall be conducted in an area separate from other workplace activities and personnel;

(2) training practice shall be conducted under the supervision of the trainer;

(3) training practice shall include the actual operation or simulated performance of all operating tasks such as load handling, maneuvering, traveling, stopping, starting, and other activities under the conditions which will be encountered in the use of the truck.

4.19.5 Testing, Retraining, and Enforcement.

(a) During training, performance and oral and/or written tests shall be given by the employer to measure the skill and knowledge of the operator in meeting the requirements of the Standard. Employers shall establish a pass/fail requirement for such tests. Employers may delegate such testing to others but shall remain responsible for the testing. Appropriate records shall be kept.

(b) Operators shall be retrained when new equipment is introduced, existing equipment is modified, operating conditions change, or an operator's performance is unsatisfactory.

(c) The user shall be responsible for enforcing the safe use of the powered industrial truck according to the provisions of this Standard.

Note: Information on operator training is available from such sources as powered industrial truck manufacturers, government agencies dealing with employee safety, trade organizations of users of powered industrial trucks, public and private organizations, and safety consultants.

(For an explanation of why OSHA decided to propose a somewhat different standard, see section entitled Summary and Explanation of the Proposed Standard, below.)

Since 1971, the consensus committee has adopted other volumes for additional types of vehicles that fall within the broad definition of a powered industrial truck. Specifically, requirements have been adopted for guided industrial vehicles, rough terrain forklift trucks, industrial crane trucks, personnel and burden carriers, operator controlled industrial tow tractors, and manually propelled high lift industrial trucks. This rulemaking would adopt training requirements for all types of powered industrial trucks regardless of their usage and the industry in which they are operating.

d. Petitions and Requests

On March 15, 1988, the Industrial Truck Association (ITA) petitioned OSHA to revise its standard requiring the training of powered industrial truck operators (Ex. 3–2). The petition contained suggested language for a proposed requirement along with a model operator training program by which compliance with the recommended requirement could be met. OSHA responded to the petition on April 8, 1988, stating that work on the revision of the OSHA powered industrial truck operator training requirement would begin as soon as other priority projects were completed.

In addition to the petition, other interested persons have frequently asked questions about training operators of powered industrial trucks, such as:

- What constitutes the necessary and sufficient training of forklift operators?
- How can one ensure that all forklift operators have been trained?
- What testing, if any, should be conducted as part of the training?
- Should the prior experience of a newly hired employee be considered as fulfilling part or all of the training requirement or totally fulfilling the employer's obligation to train that employee?

Some interested persons have suggested that OSHA develop a standardized training course or at least review and comment on or endorse various training courses, programs, agenda, or outlines. Others have suggested that OSHA license or certify all powered industrial truck operators to attest to their ability to properly operate powered industrial trucks. These concerns also were considered in the development of the proposed rulemaking. OSHA is proposing to amend the current powered industrial truck operator training requirements for general industry and to adopt the same requirement for the maritime industries.

e. *Reasons for the Proposal*

As discussed in the benefits discussed below and in the Regulatory Impact Analysis, powered industrial truck accidents cause approximately 85 fatalities and 34,900 serious injuries each year. It is estimated that approximately 20 to 25 percent are at least in part caused by inadequate training.

As just discussed, the ITA and others have requested that OSHA improve its training requirement for powered industrial truck operators. ANSI has substantially upgraded its recommended training requirements. OSHA preliminarily concludes that upgrading the training requirements for powered industrial truck operators will substantially reduce a significant risk of death and injury from untrained operators driving powered industrial trucks.

II. The Powered Industrial Truck

The term powered industrial truck is defined in the American Society of Mechanical Engineers, ASME B56.1 (formerly the ANSI B56.1 standard) as a "mobile, power propelled truck used to carry, push, pull, lift, stack, or tier material."

There are presently approximately 822,830 powered industrial trucks in use in American industry. This number was generated using the available information on truck shipments of powered industrial trucks and the percentage of market that ITA members control. This information was provided OSHA by the Industrial Truck Association.

The Industrial Truck Association stated in conversations with OSHA representatives that it considers the average useful life of a powered industrial truck to be 8 years. The 8-year life cycle has been used throughout the preparation of this proposed rule and in the formulation of the Preliminary Regulatory Impact Analysis. The vehicle manufacturers also estimate that there are, on average, 1.5 operators for each industrial truck. A search of the available literature indicates that this number has not been disputed. OSHA believes that this number is a fair assessment of the number of powered industrial operators since many employers (particularly small employers) have one operator per truck and the vehicle is used only during one shift per day whereas other vehicles are used by multiple operators during multiple shifts.

Powered industrial trucks are classified by the manufacturers according to their individual characters. There are seven classes of powered industrial trucks:

Class 1—Electric Motor, Sit-down Rider, Counter-Balanced Trucks (Solid and Pneumatic Tires).
Class 2—Electric Motor Narrow Aisle Trucks (Solid Tire).
Class 3—Electric Motor Hand Trucks or Hand/Rider Trucks (Solid Tires).
Class 4—Internal Combustion Engine Trucks (Solid Tires).
Class 5—Internal Combustion Engine Trucks (Pneumatic Tires).
Class 6—Electric and Internal Combustion Engine Tractors (Solid and Pneumatic Tires).
Class 7—Rough Terrain Fork Lift Trucks (Pneumatic Tires).

Each of these different types of powered industrial trucks has its own unique characteristics, and inherent hazards. To maximize the effectiveness of the training, it must be somewhat unique for each type vehicle. For example, an operator of a high lift rider truck must have an understanding of the basics of the vehicle's stability (including those factors which affect that stability), the need to not overload the vehicle, and the need to operate the vehicle according to established rules (such as not using the vehicle to elevate employees who are standing on its forks). On the other hand, order picker trucks elevate the operator along with a platform that is used to hold material destined for storage or retrieval from storage in high stacking racks or bins. The platforms on these trucks are not completely enclosed by railings, toe boards, or other similar fall protection devices to prevent an operator from falling off an elevated platform. To be protected, the operator must wear a body harness or belt with a lanyard affixed to the mast of the vehicle or the overhead guard. Therefore, training for employees who use order picker trucks must emphasize that the use of the body belt or harness and lanyard is essential whenever the operator is aloft.

Powered industrial trucks may be powered by gasoline, propane, diesel or liquified petroleum gas engines or by electric motors. Each of the basic powerplants (except propane) and their associated components (such as mufflers on internal combustion engines and switches and wiring on electric trucks) may be upgraded and the entire truck may be approved by a nationally recognized testing laboratory for operation in certain classified hazardous areas. These classified hazardous areas are those parts of a plant, factory or other workplace where there exists or may exist concentrations of flammable gases or vapors, combustible dust, or easily ignitible flyings or fibers so that the risk of fire or explosion is increased. The current OSHA general industry standard for powered industrial trucks contains basic descriptions of the types of approved powered industrial trucks and the various classes, divisions, and groups of classified hazardous areas and some of the materials whose presence would cause classification of those areas. However, the number of substances whose presence causes the hazards of fire and/or explosion have increased greatly since promulgation of the OSHA standards. (For additional information on the properties and classifications of materials, see the National Fire Protection Association (NFPA) 505–1992 Fire Safety Standard for Powered Industrial Trucks Including Type Designation, Areas of Use, Maintenance, and Operation.) (Ex. 3–3).

In addition to the general requirements for truck operation, such as vehicle stability and load carrying capability, training must be provided for unusual situations, such as training operators to handle asymmetrical loads when their work includes this activity. The only way that unusual loads may be moved safely with some powered industrial trucks is for the operator to understand and apply the principles of moments and stability of the vehicle. (These principles are explained in more detail in the part of this preamble entitled "Powered Industrial Truck Hazards.") With many powered industrial trucks, the capacity is given as some weight at some load center [usually 24" (61 cm)]. If the operator does not understand that the load center is the distance from the vertical face of the forks to the center of gravity of the load and that loads are usually symmetrical, then the operator may pick up a load incorrectly. If the operator understands that the capacity of the vehicle decreases as the load center increases, then some asymmetrical or off-center loads may be safely picked up and moved using a high lift truck. Other type trucks, such as low lift platform trucks, can handle asymmetrical or off-center loads with minimum danger to an employee because the load is not raised far above the ground. However, because these type trucks are unable to raise loads far above the ground, they are of little or no use when working in a workplace that has high stacking racks or bins where powered industrial trucks must be able to deposit and retrieve loads from considerable distances above the ground or floor.

Powered industrial trucks also are used to move large items or many smaller items about the workplace without the restrictions that generally exist with other mechanical material handling equipment. Other material handling equipment, like overhead cranes or conveyors, are restricted to moving material along a particular, predetermined pathway. A powered industrial truck, on the other hand, may operate along any aisleway or passageway provided it is wide enough to accommodate the vehicle and can support the vehicle and its load. Once one of these trucks has left an area, there is no remaining obstruction to the flow of employee or vehicular traffic, as would normally occur when fixed equipment is used.

Powered industrial trucks may be operated in and among employees with little or no inconvenience to the employees. Although it may be convenient to operate a powered industrial truck around employees, this can be dangerous, particularly when the employees may be hidden from view (for example, when they are working behind stored material.)

These trucks may operate on almost any type surface, from smooth and level floors to rocky, uneven ground, provided they were manufactured to operate on that type floor or ground and the surface does not have an excessive slope. Different type trucks are designed and manufactured to operate in various work environments. Not only may powered industrial trucks be used for moving material about the workplace, high lift trucks are used to raise loads up to 30 or 40 feet above the floor and deposit the material on a rack, mezzanine or other elevated location and then retrieve and lower the material. Many trucks were designed specifically to operate in restricted areas such as narrow aisles and passageways.

Because powered industrial trucks are intended to accomplish specific tasks in a particular manner, their use is restricted. For example, a powered industrial truck that was designed to operate in a restricted space (such as in a narrow aisle or passageway) must be manufactured with a narrow track (the distance between the two wheels on the same axle or at the same end of the vehicle). In many cases, the maximum width of a truck must be significantly less than the minimum width of the area in which it is operated since the vehicle will normally have to make turns so that loads may be deposited in and retrieved from racks or bins which are adjacent to the aisle or passageway. Narrow aisle trucks cannot be safely operated on a floor or the ground that is not smooth.

Another design criterion, the maximum lateral dimension of the vehicle, usually dictates where the various components of the vehicle, such as the engine or motor, the transmission and the seat for the operator, will be placed. The placement of these components may be higher or lower than their most desirable locations. The placement of the various components at a higher point of the vehicle than is desirable, which is the usual case, raises the center of gravity of the entire vehicle, thereby making the vehicle less stable. The greater the distance that the center of gravity of the vehicle and its load is above the ground, the less stable the vehicle (if all other factors remain constant). A more stable design of a powered industrial truck would require a wider track. This would allow installing the engine, transmission, and other components at a lower level of the truck, thereby lowering the center of gravity of the vehicle.

Because the powered industrial truck is a motor vehicle, its operation is similar to the automobile and some of its hazards are the same as those experienced during operation of the automobile. Like the automobile, the internal combustion engine powered industrial truck will move when the gas pedal depressed, and stop when the brake is applied. Some internal combustion engine and electric powered industrial trucks have both the accelerator and brake functions combined in one pedal or other controller providing restriction to movement of the vehicle when no pressure is applied to the pedal (or when the controller is in the neutral position). As pressure is applied to the pedal or other controller, the brake is gradually released, until at a given point of controller travel, the brake is completely disengaged. At this point, the vehicle can coast without restriction from the brake. Finally, as the pedal or other controller is actuated further, the motor or engine is engaged and the vehicle moves under the power supplied by the engine or motor. The vehicle then moves progressively faster as the pedal or controller is further actuated. Clearly good training is needed when design characteristics may reduce stability, limit vision or cause non-uniform methods of control.

Powered industrial trucks also may come equipped with, or can be modified to accept, attachments that allow movement of odd shaped materials or permit the truck to carry out tasks that may not have been envisioned when the truck was designed and manufactured. Many of these attachments may be added to or installed on the vehicle by the dealer or by the employer. For example, there are powered industrial truck attachments for grasping barrels or drums of material. Some of these attachments will not only grasp a barrel or drum but allow the vehicle operator to rotate the barrel or drum to empty the vessel or lay it on its side. Another attachment that looks like a long spike may be positioned within rolled material, such as carpeting. This attachment allows the movement of material without causing damage to the material being handled. All of these attachments may adversely effect the ability of a powered industrial truck to perform its primary function or may cause the vehicle to be used safely only under limited operating conditions, such as under reduced speed or load-carrying capacity. OSHA recognizes that certain attachments may limit the safe use of the vehicle. To ensure that modifications or additions do not adversely affect the safe use of the vehicle, OSHA requires at § 1910.178(b)(4) that:

(4) Modifications and additions which affect capacity and safe operation shall not be performed by the customer or user without the manufacturer's prior written approval. Capacity, operation, and maintenance instruction plates, tags, or decals shall be changed accordingly.

When the use of specialized attachments restricts the use of the powered industrial truck or when the truck is used to lift people, it is essential that operator training must include instruction on the safe use of the vehicle so that the operator knows and understands the restrictions or limitations that are imposed upon the operation of the vehicle by the utilization of those attachments.

Another type of attachment that alters the basic use of the vehicle and presents unique hazards is an overhead hoist attachment. It is made up of a rail (like n I-beam) that is attached to the truck and supports an overhead hoist. It is very easy for an operator to pick up a load with an overhead hoist attachment while the load is close to the vehicle and, without realizing it, exceed the moment of the vehicle by moving the load further from the body of the vehicle. In order to operate this type attachment successfully, the operator must have specific training in the use of this attachment, including training in calculating the maximum load at different points in front of the vehicle and instruction in the causes of longitudinal vehicle tipover and its prevention.

In an attempt to improve the load carrying capability of the vehicle, some people add extra counterweights to powered industrial trucks. Although this will increase the ability of the vehicle to resist longitudinal tipover when the vehicle is overloaded, additional weight imposes extra stresses on the vehicle and its components. The added stresses also can cause changes in the driving characteristics of the vehicle and premature failure of the truck and its components, sometimes with catastrophic effects. Training is needed so that operators avoid creating those hazards.

III. Powered Industrial Truck Hazards

Powered industrial trucks are used in all industries. Their principle utility lies in the fact that either a large number of objects confined in a large box, crate or other container or large objects may be moved about the workplace with relative ease. Since powered industrial truck movement is controlled by the operator and is not restricted by the frame of the machine or other impediments, virtually unrestricted movement of the vehicle about the workplace is possible.

The hazards that are commonly associated with powered industrial trucks may not exist or be as pronounced for every type, make or model vehicle. For example, the hazard tipping over the vehicle due to unstable operation does not exist (except in the most extraordinary circumstance) with the low lift platform truck, the motorized hand truck or the motorized hand/rider truck because each of these trucks does not allow the raising of the load to a point that will cause the vehicle to become unstable. On the other hand, the counterbalanced rider truck and the order picker truck allow the load to be raised very high, causing the vehicle to become less stable as the load is raised.

Each type truck has different hazards associated with its operation. For example, the chance of a falling load accident occurring when the truck is a sitdown, counterbalanced rider truck is much greater than when the vehicle is a motorized hand truck because the height that the load can be raised on the sitdown rider truck is much greater than the hand truck.

Correspondingly, the method or means to prevent the accident or to protect the employee from injury may be different with different type trucks. When a rider truck is involved in a tipover accident, the operator has the opportunity to remain in the operator's position on the vehicle during the tipover, thereby minimizing the potential for injury. In most cases, the operator of a rider truck is injured in a tipover accident when he or she attempts to jump clear of the vehicle when it begins to tip over. Because the natural tendency of the operator is to jump downward, he or she lands on the floor or ground and is then crushed by the overhead guard of the vehicle. Consequently, the operator should be trained to stay with the vehicle during a lateral tipover. On the other hand, when an order picker tips over with the platform in a raised position, generally the operator should attempt to jump clear of the vehicle, and should be trained accordingly.

Because the powered industrial truck is a motor vehicle, its operation is similar to the automobile and some of its hazards are the same as those experienced during operation of the automobile. Both the automobile and the powered industrial truck are subject to some of the same hazards such as contacting both fixed and movable objects (including employees) and tipping over.

Additionally, there are hazards associated with operating the vehicle at an excessive rate of speed and the hazard of skidding on a wet or otherwise slippery ground or floor. Driving a powered industrial truck at an excessive rate of speed may result in the loss of control of the vehicle, causing the vehicle to skid, tipover, or fall off a loading dock or other elevated walking or working surface. Failure to maintain control of the vehicle also may cause the vehicle to strike an employee or some stored material, causing the material to topple and possibly injure another employee. In these cases, training which reinforces driver training is necessary so that the operator will react properly to minimize the hazard to him or herself and to other employees.

Although there are many similarities between the automobile and the powered industrial truck, there are also many differences. Here greater training is required so that operators are aware of the differences. Some of the characteristics of a powered industrial truck that have a pronounced effect upon its operation and safety that are outside their auto driving experience are its ability to change its dynamic stability, to raise, lower and tilt loads, and to steer with the rear wheels while powered by the front wheels. The capability to move loads upwards, downwards, forwards and backwards causes a shift of the center of gravity of the vehicle and can adversely affect the overall stability. When a load is raised or moved away from the vehicle, the vehicle's longitudinal stability is decreased. When the load is lowered or moved closer to the vehicle, its longitudinal stability is increased.

To mitigate the hazards of stability caused by the movement of the material being handled, OSHA has seven provisions that address proper operation of a powered industrial truck. These provisions are § 1910.178 (n)(15), (o)(1), (o)(2), (o)(3), (o)(4), (o)(5), and (o)(6). These provisions specify:

(15) While negotiating turns, speed shall be reduced to a safe level by means of turning the hand steering wheel in a smooth, sweeping motion. Except when maneuvering at a very low speed, the hand steering wheel shall be turned at a moderate, even rate.

(O) *Loading.* (1) Only stable or safely arranged loads shall be handled. Caution shall be exercised when handling off-center loads which cannot be centered.

(2) Only loads within the rated capacity of the truck shall be handled.

(3) The long or high (including multiple-tiered) loads which may affect capacity shall be adjusted.

(4) Trucks equipped with attachments shall be operated as partially loaded trucks when not handling a load.

(5) A load engaging means shall be placed under the load as far as possible; the mast shall be carefully tilted backward to stabilize the load.

(6) Extreme care shall be used when tilting the load forward or backward, particularly when high tiering. Tilting forward with load engaging means elevated shall be prohibited except to pick up a load. An elevated load shall not be tilted forward except when the

load is in a deposit position over a rack or stack. When stacking or tiering, only enough backward tilt to stabilize the load shall be used.

Knowledge of, and adherence to these principles, as well as the other requirements of the OSHA standard, are essential for safe load handling and vehicle operation. Training is needed in these requirements.

Each powered industrial truck has a different "feel" that makes its operation slightly different from the operation of other trucks. The workplaces where these trucks are being used also present particular hazards. For these reasons, a uniform or consistent set of hazards for all industrial trucks and their operation cannot be delineated. The hazards addressed in this section relating to the use of powered industrial trucks have been generalized rather than being make or model specific. For this reason, development of a single "generic" training program which fits all powered industrial trucks and their operation is impractical. In developing an effective training program, there are three major areas of concern regarding the hazards of the operation of powered industrial trucks. The three major groups of hazards of powered industrial trucks and their operation are hazards associated with the particular make and model truck, hazards of the workplace, and general hazards that apply to the operation of all or most powered industrial trucks.

There are other hazards caused by improper operation of a powered industrial truck. Among these hazards are: Falling loads caused by overloading or improperly loading powered industrial trucks (including carrying unbalanced or unstable loads); the vehicle falling from platforms, curbs, trailers or other surfaces on which the vehicle is operating; driving the vehicle while the operator has obstructed view in the direction of travel or the operator not paying full attention to the operation of the powered industrial truck; and the vehicle being operated at an excessive rate of speed. OSHA has identified several accidents that have occurred when an employee other than the operator is "given a ride" on a powered industrial truck. Most trucks were designed and are intended to allow only the operator to ride on the vehicle. The carrying of other persons may result in an accident when that other person either falls from the vehicle or contacts some obstruction when the vehicle is driven in proximity to that obstruction. Finally, powered industrial truck accidents have occurred because the vehicle was not maintained (most commonly, employees being overcome by excessive carbon monoxide exposure) or when the powered industrial truck was not being maintained properly.

Each of these hazards may be more or less consequential based upon the method of operation of the powered industrial truck, the loads being carried, and the workplace where the vehicle is being operated. Truck operators must be trained to recognize unsafe conditions and how to react to them when they occur.

Several features of a powered industrial truck contribute either directly or indirectly to the existence or severity of the hazards of the vehicle. Some of the factors, that would either create or enhance the hazards of the particular truck, are the placement of the critical components of the vehicle, the age of the vehicle, and the manner in which the vehicle is operated and maintained.

There are other hazards related to the use of powered industrial trucks that are caused or enhanced by the characteristics of the workplace. Those hazards include the following: operating powered industrial trucks on rough, uneven or unlevel surfaces; operating powered industrial trucks with unusual loads; operations in hazardous (classified) areas; operation in areas where there are narrow aisles; where there is pedestrian traffic; or where employees are working in or adjacent to the path of travel of the powered industrial truck.

The operation of a powered industrial truck presents hazards not only to the operator, but also endangers other employees working with or around the vehicle. As explained in the section entitled "Accident, injury and other data", below, employees other than operators have been injured or killed in accidents involving powered industrial trucks. Proper training can reduce accidents resulting from the above causes.

IV. Accident, Injury and Other Data

This section of the preamble contains a discussion of the reports, studies and other sources of data and information that were analyzed to determine the magnitude and extent of the problems that powered industrial truck operator training can mitigate.

A. The Bureau of Labor Statistics (BLS) maintains a database entitled, Census of Fatal Occupational Injuries (CFOI). The CFOI is a compilation of information on fatal work injuries that occurred in the 50 States and the District of Columbia. The CFOI uses death certificates, workers compensation reports and other Federal and State records to gather pertinent information. Work relationships are verified by using at least two source documents.

The program collects information on the workers and the circumstances surrounding each fatality. The data are compiled on an annual basis.

In April, 1994, BLS published a booklet entitled, Fatal Workplace Injuries in 1992: A Collection of Data and Analysis (Ex. 3–4). In this booklet, there was an article written by Gary A. Helmer entitled, Fatalities Involving Forklifts and Other Powered Industrial Carriers, 1991–1992. This report contains information contained in the CFOI on 170 fatal powered industrial truck accidents. Table 1 lists the classifications of those powered industrial truck accidents.

TABLE 1.—CLASSIFICATION OF FORK-LIFT FATALITIES, CFOI, 1991–1992

How accident occurred	No.	Percent
Forklift overturned	41	24
Forklift struck something, or ran off dock	13	8
Worker pinned between objects	19	11
Worker struck by material	29	17
Worker struck by forklift	24	14
Worker fell from forklift	24	14
Worker died during forklift repair	10	6
Other accident	10	6
Total	170	100

Source: Bureau of Labor Statistics, Fatal Workplace Injuries in 1992, A Collection of Data and Analysis, Report 870, April 1994.

B. Measuring the Effectiveness of an Industrial Lift Truck Safety Training Program.

In 1984, H. Harvey Cohen and Roger C. Jensen, working under contract with the National Institute for Occupational Safety and Health (NIOSH), published an article in the Journal of Safety Research (Fall 1984, Vol. 15, No. 3, pps. 125–135) entitled, Measuring the Effectiveness of an Industrial Lift Truck Safety Training Program (Ex. 3–5). The article contained an analysis of two studies that were undertaken to measure objectively the effects of safety training of powered industrial truck operators.

This article detailed the results of an experiment that was conducted to evaluate the value of training powered industrial truck operators using a behavioral (work) sampling procedure to obtain objective data about work practices that correlate with injury risk. There were two separate studies conducted in this experiment, one at each of two similar warehouses. The studies that comprised the experiment were conducted to assess the value of training and the influence of post training actions on the safety performance of workers.

There were 14 criteria used in measuring the performance of the trainees. Each of the criterion was selected because it was (a) measurable, (b) frequently observable, (c) capable of being reliably observed, (d) related to accident occurrence, and (e) amenable to corrective action through training. The fourteen criteria observed were: Warns other operators, yields to trucks, warns co-workers, yields to co-workers, sounds horn at blind intersection, slows down at blind intersection, looks at blind intersection, looks in direction of travel, maintains moderate speed, avoids quick starts/changes of direction, keeps all body parts within truck, maintains forks in proper position, maintains balanced load, and drives properly in reverse. Each observation of the operation of the powered industrial trucks resulted in all criteria being evaluated (either correctly performed, incorrectly performed, or not observed). An error rate for each criterion was calculated by dividing the number of incorrect behaviors observed by the total behaviors observed.

Each of the groups of employees were subdivided into smaller groups. These groups were then given training at different times during the study and, in some cases, additional feedback following the training.

The first study was conducted in four phases. The pretraining phase was conducted with none of the operators having received special training. During the second phase, the control group remained untrained, the treatment group received training, and the treatment-plus-feedback group received training and also received performance feedback. In the third phase, the control group received training so that all three groups had received training but only the training-plus-feedback group received performance feedback. The retention phase started three months after the end of the third phase of the study and the performance of all operators was evaluated without regard to their previous categorization.

The error rates of the various groups during the different phases of the study are given in Table 2.

TABLE 2.—SUMMARY OF MEAN ERROR RATES [1]
[Warehouse 1]

Group	Pre-training	Post-training 1	Post-training 2	Retention
Control	.34	.32	.23	
Training	.33	.27	.26	
Training + Feedback	.35	.27	.25	
All operators	.34	.27	.25	.19

The mean error rate is defined in the study as the number of incorrect behaviors observed divided by the total behaviors observed.

NOTE: The mean error rate for all operators began at .34, that is, in 34 percent of the observed criteria, the tasks observed and evaluated were performed improperly.

Source: Measuring the Effectiveness of Industrial Lift Truck Safety Training Program, Journal of Safety Research, Vol. 15, No. 3, Fall 1984, pp. 125–135.

Following the initial training (post-training 1), all three groups showed a decrease in their mean error rates with the training-plus-feedback group showing the largest decrease (from .35 to .27, a 23 percent decrease) followed by the training-only (from .33 to .27, an 18 percent decrease) and the control group (from .34 to .32, a 6 percent decrease). The reduction in the error rate of the control group from the pre-training to the post-training 1 phase of the study was attributed to a peer modeling influence, i.e., the control group operators were copying the behavior of their previously trained counterparts. Toward the end of the post-training 1 phase, the error rates of the three groups converged, suggesting that the effects of the training program had begun to wear off. Observers also noted that some behaviors were being compromised when employees of different knowledge levels were required to interact, particularly in conflict avoidance situations such as signaling and yielding at blind intersections.

During the post-training 2 phase of the study, all groups improved in performance, particularly the original control group. This group's performance improved by 28 percent (from a mean error rate of .32 to .23). Additional evidence of the effect of peer modeling may be deduced from the fact that the performance of the other two groups (the training and the training and feedback groups) continued to improve although there was no additional instruction given to those groups.

The retention phase of the study was conducted three months following the completion of the post-training 2 phase of the study. It was intended to determine the longer term effects of the training. The results of this phase of the study indicate an additional improvement in the performance of the operators with the mean error rate decreasing from .25 to .19, a 24 percent improvement in their performance. The total performance gain achieved during this study was a 44 percent improvement from the pre-training (baseline) phase through the retention phase (from a mean error rate of .34 to a final error rate of .19). The data indicate that there were significantly fewer errors at each successive phase of the study.

The second study was conducted in order to verify and extend the findings of the first study. Consequently, a modified experimental design was used to eliminate the mitigating influence of the untrained control group. In the second study, all operators were trained at the same time and all received performance feedback. Comparisons were made only before and after training. The study was divided into three phases: Pre-training, post-training and retention. The retention phase of the study was again conducted three months after the conclusion of the prior phase. The mean error rates during the three phases of the study are given in Table 3.

TABLE 3.—SUMMARY OF MEAN ERROR RATES STUDY 2

Pre-training	Post-training	Retention
.23	.09	.07

Source: Measuring the Effectiveness of Industrial Lift Truck Safety Training Program, Journal of Safety Research, Vol. 15, No. 3, Fall 1984, pp. 125–135.

Following the training of the vehicle operators, there was a 61 percent improvement in performance scores (from an error rate of .23 to .09). Observation in the retention phase of this study showed an additional reduction of 22 percent in mean error rates (from .09 to .07 mean error rate). This corresponds closely to the 24 percent gain experienced in Study 1. The overall improvement in mean error rates between the pre-training error rate (.23) to that achieved during the retention phase (.07) was a reduction of 70 percent.

C. In 1987, Nancy Stout-Wiegand of the National Institute for Occupational Safety and Health (NIOSH) published an article in the Journal of Safety Research (Winter 1987, Vol 18, No. 4, pp. 179–190) entitled, Characteristics of Work-Related Injuries Involving Forklift Trucks (Ex. 3–6). This article analyzed powered industrial truck injuries reported in two occupational injury databases—the National Electronic Injury Surveillance System (NEISS) and the Bureau of Labor Statistics' Supplementary Data System (SDS).

The NEISS database is composed of records from a national sample of 200 hospital emergency rooms and burn centers handling all types of injuries. The NEISS database was originally established by the Consumer Product Safety Commission, therefore, the original intent was to gather data about accidents involving commercial products rather than industrial injuries. The hospital emergency rooms were not necessarily those located in industrial areas that would predominantly treat industrial injuries and illnesses. The data from this sample are weighted to represent the nation in numbers and characteristics of traumatic injuries treated in emergency rooms and burn centers. A subset of this database—the work related injuries—is maintained by NIOSH. Since the NEISS database records only injuries treated in emergency rooms and burn centers, traumatic work injuries treated by private practitioners or by industry or private clinics are not included in the NEISS database. Moreover, chronic injuries, such as injuries due to overexertion, are not as likely to be treated in emergency room as are acute traumatic injuries, and, therefore are probably underrepresented in the NEISS database. Other probable sources of error in the calculation of accident rates include misclassification of the sources of injury or the agent of injury. For example, if an employee fell while elevated on the forks of a powered industrial truck, the accident could be misclassified as a fall from elevation rather than a fall from a forklift. Similarly, if an employee were struck in the head by part of a load which fell from a powered industrial truck, the accident could be classified as employee struck by falling object. In either case, the accident would have involved a powered industrial truck, but in neither case would the accident have been classified as one in which a powered industrial truck was involved.

The Supplementary Data System (SDS) database is composed of workers' compensation claims for injuries involving lost workdays. There were 30 states that provided information to the SDS system. The SDS system reports the occupations of injured workers and states where the claim was filed. SDS includes only compensable injuries. The definition of a compensable injury varies from state to state, with some injuries being compensable, for example, if they result in one day or more away from work. In other states, the time away from work may be up to 7 days before the injury becomes compensable.

The SDS and NEISS data do not necessarily represent the same injuries because injuries treated in emergency rooms do not always result in lost workdays. At the same time, compensable injuries included in SDS may not have been treated in emergency rooms and thus would not be represented in NEISS. However, both of these databases represent the more serious injuries involving powered industrial trucks, that is, those requiring treatment in emergency rooms and those which result in compensable injuries.

In 1983, the SDS system identified 13,417 workers' compensation claims for lost-workday injuries involving powered industrial trucks that occurred in 30 states. Assuming that these 30 states represent an average of the whole population, then the number of accidents which occurred nationally would be five-thirds of the 13,417 accidents, or approximately 22,400 compensation claims for lost-workday injuries involving powered industrial trucks filed nationally. This number is comparable to the estimated 24,000 forklift-related injuries that were treated in U.S. emergency rooms in 1983 as reported by NIOSH from information gathered by the NEISS system. In 1985, the NEISS system figures were used to determine that about 34,000 powered industrial truck related accidents were treated in emergency rooms. This is an increase of about 39% over a three-year period of time.

This report also contained a tabulation of the occupations of the injured workers. The breakdown of the occupations of those employees and the corresponding percentage of the accidents is listed in Table 4.

TABLE 4.—PERCENTAGE DISTRIBUTION OF POWERED INDUSTRIAL TRUCK INJURIES BY OCCUPATION OF INJURED EMPLOYEE

Occupation	Percent
Professional, technical and kindred workers	0.3
Managers and administrators (except farm)	2.0
Sales workers	0.8
Clerical and kindred workers	5.0
Craftsmen and kindred workers	(15.5)
Mechanics	6.5
Foremen	3.0
Other craftsmen and kindred workers	6.0
Operatives (except transportation)	(17.5)
Assemblers	1.4
Packers/wrappers	1.1
Welders	0.9
Miscellaneous/unspecified operatives	9.2
Other operatives	4.9
Transportation equipment operatives	(20.8)
Powered industrial truck operators	12.3
Truck drivers	5.5
Motormen	1.7
Deliverymen	1.2
Other transportation equipment operators	0.1
Laborers (except farm)	(37.3)
Warehousemen	10.4
Freight and material handlers	7.3
Stock handlers	4.4
Construction laborers	2.2
Miscellaneous/unspecified laborers	8.0
Other laborers	1.6
Farmers (managers and laborers)	1.5
Service workers	1.8
Occupations unspecified	1.1

Source: Characteristics of Work-Related Injuries Involving Forklift Trucks, Journal of Safety Research, Vol. 18 No. 4, Winter 1987, pp. 179–190.

D. Industrial Forklift Truck Fatalities—A Summary.

The Office of Data Analysis (ODA) of OSHA's Directorate of Policy conducted an examination of 53 investigative case files involving powered industrial truck fatalities that occurred between 1980 and 1986 (Ex. 3–7). The results of their analysis is summarized below.

TABLE 5.—OFFICE OF DATA ANALYSIS TYPE ACCIDENTS—53 POWERED INDUSTRIAL TRUCK FATALITIES

Type accident	No.	Percent
Crushed by tipping vehicle	22	42
Crushed between vehicle and a surface	13	25

TABLE 5.—OFFICE OF DATA ANALYSIS TYPE ACCIDENTS—53 POWERED INDUSTRIAL TRUCK FATALITIES—Continued

Type accident	No.	Percent
Crushed between two vehicles	6	11
Struck or run over by vehicle	5	10
Struck by falling material	4	8
Fall from platform on forks	2	4
Accidental activation of controls	1	2

Source: *Industrial Forklift Truck Fatalities—A Summary*, Report from Office of Data Analysis, Directorate of Policy, OSHA, dated June 1990.

The single largest cause of the accidents was vehicle tipovers. These tipovers were attributed to the following: (1) The vehicle being out of control (speeding, elevated loads, mechanical problems, etc.; 7 instances—13 percent); (2) the vehicle being run off/over the edge of the surface (4 instances—8 percent); (3) attempting to make too sharp a turn (excessive speed, unbalanced load, etc.; 4 instances—8 percent); (4) employee jumped from overturning vehicle being pulled by another vehicle (2 instances—4 percent); vehicle skidded or slipped on slippery surface (2 instances—4 percent); (5) wheels on one side of vehicle ran over raised surface or object (2 instances—4 percent); and (6) vehicle tipped over when struck by another vehicle (1 instance—2 percent).

The second highest number of fatalities reported in the ODA study was caused by an employee being crushed between a vehicle and a surface. The accidents were attributed to: (1) The operator getting off the vehicle while it was running (7 instances—13 percent); (2) worker on platform being crushed between platform and overhead surface (2 instances—4 percent); (3) employees leg being caught when vehicle sideswiped metal surface (1 instance—2 percent); (4) employee attempting to prevent vehicle tipover by holding up overhead guard (1 instance—2 percent); (5) employee changing tire and vehicle fell from jack (1 instance—2 percent) and (6) empty 55 gallon drum used for support vehicle during maintenance collapsed (1 instance—2 percent).

The six accidents that were attributed to employees being crushed between two vehicles were caused by contact between two moving powered industrial trucks (4 cases) and between a powered industrial truck and a stationary vehicle in the other two instances.

Of the five accidents which were identified as an employee being struck or run over by vehicle, four were accidents where employees other than the vehicle operator were struck by the vehicle. The remaining one was an operator trying unsuccessfully to board a free rolling vehicle.

E. The OSHA Fatality/Catastrophe Reports. OSHA records a summary of the results of investigations of all accidents resulting in fatalities, catastrophes, amputations and hospitalizations of two or more days, and those accidents that have received significant publicity or property damage. These summaries are recorded on an OSHA Form 170 and include an abstract describing the activities taking place at the time of the accident and the causes of the accident. These reports are stored in a computerized database system.

OSHA queried the computer for all reports that contained the keyword "industrial truck". There were 4268 total reports in the system that resulted in 3038 fatalities, 3244 serious injuries, and 1413 non-serious injuries (many of the accidents resulted in multiple fatalities and/or injuries). The use of the keyword "industrial truck" produced a printout of 208 accidents (Ex. 3–8). These 208 accidents resulted in 147 fatalities, 115 serious injuries and 34 non-serious injuries.

By adding the number of fatalities, serious injuries and non-serious injuries and dividing by the number of accidents, it was determined that 1.4 injuries of some nature occurred per accident. OSHA also determined the percent of each of the three classes of accidents that involved powered industrial trucks. Those percentages are 4.8 percent of the fatalities, 3.5 percent of the serious injuries and 2.4 percent of the non-serious injuries were attributable to an accident that involved a powered industrial truck.

OSHA looked at the OSHA 170s to determine the causes of the accidents that were attributable to the use of powered industrial trucks in general industry. Table 6 presents a compilation of the causes of those accidents.

TABLE 6.—CAUSES OF ACCIDENTS [1]—OSHA INVESTIGATION SUMMARIES (OSHA 170s)

Cause	No. of reports
No training [2]	19
Improper equipment	10
Overturn	53
Unstable load	45
Overload, improper use	15

TABLE 6.—CAUSES OF ACCIDENTS [1]—OSHA INVESTIGATION SUMMARIES (OSHA 170s)—Continued

Cause	No. of reports
Obstructed view	10
Carrying excess passenger	8
Operator inattention	59
Falling from platform or curb	9
Falling from trailer	6
Elevated employee	26
Operator struck by load	37
Other employee struck by load	8
Accident during maintenance	14
Vehicle left in gear	6
Speeding	5
Not powered industrial truck accident	9

[1] The causes of the accidents were determined by the narrative in the accident report. In most cases, the narrative emphasized the cause of the accident, however, in a few cases, reasonable and appropriate assumptions were made. In some cases, multiple accident causes were described in the narrative portion of the report, or were assumed to have caused the accident. (See Ex. 3–8.)

[2] Of the 19 instances when the report contained the indication that a lack of training was one of the causal factors of the accident, there were 6 serious violations issued, 2 other (nonserious) violations and 11 instances where no citation was issued.

Source: Office of Electrical, Electronic and Mechanical Engineering Safety Standards, Directorate of Safety Standards Programs, OSHA.

Using the OSHA Form 170 data, OSHA also compiled a listing of the industries in which accidents occurred. Table 7 presents a tabulation of the SIC codes, the description of the industry, and the number of times that accidents were identified as having occurred in those industries. For a complete listing of the individual industries, see Ex. 3–9.

TABLE 7.—INDUSTRIES WHERE ACCIDENTS OCCURRED—OSHA INVESTIGATIVE SUMMARY (OSHA FORM 170) REPORTS

SICP division	Description	Times cited
B	Mining	4
C	Construction	25
D	Manufacturing	95
E	Transportation, communication and utilities	22
F	Wholesale trades	25
G	Retail trades	18
I	Services	7
J	Public administration	4

NOTE: The breakdown of accidents does not include agricultural accidents since establishments of 10 or less employees in this industry are exempt from OSHA jurisdiction.

Source: Office of Electrical, Electronic and Mechanical Engineering Safety Standards, Directorate of Safety Standards Programs, OSHA.

F. The OSHA Emergency Communications System Reports.

OSHA has another internal system for collecting information about serious accidents. This is a telephone system which requires that serious and/or significant accidents be telephoned into the National Office.

The telephone call system is part of the OSHA emergency communications system. Regional Administrators are required to file a first report of fatalities, catastrophes and other important events (such as those that receive significant publicity) to the National Office. The information contained in these reports is disseminated to the responsible officials in OSHA and to the directorates of the Agency. These reports are broken down within the various offices and distributed to the appropriate personnel. There are approximately 1200 reports received by the National Office yearly. See Ex. 3–10.

None of the reports are screened before the OSHA National Office receives them to eliminate those from a certain industry, occupation or because of other factors. Although these reports may not be considered statistically significant by themselves in attempting to determine the number of accidents that have occurred, the lack of prior screening indicates that they represent a reasonable sampling of the most serious type accidents and that the causes of the accidents closely parallel the distribution of the causes of all accidents.

OSHA has examined the First Report of Serious Injury reports and identified 247 that involved powered industrial trucks. These accidents occurred between 1980 and the present. OSHA looked at the number of accidents reported through its telephonic system and determined the percentage of those accidents that involved powered industrial trucks. Table 8 contains a listing of the number of First Reports of Serious Accident reports which were received from 1980 to present, the number of those accidents which involved powered industrial trucks, and the corresponding percentage.

TABLE 8.—YEARLY SUMMARY OF FIRST REPORT OF SERIOUS ACCIDENTS

Year	Total reports	Pit accidents	Percent
1980	200	2	1
1981	125	2	1.6
1982	113	0	0
1983	115	3	2.6
1984	181	1	.6
1985	456	15	3.3
1986	1,147	44	3.8
1987	1,236	38	3.1
1988	1,330	47	3.5
1989	1,150	44	3.8
1990	1,105	41	3.7
1991	[1] 215	10	4.7
Totals [2]	6,424	247	3.6

[1] These are the number of total reports received between the first of the year until March 31.
[2] The total number of reports, the number of accidents involving powered industrial trucks and the percentage were calculated using the figures from 1985–1990. The number of accidents reported during the years 1980–1984 and those reported during 1991 were too few to be representative.

Source: Office of Electrical, Electronic and Mechanical Engineering Safety Standards, Directorate of Safety Standards Programs, OSHA.

Each of these reports were examined to determine the causes of the accidents. In some instances, multiple causes were identified. Table 9 lists the causes of the accidents and the number of accidents which were attributable to that cause.

TABLE 9.—CAUSES OF ACCIDENTS (POWERED INDUSTRIAL TRUCKS) FIRST REPORTS OF SERIOUS ACCIDENT

Cause of the accident	No. Accidents
Tipover	58
Struck by powered industrial truck	43
Struck by falling load	33
Elevated employee on truck	28
Ran off loading dock or other surface	16
Improper maintenance procedures	14
Lost control of truck	10
Truck struck material	10
Employees overcome by carbon monoxide or propane fuel	10
Faulty powered industrial truck	7
Unloading unchocked trailer	7
Employee fell from vehicle	7
Improper use of vehicle	6
Electrocutions	2

Source: Office of Electrical, Electronic and Mechanical Engineering Safety Standards, Directorate of Safety Standards Programs, OSHA.

G. The OSHA General Duty Clause Citation Analysis.

The Office of Mechanical Engineering Safety Standards of OSHA, conducted an analysis of the citations which were issued between 1979 and 1984 for violations of the general duty clause (section 5(a)(1)) of the Occupational Safety and Health Act. During that period, there were a total of 3637 inspections in which at least one 5(a)(1) citation was issued. See Ex. 3–11.

Sixty-five general duty clause citations involved powered industrial truck operations. Each was examined to determine the nature of the violation. Table 10 lists the violation that was alleged to have occurred.

TABLE 10.—SUMMARY OF GENERAL DUTY CLAUSE (5(A)(1)) CITATIONS

Violation	No. instances
Employee elevated on forks	44
Improper operation of vehicle	13
Improper maintenance on vehicle	5
No vehicle operator training	2

TABLE 10.—SUMMARY OF GENERAL DUTY CLAUSE (5(A)(1)) CITATIONS—Continued

Violation	No. instances
Order picker without fall protection	1

Source: Office of Electrical, Electronic and Mechanical Engineering Safety Standards, Directorate of Safety Standards Programs, OSHA.

V. Basis for Agency Action

OSHA believes that, as the above discussion indicates, that there is a sufficient body of data and information on which to base a revision of the existing standard for powered industrial truck operator training and the promulgation of the same requirement for powered industrial truck operator training in the construction, maritime and agriculture industries. These requirements would reduce the number of fatalities and injuries resulting from accidents involving powered industrial trucks operated by untrained or insufficiently trained employees.

According to OSHA's data and information, powered industrial truck accidents account for approximately 4.8 percent of the fatalities, 3.5 percent of the serious injuries and 2.4 percent of the non-serious injuries that occur in general industry each year. These accidents resulted in an average of 107 fatalities, 33,800 serious injuries, and 61,800 non-serious injuries per year from 1981 through 1990.

In analyzing its accident data, OSHA has derived two separate estimates of the number of fatalities and serious injuries that occur to employees due to powered industrial truck accidents. Because the two set of numbers are in the same range, the Agency has presented both. It should be noted that the number of fatalities is virtually identical using either method of derivation. However, slightly different definitions are used for estimating injuries. The other set of estimates are presented in the Preliminary Regulatory Impact Analysis, below.

There are approximately 68,400 accidents involving powered industrial trucks in general industry per year. This figure was arrived at by totaling the fatalities, serious, and non-serious injuries and dividing this result by 1.4 (the number of injuries per accident determined from the OSHA Fatality/Catastrophe Reports). According to the Industrial Truck Association (ITA), there are currently approximately 855,900 powered industrial trucks in the United States, therefore approximately 8 percent of the powered industrial trucks will be involved in an accident this year (this assumes a truck is involved in only one accident this year). Since the ITA has stated that the useful life of a powered industrial truck is 8 years, that means that at some point during its useful life, almost two-thirds of the powered industrial trucks will be involved in some type accident (again, assuming there is only one accident per truck).

OSHA also looked at the type accidents that were described in the section of this preamble entitled "Accident, injury and other data." The three reports that contained that information were the "Industrial Forklift Truck Fatalities—A Summary" (ODA Study); "The OSHA Fatality/Catastrophe Reports" (Fat/Cat Study); and the "OSHA Emergency Communications System Reports, First Reports." The number of different types of accidents are given in Table 12, below. Since the Industrial Forklift Truck Fatalities report was the only one that used a single causation methodology for categorizing the accidents, this is the only study for which percentages of the accidents were calculated. These percentages appear in parentheses following the numbers.

TABLE 11.—CAUSES OF POWERED INDUSTRIAL TRUCK ACCIDENTS

Cause	Study		
	ODA study	Fat/cats	First reports
Tipovers	22 (42%)	53	58
Struck by vehicle	24 (46%)		43
Struck by falling material	4 (8%)	90	[1] 43
Elevated employees	2 (4%)	26	28
Control activation	1 (2%)	[2] 6	
Improper equipment or usage		10	[3] 13
Vehicle overloaded		15	
Obstructed view		10	
Maintenance acc		14	14
Speeding		5	
Fell from platform			[4] 23
Lost control			10
Overcome by CO			10

TABLE 11.—CAUSES OF POWERED INDUSTRIAL TRUCK ACCIDENTS—Continued

Cause	Study		
	ODA study	Fat/cats	First reports
Employee fell from vehicle			7
Electrocution			2

[1] This number represents the accidents due to material that was in the powered industrial truck (a portion of the load) falling on an employee-33 cases, and stacked material falling on an employee when struck by a powered industrial truck-10 cases.
[2] This number represents the accidents due to the operator leaving the vehicle in gear, dismounting the vehicle and being struck when the vehicle moved.
[3] This number represents the number of accidents when either the vehicle was used improperly (6 instances) or the vehicle was defective (7 instances).
[4] This number represents the number of accidents when the operator drove the vehicle off an elevated dock (16 instances) or fell against the face of the dock when an unchocked trailer rolled away from the dock when being loaded or unloaded.

Sources: "The Forklift Truck Fatalities—A Summary Report" (ODA Study); "The OSHA Fatality/Catastrophe Reports" (Fat/Cats); and "The OSHA Emergency Communications System Reports (First Reports)".

In 9 percent of the accident investigations in which an OSHA 170 was prepared (19 of 208), lack of training was identified as a causal factor. In more than half of these accident investigations (11 of 19), lack of training was not cited by OSHA compliance officers. However, OSHA's standard specifies that only trained and authorized operators are allowed to operate powered industrial trucks. Absence of a citation when lack of training was identified as a causal factor in the accident can only be attributed to the fact that many compliance officers believe that the powered industrial truck training requirement (29 CFR 1910.178(l)) is vague and unenforceable in its present form.

In addition, most of the accidents where lack of training was not mentioned, clearly could have been avoided through better training. When OSHA completes this rulemaking, in light of the large number of industrial truck accidents, based on priorities and resources, it will consider whether to revise the entire powered industrial truck standard. Persons also may wish to comment on whether OSHA should revise the entire standard in the future.

VI. The Need for Training

Training is generally defined as making a person proficient through the use of specialized instruction and practice. Training is the means by which an employer ensures that employees have the knowledge, skills, and abilities that are necessary for the employees to do their jobs correctly.

Once an employee acquires the basic knowledge, skills, and abilities, refresher or remedial training may be used to reinforce or improve those attributes, to provide new material, to provide material that was previously discussed in a new manner, or to simply maintain an awareness of the material that had previously been taught. Refresher or remedial training is normally conducted on a predetermined periodic basis, that is, on a monthly, semi-annual, or annual basis.

Training may be as simple and informal as a supervisor pointing out either an error in the manner in which an employee is doing a job (making an on-the-spot correction) or showing an employee how to do a particular task (demonstrating the proper method to do the job). On the other end of the spectrum is the detailed, structured instruction that uses the classical methods of training (lectures, conferences, formal demonstrations, practical exercises, examinations, etc.). Formal training is usually used to impart a greater amount of, more complicated, or more detailed information to a trainee.

For the most part, employees do not start out with the innate knowledge, skills, and abilities to perform many of the complicated or difficult practices and procedures that occur commonly in the workplace. For example, many states require potential car drivers to pass either driver training and/or driver education programs to qualify for a drivers license. Even with this training, young drivers are involved in a disproportionate number of accidents. It is only after the drivers have more experience that the number of accidents decreases. Although many employees who are selected or assigned to drive powered industrial trucks are licensed to drive automobiles, there are enough dissimilarities between these two types of vehicles and their operation to require additional knowledge, skills, and abilities to operate a powered industrial trucks safely. Operational characteristics of powered industrial trucks, such as using vehicles equipped with rear-wheel steering and front-wheel drive and the hoisting—moving—lowering of loads, require operator training and practice to master the different driving skills that must be used when an employee operates powered industrial trucks.

Many of these accidents either can be prevented, or the seriousness of the injury to the employee can be mitigated by training employees. Effective training and supervision also can prevent the occurrence of unsafe acts such as speeding, failing to look in the direction of travel, and failing to slow down or stop and sound the vehicle's horn at blind intersections and other areas where pedestrian traffic may not be observable. Another example in which training can prevent or lessen the severity of an accident of this kind is directly related to the stability of powered industrial trucks when traveling with an elevated load. Effective operator training should include the admonition that the vehicle can only be moved when the load is at its lowest point. Even if this admonition is ignored and the vehicle tips over, the injury to the operator is usually minimal if the he or she stays with the vehicle. As previously discussed, the usual injury in a powered industrial truck tipover occurs when the operator attempts to jump off the vehicle when it is tipping over. Since the normal tendency is for a person to jump downward, the operator lands on the floor or ground in the path of the overhead guard and the usual injury is a crushing injury of the head, neck or back when the overhead guard contacts the employee. Training an employee to stay with the vehicle will reduce the severity of some of these injuries.

In 1990, the Office of Technology Assessment of the U.S. Congress published a book at the request of the Senate Labor and Human Resources Committee, the House Education and Labor Committee, and the Senate Finance Committee. This book is entitled, Worker Training: Competing in the New International Economy, OTA-ITE-457 (Washington, DC: U.S. Government Printing Office, September 1990; Ex. 3-12) Although this book addresses the need for training so that American industry can remain competitive in the world marketplace, there were many salient facts presented, both about the state of training in the workplace and the need for additional training.

To be effective, training must impart appropriate skills, must not include irrelevant information and must accommodate varying employee backgrounds and learning styles. Training is most effective when it is quickly reinforced on the job. Poor timing of training, lack of reinforcement at work, and other factors prevent effective transfer of knowledge to the job.

The book also pointed out that small business access to new employees with good skills is limited. Employees hired by companies reflect the labor pool available and is dependent upon the size of the company. Small companies must draw their employees from the locally available talent pool whereas larger companies can attract prospective employees from a much larger geographical area. In order to make up for the limitations of the limited talent pool, small employers usually must provide additional training and education to achieve comparable employee performance.

The OTA book pointed out that inadequate training costs firms and employers not only in health and safety risks, but also downtime, defective parts and equipment, wasted material, late deliveries, inferior quality products and poor customer service. To maximize its effectiveness, training must be focused on workplace problems because simply providing more generalized, non-directed training will not promote industrial competitiveness. If the work is not organized to tap employee skills, the training investment will be wasted.

Finally, the book emphasized that employers historically have not trained their workers for several reasons. First, high labor turnover has mistakenly led employers to believe that skilled workers will leave so their companies will not recoup their training investment. Second, many employers believe that an increase in productivity will not offset the cost of training employees. As the book points out, that is not the case.

The studies conducted by Cohen and Jensen, discussed under Accident, injury and other data earlier in this preamble, found a reduction in operator error rate of up to 70 percent. Although a 70 percent error rate reduction can not be directly equated to a corresponding reduction in the number of accidents that this or any other group of operators will experience, improper or unsafe operation of a powered industrial truck is the major cause of the accidents and their resultant fatalities and injuries. Therefore, a reduction in the unsafe operation of a powered industrial truck will reduce the number of accidents, and the resultant fatalities and injuries.

Many standards promulgated by OSHA explicitly require the employer to train employees in the safety and health aspects of their jobs. These requirements reflect OSHA's belief that training is an essential part of an effective employer's program for protecting workers from accidents and illnesses. (See Ex. 3-13 for a complete list of the OSHA standards that require training.)

Although not all powered industrial truck accident reports spell out the lack of training as a causal factor of the accidents, each accident can, in part, be attributed to either being caused or worsened by the actions or inactions of the operator. For example, when a powered industrial truck tips over, the accident is caused by one or more of several factors, including speeding, traveling with the load in an elevated position, or improperly negotiating a turn. Training can minimize the times that these events occur.

Proper training of an employee must take into account the fact that different operating conditions (including the type and size of the load, the type and condition of the surface on which the vehicle is being operated, and other factors) can adversely affect vehicle operation. Operator training must emphasize two points regarding any potential accident scenario. These two factors are: (1) The employee should not engage in activities that may cause an accident, and (2) the employee should minimize the potential for injury (either to himself or herself or to other employees) by taking appropriate actions.

OSHA is not proposing a program of licensing or certification of powered industrial truck operators either by itself or as an adjunct to operator training. OSHA does not have the resources to conduct such a program since there are close to 1.5 million employees who operate powered industrial trucks.

VII. Summary and Explanation of the Proposed Rule

OSHA is proposing to revise the training requirement for powered industrial truck operators, 29 CFR 1910.178(l), contained in the general industry standards, and to add equivalent training requirements for the maritime industries. This proposal is intended to enhance the safe operation of powered industrial trucks in the workplace.

On February 27, 1995, OSHA submitted to the Advisory Committee on Construction Safety and Health (AC) a draft of this document. The ACCOSH recommended to OSHA that the Agency not proceed with rulemaking for that industry until the Advisory Committee had sufficient time to completely study the document and provide further recommendations. Consequently, this rulemaking is limited to general industry and the maritime industries. The Agency intends to propose to adopt for the construction industry similar requirements for training the operators of powered industrial trucks after receiving and taking into account the recommendations of the ACCOSH.

In developing this proposal, OSHA looked at the training requirements of the existing national consensus standard for powered industrial trucks, ANSI B56.1–1993, as well as training requirements from other standards (both industry and government). The non-training related requirements of those standards are beyond the scope of this proposal.

OSHA has not included suggestive language contained at paragraph 4.19.2 of the consensus standard because other enforceable language in the proposed standard covers the issue. This paragraph states, "The operator training program should include the user's policies for the site where the trainee will operate the truck, the operating conditions for that location, and the specific truck the trainee will operate. The training program shall be presented to all new operators regardless of previous experience."

The Agency has not adopted the language contained in 4.19.3(a) of the consensus standard because the responsibility for providing a safe workplace (including the use of a powered industrial truck) is vested with the employer under the OSH Act. This paragraph specifies, "The primary responsibility of the operator is to use the powered industrial truck safely following the instructions given in the training program."

The consensus standard, at 4.19.4(e) and 4.19.5 specifies the type of training and the testing that should be conducted, whereas the OSHA standard leaves the methods of training up to the employer. As explained elsewhere in this preamble, the employer is responsible for selecting the methods that are employed to train the operators. In some circumstances, the employee may be able to gain valuable information from reading the operators manual for the vehicle. In other circumstances, the employee may not be able to read and comprehend the contents of the manual and may have to be shown how to operate the truck safely.

Many of the other OSHA standards and the consensus standards specify that some means be used to verify that training was conducted. Examples of such verification include: (1) Requiring documentation of the training, (2) the production and retention of lesson plans, (3) attendance rosters, and (4) the issuance of training certificates. When refresher or remedial training is specified, these other rules usually require that a set amount of training be conducted at a regular interval (for example, a certain number of hours of refresher training be conducted annually). OSHA is including evaluation by a designated person and certification that the employee has taken the training and can competently operate the truck. Course materials also must be kept. OSHA believes that this is the appropriate method of verification. As operators vary greatly in the experience and backgrounds and they will be required to operate different types of vehicles, different types and amounts of training are necessary and OSHA does not believe it can specify a rigid curriculum.

This proposed revision of the training requirement found in § 1910.178(l) for operators of powered industrial trucks and the imposition of the same requirement for operators of powered industrial trucks in other industries (construction and maritime) specifies that the employer develop a complete training program. This program consists of an evaluation of each potential truck operator and the training of the potential operator in those subject matters relating to the operation of the truck, the work environment in which the truck will be operated and the requirements of the OSHA standard. This training program also must include a periodic evaluation of the performance of the operator and refresher or remedial training as necessary. To maximize the effectiveness of the training, OSHA is proposing to allow the employer to avoid having to conduct training that is duplicative of other training the employee has previously received. Finally, the training provisions would require that the employer certify that the training and evaluations have been conducted.

At paragraph (1)(i), OSHA specifies that each potential operator of a powered industrial truck must be capable of performing the duties that are required of the job after training and appropriate accommodation. This would include being able to climb onto and off of a truck, to sit on the vehicle for extended periods of time, and to turn his or her body to be able to look in the direction of travel when driving in reverse. Elements of this evaluation may include the employee having the physical and mental abilities to perform the job. Information obtained during the initial employee evaluation can be used to, among other things, determine how best to train the employee. For example, if the employee cannot read and comprehend the operator's manuals for the type trucks that the employee will operate, then this information would have to taught by means other than having the employee try to read the truck manuals. The initial evaluation can be useful for the avoidance of duplicative training.

Paragraph (1)(ii) provides that the employer shall assure that the employee has received required training, that the employee has been evaluated and that the potential operator can perform the job competently. The evaluation must be carried out after the training by a designated person so that the employer can assure that the potential operator can perform the duties required of an operator in a competent manner. The conduct of this evaluation during the training is known as a practical exercise or a performance test. OSHA believes that only through evaluation by a knowledgeable person after training can an employer know that the employee has been adequately trained and can safely perform the job.

The designated person may be the employer if qualified. A small business person who has employees may send the employees to an outside training organization. Alternately, the employer may take or have training so that the employer is qualified as a designated person.

At paragraph (2), OSHA is proposing to require that the employer implement a training program for all powered industrial truck operators. This program would ensure that only trained drivers who have successfully completed the training program would be allowed to operate these vehicles. An exception to the rule would allow trainees to operate powered industrial trucks provided the operation is under the direct supervision of a designated person and the operation is conducted where is minimum danger to the trainee or other employees.

OSHA is proposing at paragraph (2)(ii) that the training consist of a combination of classroom instruction and practical training. The Agency believes that only by the use of a combination of training methods will the employee be adequately trained. Although classroom training is invaluable for the teaching of the principles of vehicle operation, it is the hands-on training and the evaluation of the operation of the vehicle that finally proves the adequacy of the training and the ability of the employee to use that training to successfully operate a powered industrial truck.

At paragraph (2)(iii), OSHA is proposing to require that all training be conducted by a designated person. OSHA defines a designated person as one who has the requisite knowledge, training and experience to train powered industrial truck operators. As discussed elsewhere in this preamble, the employer may have the necessary prerequisites to qualify as a designated person or he or she may assign the training responsibility to another person (either a knowledgeable employee or an trainer from outside the company).

To ensure that the training contains the appropriate information for the operator, OSHA has provided a list of subjects at paragraph (3). Under this rule, it is the responsibility of the employer to select the particular items that are pertinent to the type trucks that the employee will be allowed to operate and the work environment in which the vehicle will be operated. For example, if the employee will be allowed to operate an order picker, it is essential that he or she understand the location and function of the controls, the location and operation of the powerplant, steering and maneuvering, visibility, inspection and maintenance and other general operating functions of the vehicle. Additionally, it is essential that the employee know and understand that he or she must be restrained from falling when the platform of the truck is in an elevated position and that the truck must never be driven when the platform is elevated. Under this proposed requirement, it is the responsibility of the employer to select those elements of the training that are necessary for the type vehicle to be used and the workplace in which that vehicle will be operated. The employer may leave out elements if the employer can demonstrate that they are not relevant to safe operation in the employer's workplace.

An additional component of the training program is a continuing evaluation of the operator. At paragraph (4), OSHA specifies that this evaluation be conducted on a periodic basis so that the employee retains and uses the knowledge, skills and abilities that are necessary for the safe operation of the vehicle. This evaluation need not be conducted continuously, however, the employer should conduct these evaluations at intervals that will ensure that the operators have not forgotten or chosen to disregard their training. This evaluation does not have to be formalized but must consist of a designated person observing the operation to ensure that the use of the powered industrial truck is being conducted safely. OSHA requires that this evaluation be carried out at least annually.

OSHA is requiring at paragraph (5) that the employer certify that the required training and evaluations have been conducted. To minimize the paperwork burden on the employer, OSHA is specifying that the certification consist of the name of the employee, the date of the training or evaluation and the signature of the person conducting the training or evaluation.

Under this paragraph, OSHA also specifies that all the current training materials used in the conduct of training or the name and address of the outside trainer, if one is used, be maintained.

At paragraph (6), OSHA is proposing to allow the employer to forgo that portion of the training that an employee has previously received. The intent of these provisions is to allow the employer to not have to train an employee in those phases of the operation of a powered industrial truck if the employee knows the necessary information and has been evaluated and has proven to be competent to perform those duties.

As previously discussed, there are three major areas of consideration that must be emphasized when conducting a powered industrial truck training program. These three areas are: (1) The characteristics, operation and limitations of the vehicles that the trainee will be authorized to operate, (2) the hazards due to the characteristics of the workplace in which these vehicles will operate, and (3) the general safety rules that apply to these vehicles and their operation.

This proposed rule has been drafted in performance language to allow reasonable flexibility to the employer for developing the training program and conducting the training. OSHA recognizes the inherent differences in the capabilities and limitations of employees, both to assimilate the training and then to utilize the knowledge that has been gained. Therefore, the proposed regulation does not limit the employer by specifying the manner in which the training must be conducted. Similarly, the specific content of the training course has not been stated because there are different topics which must be taught due to variances in the operation of the many makes and models of vehicles and because there are different hazards in each workplace. However, OSHA has proposed the various subject matters that should be covered unless the employer determines they are not relevant to the employer's vehicle and workplace. Although some areas of concern may not be pertinent to any one workplace and vehicle, other areas are pertinent to all vehicles and workplaces.

OSHA believes that a training program needs to be conducted before the employee begins to operate a vehicle. To this end, OSHA has required initial training of employees so that they will acquire the knowledge and skills are necessary for the safe operation of the powered industrial truck before being allowed to operate the vehicle without close supervision.

OSHA has left the particulars of the type of training (lecture, conference, demonstration, practical exercise, test or examination, etc.) to the employer. The length of the training and other variables must be based on the employee's experience and other qualifications and the nature of the work environment. The training must be based upon the type of vehicles the employee will be allowed to operate, the conditions that exist in the workplace, the general safety rules from this OSHA standard, the ability of the trainer to teach, and the ability of the trainee to learn. The ability of the employee to assimilate the information presented in the training must be used as the primary criterion for the length, type and other details of the training. Since each employee is different in his or her ability to comprehend, assimilate and use the information received in the training, OSHA believes that one standardized training course will not suffice for all employees.

The employer may choose the training provider. This could include contracting with an outside professional training company to come into the company and train the powered industrial truck operators or the employer developing and conducting the training program. In either case, the employer can choose the method or methods by which the employees will be trained and when the training is conducted.

The standard requires not only appropriate training but evaluation of the operators competency by a designated person with the knowledge to make that evaluation. This is the method that will most accurately prove that the operator has been trained and that the training has been, and continues to be, effective. Through observation of the operation of the vehicle, these questions can be answered.

When a new employee claims prior experience in operating a powered industrial truck, the employer must ensure that the employee knows how to operate the vehicle safely. This can be ascertained by questioning the employee on various aspects of the operation of the truck and by requiring the operator to demonstrate his or her ability to operate the vehicle safely through the conduct of a practical exercise.

In making a determination of an employee's claim of sufficient prior experience, the employer must consider the type of equipment that this employee professes to have operated, how long ago this experience was gained, and the type work environment in which the employee worked. Written documentation of the earlier training is also necessary to determine that proper training has been given. In addition, the competency of the employee must be evaluated. Based on the resolution of these issues, the employer can determine whether the experience is recent and thorough enough, the documentation complete, and the competency sufficient to forgo some or much of the initial training. Some training on the specific factors of the new employees workplace is always going to be necessary. Again, the major criterion of evaluation of the employee is: Does the person know how to do the job and does the vehicle operator use those knowledge, skills and abilities to do the job safely?

OSHA also is proposing to add two non-mandatory appendices. These appendices are intended to provide guidance to employers in establishing a training program (Appendix A) and in understanding to basic principles of stability (Appendix B). In neither case is the information contained in these appendices intended to provide a exhaustive explanation of the techniques of conducting training or of understanding the principles of stability, but each appendix is intended to introduce the basic concepts so that the employer can utilize the material to provide basic training.

VIII. Statutory Considerations

A. Introduction

Section 2(b)(3) of the Occupational Safety and Health Act authorizes "the Secretary of Labor to set mandatory *occupational safety and health standards* applicable to businesses affecting interstate commerce", and section 5(a)(2) provides that "[e]ach employer shall comply with *occupational safety and health standards* promulgated under this Act" (emphasis added). Section 3(8) of the OSH Act (29 U.S.C. 652(8)) provides that "the term 'occupational safety and health standard' means a standard which requires conditions, or the adoption or use of one or more practices, means, methods, operations, or processes, reasonably necessary or appropriate to provide safe or healthful employment and places of employment."

OSHA considers a standard to be "reasonably necessary or appropriate" within the meaning of section 3(8) if it meets the following criteria:

(1) The standard will substantially reduce a significant risk of material harm;
(2) Compliance is technologically feasible in the sense that the protective measures being required already exist, can be brought into existence with available technology, or can be created with technology that can reasonably be developed;
(3) Compliance is economically feasible in the sense that industry can absorb or pass on the costs without major dislocation or threat of instability; and
(4) The standard is cost effective in that it employs the least expensive protective measures capable of reducing or eliminating significant risk. Additionally, safety standards must better effectuate the Act's protective purpose than any applicable national consensus standard, must be compatible with prior agency action, must be responsive to significant comment in the record, and, to the extent allowed by statute, must be consistent with applicable Executive Orders. OSHA believes that application of these criteria results in standards that provide a high degree of worker protection without undue burden on employers.

OSHA has long interpreted section 3(8) of the OSH Act to require that, before it promulgates "a health or safety standard, it must find that a place of employment is unsafe—in the sense that significant risks are present and can be eliminated or lessened by a change in practices [See *Industrial Union Dep't, AFL-CIO* v. *American Petroleum Inst.*, 448 U.S. 607, 642 (1980) (plurality) (Benzene)." When, as frequently happens in safety rulemaking, OSHA promulgates standards that differ from existing national consensus standards, it must explain "why the rule as adopted will better effectuate the purposes of this Act than the national consensus standard [29 U.S.C. 655(b)(8)]." Thus, national consensus standards provide the minimum level of effectiveness for standards which OSHA may adopt (29 U.S.C. 655(a)).

As a result, OSHA is precluded from regulating insignificant safety risks or from issuing safety standards that do not lessen risk in a significant way.

The OSH Act also limits OSHA's discretion to issue overly burdensome rules, as the agency also has long recognized that "any standard that was not economically or technologically feasible would *a fortiori* not be 'reasonably necessary or appropriate' under the Act. See *Industrial Union Dep't* v. *Hodgson*, [499 F.2d 467, 478 (D.C. Cir. 1974)] ('Congress does not appear to have intended to protect employees by putting their employers out of business.") [*American Textile Mfrs. Inst. Inc.*, 452 U.S. at 513 n. 31 (a standard is economically feasible even if it portends "disaster for some marginal firms," but it is economically infeasible if it "threaten[s] massive dislocation to, or imperil[s] the existence of," the industry)]."

By stating the test in terms of "threat" and "peril," the Supreme Court made clear in *ATMI* that economic infeasibility begins short of industry-wide bankruptcy. OSHA itself has placed the line considerably below this level. (See for example, *ATMI*, 452 U.S. at 527 n. 50; 43 FR 27,360 (June 23, 1978). Proposed 200 µg/m3 PEL for cotton dust did not raise serious possibility of industry-wide bankruptcy, but impact on weaving sector would be severe, possibly requiring reconstruction of 90 percent of all weave rooms. OSHA concluded that the 200 µg/m3 level was not feasible for weaving and that 750 µg/m3 was all that could reasonably be required). See also 54 FR 29,245–246 (July 11, 1989); *American Iron & Steel Institute*, 939 F.2d at 1003. OSHA raised the engineering control level for lead in small nonferrous foundries to avoid the possibility of bankruptcy for about half of small foundries even though the industry as a whole could have survived the loss of small firms.)"Although the cotton dust and lead rulemakings involved health standards, the economic feasibility ceiling established therein applies equally to safety standards. Indeed, because feasibility is a necessary element of a "reasonably necessary or appropriate" standard, this ceiling boundary is the same for health and safety rulemaking since it comes from section 3(8), which governs all permanent OSHA standards.

All OSHA standards must also be cost-effective in the sense that the protective measures being required must be the least expensive measures capable of achieving the desired end (*ATMI*, at 514 n. 32; *Building and Constr. Trades Dep't AFL-CIO v. Brock*, 838 F.2d 1258, 1269 (D.C. Cir. 1988)). OSHA gives additional consideration to financial impact in setting the period of time that should be allowed for compliance allowing as much as ten years for compliance phase-in. (See *United Steelworkers of Am. v. Marshall*, 647 F.2d 1189, 1278 (D.C. Cir. 1980), *cert. denied*, 453 U.S. 913 (1981).)

Additionally, OSHA's enforcement policy takes account of financial hardship on an individualized basis. OSHA's Field Operations Manual provides that, based on an employer's economic situation, OSHA may extend the period within which a violation must be corrected after issuance of a citation (CPL. 2.45B, Chapter III, paragraph E6d(3)(a), Dec. 31, 1990).

To reach the necessary findings and conclusions that a safety standard substantially reduces a significant risk of harm, is both technologically and economically feasible, and is cost effective, OSHA must conduct rulemaking in accord with the requirements of section 6 of the OSH Act. The regulatory proceeding allows it to determine the qualitative and, if possible, the quantitative nature of the risk with and without regulation, the technological feasibility of compliance, the availability of capital to the industry and the extent to which that capital is required for other purposes, the industry's profit history, the industry's ability to absorb costs or pass them on to the consumer, the impact of higher costs on demand, and the impact on competition with substitutes and imports. (See *ATMI* at 2501–2503; *American Iron & Steel Institute* generally.)

Finally, general principles of administrative law require the Agency to justify significant departures from prior practice. (See *International Union, UAW v. Pendergrass*, 878 F.2d 389, 400 (D.C. 1989)). In the twenty years since enactment of the OSH Act, OSHA has promulgated numerous safety standards—standards that provide benchmarks for judging risks, benefits, and feasibility of compliance in subsequent rulemakings. (OSHA's Hazardous Waste Operations and Emergency Response Standard, for example, required use of existing technology and well accepted safety practices to eliminate at least 32 deaths and 18,700 lost workday injuries at a cost of about $153 million per year (54 FR 9311–9312; March 6, 1989). The Excavation standard also drew on existing technology and recognized safety practices to save 74 lives and over 800 lost workday injuries annually at a cost of about $306 million. (54 FR 45,954; Oct. 31, 1989). OSHA's Grain Handling Facilities standard relied primarily on simple housekeeping measures to save 18 lives and 394 injuries annually, at a total net cost of $5.9 to $33.4 million (52 FR 49,622; Dec. 31, 1991).)

B. The proposed amendment to the standard for the training of powered industrial truck operators and the promulgation of like requirements for the construction and maritime industries complies with the statutory criteria described above.

As explained in Section I, *Background*, Section II, *The Powered Industrial Truck*, Section III, *Powered Industrial Truck Hazards*, Section IV, *Accident, Injury and Other Data*, and Section V, *Basis for Agency Action*, earlier in this preamble, and in Section IX, *Summary of the Regulatory Impact and Regulatory Flexibility Analysis and Environmental Impact Assessment*, later in this preamble, OSHA has determined that the operation of powered industrial trucks by untrained or inadequately trained operators pose significant risks to employees. There have been on average 85 fatalities, 34,900 serious injuries and 61,800 non-serious injuries annually since 1981 due to unsafe powered industrial truck operation. OSHA estimates that compliance with the revised training requirement for powered industrial truck operator will reduce the risk of hazards to those operators and other employees by 25 percent (preventing 17 to 22 fatalities, 10,898 to 14,118 serious injuries and 15,450 non-serious injuries annually). This constitutes a substantial reduction of significant risk of material harm.

The Agency believes that compliance is technologically feasible because there exists a current rule for the training of powered industrial truck operators and the revised regulation specifies in more detail what is to be taught to those operators, and requires the employer to institute effective supervisory measures to ensure continued safe operation of those vehicles. In many companies, the training of vehicle operators and the subsequent supervisory measures required by the standard have already been implemented.

Additionally, OSHA believes that compliance is economically feasible, because, as documented by the Regulatory Impact Analysis, all regulated sectors can readily absorb or pass on compliance costs.

The standard's costs, benefits, and compliance requirements are reasonable, amounting to approximately 34.9 million in the first year and 19.4 million per year thereafter, preventing 17 to 22 fatalities, 10,898 to 14,118 serious injuries and 15,450 non-serious injuries per year. As explained above, using another definition, OSHA estimates that it will eliminate between 11,968 and 15,504 lost workday injuries in addition to the fatalities prevented. These percentages are consistent with those of other OSHA safety standards.

C. The requirement for the training of powered industrial truck operators is necessary to address the significant risks of material harm posed by the operation of those vehicles.

OSHA believes that Section I, *Background*, Section II, *The Powered Industrial Truck*, Section III, *Powered*

Industrial Truck Hazards, Section IV, *Accident, Injury and Other Data*, and Section V, *Basis for Agency Action*, earlier in this preamble have clearly and comprehensively set out the Agency's bases for concluding that the operation of powered industrial trucks by untrained or inadequately trained employees pose significant risks and that the training of those operators is reasonably necessary to protect affected employees from those risks. In particular, as detailed in Section IX, *Preliminary Regulatory Impact and Regulatory Flexibility Analysis and Environmental Impact Assessment*, later in this preamble, OSHA estimates that the improper operation of powered industrial trucks causes 85 fatalities, 34,902 serious injuries, and 61,800 non-serious injuries annually, and that revision of and compliance with the requirements of the OSHA standard for the training of powered industrial truck operators will reduce the risk of fatality and injury by 25 percent (preventing 17 to 22 fatalities, 10,898 to 14,118 serious injuries and 15,450 non-serious injuries).

OSHA emphasizes that its risk assessment is based on employee exposure to the hazards of the operation of powered industrial trucks, hazards that exists in a large range of industries. Although Section IX, *Preliminary Regulatory Impact and Regulatory Flexibility Analysis and Environmental Impact Assessment*, later in this preamble, presents OSHA's estimate of the costs and benefits of the revision of the training requirement in terms of the Standard Industrial Classification (SIC) codes for the industries regulated, OSHA does not believe that the risk associated with these hazards vary according to what SIC code a vehicle may be operated in. Thus, some of the industry categories within the scope of the final rule that will have compliance costs have had few or no documented powered industrial truck accidents or injuries or fatalities during the period covered by the PRIA. In this case, OSHA has considered developing a scope of the rule to cover those situations it has determined to be hazardous. As explained more fully below, OSHA has determined that the lack of prior documented injuries and deaths in some SIC Codes does not indicate that the employees in those industries are not exposed to significant risks from the unsafe operation of powered industrial trucks. As the summary of the PRIA explains in detail, OSHA has determined that it is appropriate to include those industries within the scope of the standard because employees in those industries are exposed to the same kinds of hazards as employees in industries for which there are reported injuries and fatalities.

Even in industry sectors in which no injuries or fatalities have been reported, the Agency believes there is sufficient information for OSHA to determine that employees who work in areas in which powered industrial trucks are operated or operate those vehicles face significant risks, based on analysis of the elements of the hazards identified and of the similarity of hazard elements between industry sectors. Therefore, the Agency has determined that all employees who operate those vehicles or work in areas in which those vehicles are operated face a significant risk of material harm and that compliance with the powered industrial truck standard is reasonably necessary to protect affected employees from those risks, regardless of the number of accidents and injuries reported for the SIC code to which the employer has been assigned.

Also, because of the difficulties the Agency has experienced in compiling a database for powered industrial truck accidents, injuries or fatalities may have occurred in industries, including those for which no incidents have been documented, without being recorded. In addition, the SIC code-based organization of incident data may mask actual or potential hazards of the operation of powered industrial trucks because, while a business is classified for SIC purposes according to its principal activity, the workplace may also contain warehousing areas where materials are stored as a "secondary" purpose, that have necessitated the use of powered industrial trucks with their resultant injuries or fatalities. For example, a new car dealer would be classified under the new car dealer SIC, even though the dealer may store a large number of auto accessories, such as tires and batteries. In many instances, large quantities of items like batteries are palletized for ease of handling. When these pallets of material are delivered to the dealer, the items are either removed from the pallet and handling manually, or the pallet and the material are moved with some type of powered industrial truck, such as a pallet jack. Although the workplace is a new car dealer, a powered industrial truck is in use and an accident would have nothing to do with selling new cars. Therefore, OSHA believes, based on the limitations of the accident data and the circumstantial nature of many vehicle accidents, that it is appropriate to require that employers protect affected employees from the hazards of vehicle operations in all workplaces where powered industrial trucks are used, rather than to characterize workplaces according to the injury or fatality experience of the SIC codes in which they have been classified.

The Agency also notes that many accidents that occur as a result of powered industrial truck operations are not classified as an accident involving a truck. For example, if a powered industrial truck is used to lift an employee who is standing on the forks of the vehicle and the employee falls from those forks while aloft, the accident could be classified as a fall from height or a fall from an elevated platform. In both instances, the fact that the employee was unsafely taken aloft on the forks of a powered industrial truck and fell from those forks is not transferred to the accident report because the accident was attributed to other causes.

Finally, it is well established in the OSH Act enforcement context that the lack of injuries or deaths to a particular employer's employees does not establish that the employees are not exposed to a hazard. In a frequently quoted passage, the Fifth Circuit long ago observed that "the goal of the Act is to prevent the first accident, not to serve as a source of consolation for the first victim or his survivors" (*Mineral Industries & Heavy Construction Group v. OSHRC*, 639 F.2d 1289, 1294 (5th Cir. 1981)). This principle applies to regulatory actions as well. Once the agency determines that exposure to a particular condition constitutes a significant risk, it need not repeat that analysis for every situation or type of workplace in which the condition is found.

In addition, those segments with fewer trucks and, consequently fewer accidents, will have lower costs for training and evaluation. However, the risk to each individual operator for each year of operation is approximately the same as in industries with more trucks and operators. This approach was upheld in *International Union, UAW, v. OSHA*, —F. 2d—, (D.C. Circ., October 21, 1994)

For all of the foregoing reasons, OSHA has determined that it is inappropriate to exclude any of the SICs merely because they have not recently had documented powered industrial truck injuries or fatalities, insofar as those SICs contain workplaces where those vehicles are operated.

D. Conclusion

OSHA has determined that the powered industrial truck standard, like other safety standards, is subject to the constraints of section 3(8) of the OSH

Act, that the standard is "reasonably necessary or appropriate to provide safe or healthful employment and places of employment." But the standard is not subject to the section 6(b)(5) requirement that it limit significant risk "to the extent feasible."

The Agency believes that the use of powered industrial trucks in the workplace by untrained or poorly trained employees poses significant risks and that the need to require that only properly trained employees operate those vehicles is reasonably necessary to protect affected employees from those risks. OSHA also has determined that compliance with the standard for the training of those operators is technologically feasible because many companies offer the type training that the standard would require. In addition, OSHA believes that compliance is economically feasible, because, as documented by the Preliminary Regulatory Impact Analysis (Ex. 2), all regulated sectors can readily absorb or pass on initial compliance costs and economic benefits will ultimately exceed compliance costs. In particular, the Agency believes that compliance with the powered industrial truck training requirement will result in substantial cost savings and productivity gains at facilities that utilize powered industrial trucks that might otherwise be disrupted by accidents and injuries.

As detailed in the Summary of the Preliminary Regulatory Impact Analysis, the standard's costs, benefits, and compliance requirements are consistent with those of other OSHA safety standards. For example, the Hazardous Waste Operations and Emergency Response standard (29 CFR 1910.120) requires the use of existing technology and well accepted safety practices to eliminate at least 32 deaths and 18,700 lost workday injuries at a cost of about $153 million per year (54 FR 9311–9312; March 6, 1989). The Excavations standard (29 CFR 1926, Subpart P) also drew on existing technology and recognized safety practices to save 74 lives and over 800 lost workday injuries annually at a cost of about $306 million (54 FR 45,954; Oct. 31, 1989). Additionally, the Grain Handling Facilities standard (29 CFR 1910.272) relied primarily on simple housekeeping measures to save 18 lives and 394 injuries annually, at a total net cost of between $5.9 million and $33.4 million (52 FR 49,622; Dec. 31, 1987). Also, compliance with the planning, work practice, and training provisions of the Process Safety Management standard (29 CFR 1910.119) will reduce the risk of catastrophic fire and explosion (330 fatalities and 1917 injuries and illnesses annually) by 80 percent, at an annualized cost of $888.7 million in the first five years and at an annualized cost of $470.8 million in the following five years.

IX. Summary of the Preliminary Economic, Feasibility and Regulatory Flexibility Analyses and Environmental Impact Assessment

A. Introduction

Executive Order 12866 and the Regulatory Flexibility Act require Federal Agencies to analyze the costs, benefits and other consequences and impacts of proposed standards and final rules. Consistent with these requirements, OSHA has prepared a preliminary economic analysis for the proposed revisions to and adoption of the powered industrial truck operator training provisions which are proposed in this document.

This analysis includes a description of the industries that would be affected by the regulation, an assessment of the benefits attributable to adoption of the proposal, a determination of the technological feasibility of the proposed revisions, estimation of the costs of compliance, a determination of the economic feasibility of compliance with the proposed provisions, and an analysis of the economic and other impacts of this rulemaking. The Advisory Committee on Construction Safety and Health is currently reviewing the proposed rule for applicability to the construction industry and based on the Advisory Committee's recommendations, OSHA may extend the coverage of the proposed rule to this sector in the future.

Affected Industries

Using powered industrial truck sales data provided by the Industrial Truck Association (ITA), OSHA estimates that there are 822,831 industrial trucks in use in industries covered by the proposed standard. Industries with the largest number of powered industrial trucks include wholesale trade-non-durable goods (SIC 51) with an estimated 109,232 powered industrial trucks, and food and kindred products (SIC 20) with an estimated 71,275 such trucks.

The proposed OSHA revisions will cover workers who operate powered industrial trucks. This includes operators using these vehicles in the general industry and maritime sectors. The population-at-risk in powered industrial truck accidents consists primarily of the operators of these trucks. Operators of powered industrial trucks include workers employed as designated truck operators as well as those who might operate powered industrial truck as part of another job. These alternate users of powered industrial trucks include shipping and receiving clerks, order pickers, maintenance personnel, and general temporary workers. Non-driving workers such as warehousemen, materials handlers, laborers and pedestrians who work on or are present in the vicinity of powered industrial trucks are also injured or killed in powered industrial truck accidents. Estimates of the number of non-driving employees are not included in the population-at-risk numbers presented in this economic analysis. However, non-driving employees are included in the number of preventable fatal and non-fatal injuries estimated to be associated with compliance with the proposed rule.

OSHA estimates that approximately 1.2 million workers are employed as industrial truck operators in industries regulated by OSHA. Industries with the largest number of operators include wholesale trade (SIC 51) with 163,848 operators, and food and kindred products (SIC 20) with 106,913 operators.

Technological Feasibility

OSHA could not identify any requirement in the proposed standard that raises technological feasibility problems for establishments that use industrial trucks. On the contrary, there is substantial evidence that establishments can achieve compliance with all requirements using existing methods and equipment. In addition, the standard introduces no technological requirements of any type. Therefore, OSHA has preliminarily concluded that technological feasibility is not an issue for the proposed standard.

Costs of Compliance

The proposed OSHA industrial truck operator training standard would expand the initial training required by the existing standard to include information on the operating instructions and warnings appropriate to the type of truck used, the specific hazards in the workplace where the truck will be operated, and instructions pertaining to the requirements of the OSHA standard. Additionally, the proposed standard requires employers to monitor the performance of industrial truck operators through an annual evaluation and to provide remedial training when this evaluation suggests that such training is needed.

OSHA estimates that the first year cost of compliance with the proposed standard will be $34.9 million and that the annual cost of compliance thereafter will be $19.4 million. Table 12 outlines the annual costs by each sector affected by the proposed standard. Industry sectors with the highest estimated annualized compliance costs are manufacturing, with $9.8 million, and wholesale and retail trade with $5.6 million. Existing industry practice was taken into consideration when calculating costs, i.e., where employers have already voluntarily implemented practices that would be required by the proposed standard, no cost is attributed to the standard. OSHA welcomes comments on the preliminary costs and assumptions presented in this Preliminary Economic Analysis.

TABLE 12.—ESTIMATED ANNUALIZED COMPLIANCE COSTS FOR THE PROPOSED INDUSTRIAL TRUCK OPERATOR TRAINING STANDARD

Sector	Initial evaluation	Initial training	Monitoring	Remedial training	Total
Agriculture	$2,457	$28,637	$39,404	$2,251	$72,749
Mining [a]	1,109	12,923	17,778	1,016	32,825
Manufacturing	332,222	3,872,651	5,327,726	304,441	9,837,040
Transportation and Utilities	91,344	1,064,777	1,464,847	83,706	2,704,674
Wholesale and Retail Trade	189,193	2,205,396	3,034,034	173,373	5,601,996
Finance, Insurance, & Real Estate	2,607	30,389	41,807	2,389	77,192
Services	37,477	436,859	601,001	34,343	1,109,679
Total	656,408	7,651,632	10,526,595	601,519	19,436,154

[a] Oil and gas extraction.
Note: Costs are annualized over 10 years at a 7 percent interest rate (annualization factor 0.1424).
Source: US Department of Labor, OSHA, Office of Regulatory Analysis, based on ERG [1, Section 3].

Benefits

An estimated 85 fatalities and 34,902 injuries result annually from industrial truck-related accidents. As presented in Table 13, OSHA estimates that full compliance with the proposed standard will prevent between 17 and 22 of these fatalities per year and between 10,898 and 14,118 lost workday injuries. These preventable fatalities and injuries are in addition to lives saved and injuries prevented by OSHA's existing standard.

The proposed standard will also reduce property damage and training-related litigation. OSHA's preliminary analysis of the impacts of improved training show reductions in property damage valued at an estimated $8 million to $42 million annually. In addition, OSHA estimates that approximately $770,018 will be saved annually in damages and settlements in court cases that would have been awarded as a result of injuries caused by deficiencies in industrial truck operator training.

TABLE 13.—NUMBER OF FATALITIES AND INJURIES PREVENTED BY COMPLIANCE WITH THE PROPOSED POWERED INDUSTRIAL TRUCK TRAINING STANDARD

Industry group	Total number of industrial truck fatalities	Preventable fatalities under proposed standard		Total number of industrial truck injuries	Preventable injuries under proposed standard	
		Low	High		Low	High
Forestry, Fishing and Agricultural Services	0	0	0	219	68	88
Mining—oil and gas extraction	1	0.2	0.3	84	26	34
Manufacturing	30	5.9	7.7	14,895	4,651	6,025
Transportation, communication, and utilities	20	3.9	5.1	4,265	1,332	1,725
Wholesale and retail trade	25	4.9	6.4	12,012	3,751	4,859
Finance, insurance, and real estate	0	0	0	212	66	86
Services	9	1.8	2.3	3,215	1,004	1,300
All industries	85	17	22	34,902	10,898	14,118

Source: U.S. Department of Labor, OSHA, Office of Regulatory Analysis, based on ERG Report (1, Section 4).

Economic Impacts and Regulatory Flexibility Analysis

OSHA assessed the potential economic impacts of compliance with the proposed standard and has preliminarily determined that the standard is economically feasible for all industry groups. Detailed information at the three-digit SIC level is presented in OSHA's Preliminary Economic Analysis. When an industry enjoys an inelastic demand for its products, an increase in operating costs can ordinarily be passed on to consumers. In this case, the maximum expected price increase is calculated by dividing the average estimated compliance cost in each industry by the average revenue for that industry. OSHA estimates that the average price increase would be negligible, about 0.0002 percent. Table 14 shows that the average price increase at the two-digit SIC level would be extremely small. (For impacts at the three-digit SIC level, see economic analysis, Table V-1). These estimates indicate that even if all costs were passed on to consumers through price increases, the proposed standard would have a negligible impact on prices overall.

Given the minuscule price increases necessary to cover the cost of the proposed training requirements, employers should be able to pass along compliance costs to customers. However, even if all costs were absorbed by the affected firms, the average reduction in profits would be only 0.007 percent. As presented in Table 14, the largest potential decrease in profits—0.038 percent—would occur in SIC 51, Nondurable Goods. Because most firms will not find it necessary to absorb all of the costs from profits and should be able to pass most if not all of the standard's costs on to consumers, average profits are not expected to decline to the extent calculated here. OSHA, therefore, does not expect the revised standard to have a significant economic impact on affected firms or industries.

TABLE 14.—ECONOMIC IMPACT OF THE PROPOSED POWERED INDUSTRIAL TRUCKS OPERATOR TRAINING STANDARD

SIC/Industry sector	Value of industry shipments, receipts or sales ($ millions)	Annualized compliance costs	Compliance costs as a percent of sales	Pre-tax income ($ millions)	Compliance costs as a percent of pre-tax income
07 Agricultural services	NA	$72,749			
13 Mining—oil and gas extraction	$48,178	32,825	Negligible		
20 Food and kindred products	387,601	1,774,023	0.0005	36,213	0.005
21 Tobacco products	32,032	43,951	0.0001	(¹)	(¹)
22 Textile mill products	65,706	384,461	0.0006	5,102	0.008
23 Apparel and other textile products	65,345	109,656	0.0002	3,548	0.003
24 Lumber and wood products	70,569	415,093	0.0006	2,881	0.014
25 Furniture and fixtures	40,027	194,006	0.0005	1,942	0.010
26 Paper and allied products	128,824	760,042	0.0006	7,307	0.010
27 Printing, publishing, and allied industries	156,685	435,959	0.0003	13,171	0.003
28 Chemicals and allied products	292,326	931,407	0.0003	24,169	0.004
29 Petroleum refining and related industries	158,076	92,786	0.0001	11,193	0.001
30 Rubber and miscellaneous plastics products	100,668	522,973	0.0005	5,366	0.010
31 Leather and leather products	9,142	47,059	0.0005	(²)	(²)
32 Stone, clay, glass, and concrete products	59,611	396,003	0.0007	2,664	0.015
33 Primary metal industries	132,837	567,368	0.0004	3,133	0.018
34 Fabricated metal products	157,077	717,423	0.0005	7,660	0.009
35 Industrial and commercial machinery and computer equip	243,479	900,774	0.0004		
36 Electric and electronic equipment	197,880	492,784	0.0002	15,378	0.003
37 Transportation equipment	364,032	691,674	0.0002	1,916	0.036
38 Instruments and related equipment	127,160	141,176	0.0001	8,326	0.002
39 Miscellaneous manufacturing industries	37,131	218,423	0.0006	2,418	0.009
40 Railroad transportation	44,422	69,042	0.0002		
41 Local, suburban, and interurban passenger transit	8,094	51,782	0.0006		
42 Trucking and warehousing	110,103	1,800,849	0.0016		
44 Water transportation	18,336	105,655	0.0006		
45 Transportation by air	82,055	188,820	0.0002		
46 Pipelines, except natural gas	2,098	4,707	0.0002		
47 Transportation services	54,432	156,391	0.0003		
48 Communications	232,257	60,673	Negligible		
49 Electric, gas and sanitary services	292,280	266,754	0.0001		
50 Durable goods	981,208	1,335,982	0.0001	4,880	0.027
51 Nondurable goods	943,174	2,201,118	0.0002	5,831	0.038
52 Building materials and garden supplies	115,855	426,997	0.0004		
53 General merchandise stores	266,991	683,253	0.0003		
54 Food stores	392,400	690,815	0.0002		
55 Automatic dealers and service stations	587,890	67,212	Negligible		
56 Apparel and accessory stores	106,128	39,537	Negligible		
57 Furniture and home furnishings stores	113,673	136,765	0.0001		
58 Eating and drinking places	211,036	28,035	Negligible		
59 Miscellaneous retails	249,463	265,974	0.0001		
60 Banking	48,477	15,103	Negligible		
61 Credit agencies other than banks	69,148	6,293	Negligible		
62 Security and commodity brokers and services	41,226	5,034	Negligible		
63 Insurance carriers	521,036	27,269	Negligible		
64 Insurance agents, brokers, and services	31,623	2,937	Negligible		
65 Real estate	96,942	13,425	Negligible		
67 Holding and other investment offices	47,301	7,132	Negligible		
70 Hotels and other lodging places	64,630	13,486	Negligible		
72 Personal services	59,052	13,486	Negligible		
78 Motion pictures	43,838	17,164	Negligible		
79 Amusement and recreation services	51,107	25,746	Negligible		
80 Health services	285,040	72,743	Negligible		
81 Legal services	96,179	4,495	Negligible		
82 Educational services	4,617	64,569	0.0014		
83 Social services	68,312	22,068	Negligible		
84 Museums, art galleries, botanical and zoological gardens	3,551	1,226	Negligible		
86 Membership organizations	39,118	7,765	Negligible		
87 Engineering, accounting, research and management svcs	224,238	52,309	Negligible		
89 Miscellaneous services, n.e.c.	23,871	15,938	0.0001		

TABLE 14.—ECONOMIC IMPACT OF THE PROPOSED POWERED INDUSTRIAL TRUCKS OPERATOR TRAINING STANDARD—Continued

SIC/Industry sector	Value of industry shipments, receipts or sales ($ millions)	Annualized compliance costs	Compliance costs as a percent of sales	Pre-tax income ($ millions)	Compliance costs as a percent of pre-tax income
Totals	....	19,436,154	0.0002	....	0.007

1 =included under SIC 20.
2 =included under SIC 23.
Negligible denotes less than 0.00001 percent.
Source: US Department of Labor, OSHA, Office of Regulatory Analysis, based on ERG Report (1, Chapter 6).

In accordance with the Regulatory Flexibility Act of 1980 (5 U.S.C. 601 et seq.), OSHA has also analyzed the economic impact of the proposed standard on small establishments (19 or fewer employees), looking particularly for evidence that the rule would have a significant impact on a substantial number of small entities. Small businesses will incur lower compliance costs than larger businesses because the compliance costs depend directly on the number of industrial truck operators in a given facility. OSHA has preliminarily concluded that it would not have a significant impact upon a substantial number of small entities. Assuming a 15 percent turnover rate, compliance costs for a typical small business in public warehousing and storage (SIC 422) will be $1,188 in the first year and $280 annually thereafter. OSHA estimates that the average price impact for small establishments will not exceed 0.12 percent. Similarly, OSHA estimates that, if the average establishment could not pass any of these costs to its customers through this very small price increase (a highly unlikely scenario), the costs would impact average profits by less than 1.2 percent. These impacts are judged to be relatively minor; therefore, the proposed standard is economically feasible for small establishments.

XI. Environmental Assessment

The proposed rules have been reviewed in accordance with the requirements of the National Environmental Policy Act (NEPA) of 1969 (42 U.S.C. 4321 et seq.), the regulations of the Council of Environmental Quality (CEQ) (40 CFR part 1500), and DOL NEPA procedures (29 CFR part 11). The provision of the standard focuses on the reduction and avoidance of incidents involving powered industrial trucks. Consequently, no major negative impact is foreseen on air, water or soil quality, plant or animal life, the use of land or other aspects of the environment. Therefore, this revision is categorized as an excluded action according to subpart B, § 11.10 of the DOL NEPA regulations.

X. International Trade

This revision of the OSHA standards on powered industrial trucks and the promulgation of the same standard for other industries is not likely to have a significant effect on international trade because of the small magnitude of any price increase that would be required for passing forward compliance costs. As shown above, the maximum price increases generated from the proposed rule would be less that 1.0 percent for the majority of affected establishments. Further, none of the compliance requirements affect the demand for foreign-made safety equipment. It can be concluded, therefore, that there will be no measurable impacts on foreign trade.

XII. Federalism

This proposed regulation has been reviewed in accordance with Executive Order 12612 (52 FR 41685, October 30, 1987), regarding Federalism. This Order requires that agencies, to the extent possible, refrain from limiting state policy options, consult with states prior to taking any actions which would restrict state policy options, and take such actions only when there is clear constitutional authority and the presence of a problem of national scope. The Order provides for preemption of state law only if there is a clear Congressional intent for the Agency to do so. Any such preemption is to be limited to the extent possible.

Section 18 of the Occupational Safety and Health Act (OSH Act) expresses Congress' intent to preempt state laws relating to issues on which Federal OSHA has promulgated occupational safety and health standards. Under the OSH Act, a state can avoid preemption in issues covered by Federal standards only if it submits, and obtains Federal approval of, a plan for the development of such standards and their enforcement. Occupational safety and health standards developed by such Plan states must, among other things, be at least as effective in providing safe and healthful employment and places of employment as the Federal standards. When such standards are applicable to products distributed or used in interstate commerce they may not unduly burden commerce and must be justified by compelling local conditions.

The Federal proposed standard on powered industrial truck operator training addresses hazards that are not unique to any one state or region of the country. Nonetheless, states with occupational safety and health plans approved under section 18 of the OSH Act will be able to develop their own state standards to deal with any special problems which might be encountered in a particular state. Moreover, because this standard is written in general, performance-oriented terms, there is considerable flexibility for state plans to require, and for affected employers to use, methods of compliance which are appropriate to the working conditions covered by the standard.

In brief, this proposed rule addresses a clear national problem related to occupational safety and health in general industry. Those states which have elected to participate under section 18 of the OSH Act are not preempted by this standard, and will be able to address any special conditions within the framework of the Federal Act while ensuring that the state standards are at least as effective as their standard. State comments are invited on this proposal and will be fully considered prior to promulgation of a final rule.

XIII. Public Participation

Interested persons are requested to submit written data, views and arguments concerning this proposal. These comments must be postmarked by July 12, 1995, and submitted in quadruplicate to the Docket Office; Docket No. S–008, Room N2624; U.S. Department of Labor, Occupational Safety and Health Administration; 200

Constitution Ave., NW., Washington, DC 20210.

All written comments received within the specified comment period will be made a part of the record and will be available for public inspection and copying at the above Docket Office address.

Additionally, under section 6(b)(3) of the OSH Act and 29 CFR 1911.11, interested persons may file objections to the proposal and request an informal hearing. The objections and hearing requests should be submitted in quadruplicate to the Docket Office at the above address and must comply with the following conditions:

1. The objection must include the name and address of the objector;
2. The objections must be postmarked by July 12, 1995;
3. The objections must specify with particularity grounds upon which the objection is based;
4. Each objection must be separately numbered; and
5. The objections must be accompanied by a detailed summary of the evidence proposed to be adduced at the requested hearing.

Interested persons who have objections to various provisions or have changes to recommend may of course make those objections and their recommendations in their comments and OSHA will fully consider them. There is only need to file formal "objections" separately if the interested person requests a public hearing.

OSHA recognizes that there may be interested persons who, through their knowledge of safety or their experience in the operations involved, would wish to endorse or support certain provisions in the standard. OSHA welcomes such supportive comments, including any pertinent accident data or cost information which may be available, in order that the record of this rulemaking will present a balanced picture of the public response on the issues involved.

XIV. State Plan Standards

The 25 States with their own OSHA approved occupational safety and health plans must adopt a comparable standard within six months of the publication date of the final standard. These States are: Alaska, Arizona, California, Connecticut (for State and local government employees only), Hawaii, Indiana, Iowa, Kentucky, Maryland, Michigan, Minnesota, Nevada, New Mexico, New York (for State and local government employees only), North Carolina, Oregon, Puerto Rico, South Carolina, Tennessee, Utah, Vermont, Virginia, Virgin Island, Washington, and Wyoming. Until such time as a State standard is promulgated, Federal OSHA will provide interim enforcement assistance, as appropriate, in those States.

List of Subjects

29 CFR Part 1910

Motor vehicle safety, Occupational safety and health, Transportation.

29 CFR Part 1915

Motor vehicle safety, Occupational safety and health, Transportation, Vessels.

29 CFR Part 1917

Marine terminals, Motor vehicle safety, Occupational safety and health, Vessels.

29 CFR Part 1918

Longshoring, Motor vehicle safety, Occupational safety and health, Vessels.

XV. Authority

This document was prepared under the direction of Joseph A. Dear, Assistant Secretary of Labor for Occupational Safety and Health, U.S. Department of Labor, 200 Constitution Avenue, NW., Washington, DC 20210.

Accordingly, pursuant to section 4, 6(b), 8(c) and 8(g) of the Occupational Safety and Health Act of 1970 (29 U.S.C. 653, 655, 657), Secretary of Labor's Order No. 1–90 (55 FR 9033), and 29 CFR part 1911, it is proposed to amend 29 CFR parts 1910, 1915, 1917, 1918 and 1926 as set forth below.

Signed at Washington, DC, this 24th day of February, 1995.

Joseph A. Dear,
Assistant Secretary of Labor.

PART 1910—OCCUPATIONAL SAFETY AND HEALTH STANDARDS

1. The authority citation for subpart N of part 1910 would be revised to read as follows:

Authority: Secs. 4, 6, 8 of the Occupational Safety and Health Act of 1970 (29 U.S.C. 653, 655, 657); Secretary of Labor's Order No. 12–71 (36 FR 8754), 8–76 (41 FR 25059), 9–83 (48 FR 35736) or 1–90 (55 FR 9033), as applicable.

Section 1910.177 also issued under 5 U.S.C. 553 and 29 CFR part 1911.

Sections 1910.176, 1910.178, 1910.179, 1910.183, 1910.184, 1910.189, and 1910.190 also issued under 29 CFR part 1911.

2. Section 1910.178 would be amended by revising paragraph (l) and by adding appendices A and B at the end of the section to read as follows:

§ 1910.178 Powered industrial trucks.

* * * * *

(l) *Operator training.*

(1) *Operator qualifications.* (i) The employer shall ensure that each potential operator of a powered industrial truck is capable of performing the duties that are required of the job.

(ii) In determining operator qualifications, the employer shall ensure that each potential operator has received the training required by this paragraph (l), that each potential operator has been evaluated by a designated person while performing the required duties, and that each potential operator performs those operations competently.

(2) *Training program implementation.*

(i) The employer shall implement a training program and ensure that only trained drivers who have successfully completed the training program are allowed to operate powered industrial trucks. Exception: Trainees under the direct supervision of a designated person shall be allowed to operate a powered industrial truck provided the operation of the vehicle is conducted in an area where other employees are not near and the operation of the truck is under controlled conditions.

(ii) Training shall consist of a combination of classroom instruction (Lecture, discussion, video tapes, and/or conference) and practical training (demonstrations and practical exercises by the trainee).

(iii) All training and evaluation shall be conducted by a designated person who has the requisite knowledge, training and experience to train powered industrial truck operators and judge their competency.

(3) *Training program content.* Powered industrial truck operator trainees shall be trained in the following topics unless the employer can demonstrate that some of the topics are not needed for safe operation.

(i) Truck related topics.

(A) All operating instructions, warnings and precautions for the types of trucks the operator will be authorized to operate;

(B) Similarities to and differences from the automobile;

(C) Controls and instrumentation: location, what they do and how they work;

(D) Power plant operation and maintenance;

(E) Steering and maneuvering;

(F) Visibility (including restrictions due to loading);

(G) Fork and attachment adaption, operation and limitations of their utilization;

(H) Vehicle capacity;

(I) Vehicle stability;

(J) Vehicle inspection and maintenance;

(K) Refueling or charging, recharging batteries;
(L) Operating limitations; and
(M) Any other operating instruction, warning or precaution listed in the operator's manual for the type vehicle which the employee is being trained to operate.

(ii) Workplace related topics.
(A) Surface conditions where the vehicle will be operated;
(B) Composition of probable loads and load stability;
(C) Load manipulation, stacking, unstacking;
(D) Pedestrian traffic;
(E) Narrow aisles and other restricted places of operation;
(F) Operating in hazardous classified locations;
(G) Operating the truck on ramps and other sloped surfaces that could affect the stability of the vehicle;
(H) Other unique or potentially hazardous environmental conditions that exist or may exist in the workplace; and
(I) Operating the vehicle in closed environments and other areas where insufficient ventilation could cause a buildup of carbon monoxide or diesel exhaust.

(iii) The requirements of this section.
(4) *Evaluation and refresher or remedial training.*
(i) Sufficient evaluation and remedial training shall be conducted so that the employee retains and uses the knowledge, skills and ability needed to operate the powered industrial truck safely.
(ii) An evaluation of the performance of each powered industrial truck operator shall be conducted at least annually by a designated person.
(iii) Refresher or remedial training shall be provided when there is reason to believe that there has been unsafe operation, when an accident or a near-miss occurs or when an evaluation indicates that the operator is not capable of performing the assigned duties.
(5) *Certification.*
(i) The employer shall certify that each operator has received the training, has been evaluated as required by this paragraph, and has demonstrated competency in the performance of the operator's duties. The certification shall include the name of the trainee, the date of training, and the signature of the person performing the training and evaluation.
(ii) The employer shall retain the current training materials and course outline or the name and address of the person who conducted the training if it was conducted by an outside trainer.
(6) *Avoidance of Duplicative Training.*

(i) Each current truck operator who has received training in any of the elements specified in paragraph (l)(3) of this section for the types of trucks the employee is authorized to operate and the type workplace that the trucks are being operated in need not be retrained in those elements if the employer certifies in accordance with paragraph (l)(5)(i) of this section that the operator has been evaluated to be competent to perform those duties.

(ii) Each new truck operator who has received training in any of the elements specified in paragraph (l)(3) of this section for the types of trucks the employee will be authorized to operate and the type of workplace in which the trucks will be operated need not be retrained in those elements before initial assignment in the workplace if the employer has written documentation of the training and if the employee is evaluated pursuant to paragraph (l)(4) of this section to be competent.

Note to paragraph (l): Appendices A and B at the end of this section provide non-mandatory guidance to assist employers in implementing this paragraph (l).

* * * * *

Appendixes to 31910.178

Appendix A—Training of Powered Industrial Truck Operators

(Non-mandatory appendix to paragraph (l) of this section)

A-1. Operator Selection

A-1.1. Prospective operators of powered industrial trucks should be identified based upon their ability to be trained and accommodated to perform job functions that are essential to the operation of a powered industrial truck. Determination of the capabilities of a prospective operator to fulfill the demands of the job should be based upon the tasks that the job demands.

A-1.2. The employer should identify all the aspects of the job that the employee must meet/perform when doing his or her job. These aspects could include the level at which the employee must see and hear, the physical demands of the job, and the environmental extremes of the job.

A-1.3. One factor to be considered is the ability of the candidate to see and hear within reasonably acceptable limits. Included in the vision requirements are the ability to see at distance and peripherally. In certain instances, there also is a requirement for the candidate to discern different colors, primarily red, yellow and green.

A-1.4. The environmental extremes that might be demanded of a potential powered industrial truck operator include that ability of the person to work in areas of excessive cold or heat.

A-1.5. After an employee has been trained and appropriate accommodations have been made, the employer needs to determine whether the employee can safely perform the job.

A-2. The Method(s) of Training

A-2.1. Among the many methods of training are the lecture, conference, demonstration, test (written and/or oral) and the practical exercise. In most instances, a combination of these methods have been successfully used to train employees in the knowledge, skills and abilities that are essential to perform the job function that the employee is being trained to perform. To enhance the training and to make the training more understandable to the employee, employers and other trainers have used movies, slides, video tapes and other visual presentations. Making the presentation more understandable has several advantages including:
(1) The employees being trained remain more attentive during the presentation if graphical presentation are used, thereby increasing the effectiveness of the training;
(2) The use of visual presentations allows the trainer to ensure that the necessary information is covered during the training;
(3) The use of graphics makes better utilization of the training time by decreasing the need for the instructor to carry on long discussions about the instructional material; and
(4) The use of graphics during instruction provides greater retention by the trainees.

A-3. Training Program Content

A-3.1. Because each type (make and model) powered industrial truck has different operating characteristics, limitations and other unique features, an optimum employee training program for powered industrial truck operators must be based upon the type vehicles that the employee will be trained and authorized to operate. The training must also emphasize the features of the workplace which will affect the manner in which the vehicle must be operated. Finally, the training must include the general safety rules applicable to the operation of all powered industrial trucks.

A-3.2. Selection of the methods of training the operators has been left to the reasonable determination of the employer. Whereas some employees can assimilate instructional material while seated in a classroom, other employees may learn best by observing the conduct of operations (demonstration) and/or by having to personally conduct the operations (practical exercise). In some instances, an employee can receive valuable instruction through the use of electronic mediums, such as the use of video tapes and movies. In most instances, a combination of the different training methods may provide the mechanism for providing the best training in the least amount of time. OSHA has specified at paragraph (l)(2)(ii) of this section that the training must consist of a combination classroom instruction and practical exercise. The use of both these modes of instruction is the only way of assuring that the trainee has received and comprehended the instruction and can utilize the information to safely operate a powered industrial truck.

A-4. Initial Training

A-4.1. The following is an outline of a generalized forklift operator training program:

(1) Characteristics of the powered industrial truck(s) the employee will be allowed to operate:

(a) Similarities to and differences from the automobile;
(b) Controls and instrumentation: location, what they do and how they work;
(c) Power plant operation and maintenance;
(d) Steering and maneuvering;
(e) Visibility;
(f) Fork and/or attachment adaption, operation and limitations of their utilization;
(g) Vehicle capacity;
(h) Vehicle stability;
(i) Vehicle inspection and maintenance;
(j) Refueling or charging, recharging batteries.
(k) Operating limitations.
(l) Any other operating instruction, warning or precaution listed in the operator's manual for the type vehicle which the employee is being trained to operate.

(2) The operating environment:
(a) Floor surfaces and/or ground conditions where the vehicle will be operated;
(b) Composition of probable loads and load stability;
(c) Load manipulation, stacking, unstacking;
(d) Pedestrian traffic;
(e) Narrow aisle and restricted place operation;
(f) Operating in classified hazardous locations;
(g) Operating the truck on ramps and other sloped surfaces which would affect the stability of the vehicle;
(h) Other unique or potentially hazardous environmental conditions which exist or may exist in the workplace.
(i) Operating the vehicle in closed environments and other areas where insufficient ventilation could cause a buildup of carbon monoxide or diesel exhaust.

(3) The requirements of this OSHA Standard.

A-5. Trainee Evaluation

A-5.1. The provisions of these proposed requirements specify that an employee evaluation be conducted both as part of the training and after completion of the training. The initial evaluation is useful for many reasons, including:

(1) the employer can determine what methods of instruction will produce a proficient truck operator with the minimum of time and effort;
(2) the employer can gain insight into the previous training that the trainee has received; and
(3) a determination can be made as to whether the trainee will be able to successfully operate a powered industrial truck. This initial evaluation can be completed by having the employee fill out a questionnaire, by an oral interview, or by a combination of these mechanisms. In many cases, answers received by the employee can be substantiated by contact with other employees or previous employers.

A-6. Refresher or Remedial Training

A-6.1. (The type information listed at paragraph A-6.2 of this appendix would be used when the training is more than an on-the-spot correction being made by a supervisor or when there have been multiple instances of on-the-spot corrections having to be made.) When an on-the-spot correction is used, the person making the correction should point out the incorrect manner of operation of the truck or other unsafe act being conducted, tell the employee how to do the operation correctly, and then ensure that the employee does the operation correctly.

A-6.2. The following items may be used when a more general, structured retraining program is utilized to train employees and eliminate unsafe operation of the vehicle:

(1) Common unsafe situations encountered in the workplace;
(2) Unsafe methods of operating observed or known to be used;
(3) The need for constant attentiveness to the vehicle, the workplace conditions and the manner in which the vehicle is operated.

A-6.3. Details about the above subject areas need to be expanded upon so that the operator receives all the information which is necessary for the safe operation of the vehicle. Insight into some of the specifics of the above subject areas may be obtained from the vehicle manufacturers' literature, the national consensus standards [e.g. the ANSI B56 series of standards (current revisions)] and this OSHA Standard.

Appendix B—Stability of Powered Industrial Trucks

(Non-mandatory appendix to paragraph (l) of this section)

B-1. Definitions

To understand the principle of stability, understanding definitions of the following is necessary:

Center of gravity is that point of an object at which all of the weight of an object can be considered to be concentrated.

Counterweight is the weight that is a part of the basic structure of a truck that is used to offset the weight of a load and to maximize the resistance of the vehicle to tipping over.

Fulcrum is the axis of rotation of the truck when it tips over.

Grade is the slope of any surface that is usually measured as the number of feet of rise or fall over a hundred foot horizontal distance (this measurement is designated as a percent).

Lateral stability is the resistance of a truck to tipping over sideways.

Line of action is an imaginary vertical line through the center of gravity of an object.

Load center is the horizontal distance from the edge of the load (or the vertical face of the forks or other attachment) to the line of action through the center of gravity of the load.

Longitudinal stability is the resistance of a truck to overturning forward or rearward.

Moment is the product of the weight of the object times the distance from a fixed point. In the case of a powered industrial truck, the distance is measured from the point that the truck will tip over to the line of action of the object. The distance is always measured perpendicular to the line of action.

Track is the distance between wheels on the same axle of a vehicle.

Wheelbase is the distance between the centerline of the front and rear wheels of a vehicle.

B-2. General

B-2.1. Stability determination for a powered industrial truck is not complicated once a few basic principles are understood. There are many factors that influence vehicle stability. Vehicle wheelbase, track, height and weight distribution of the load, and the location of the counterweights of the vehicle (if the vehicle is so equipped), all contribute to the stability of the vehicle.

B-2.2. The "stability triangle", used in most discussions of stability, is not mysterious but is used to demonstrate truck stability in rather simple fashion.

B-3. Basic Principles

B-3.1. The determination of whether an object is stable is dependent on the moment of an object at one end of a system being greater than, equal to or smaller than the moment of an object at the other end of that system. This is the same principle on which a see saw or teeter-totter works, that is, if the product of the load and distance from the fulcrum (moment) is equal to the moment at the other end of the device, the device is balanced and it will not move. However, if there is a greater moment at one end of the device, the device will try to move downward at the end with the greater moment.

B-3.2. Longitudinal stability of a counterbalanced powered industrial truck is dependent on the moment of the vehicle and the moment of the load. In other words, if the mathematic product of the load moment (the distance is from the front wheels, the point about which the vehicle would tip forward) the system is balanced and will not tip forward. However, if the load-moment is greater than the vehicle-moment, the greater load-moment will force the truck to tip forward.

B-4. The Stability Triangle

B-4.1. Almost all counterbalanced powered industrial trucks have a three point suspension system, that is, the vehicle is supported at three points. This is true even if it has four wheels. The steer axle of most trucks is attached to the truck by means of a pivot pin in the center of the axle. This three point support forms a triangle called the stability triangle when the points are connected with imaginary lines. Figure 1 depicts the stability triangle.

BILLING CODE 4510-26-P

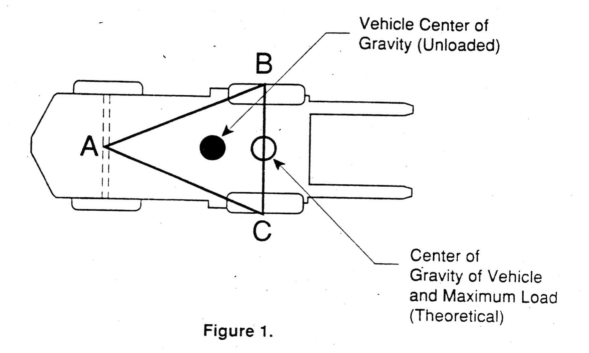

Figure 1.

NOTES:

1. When the vehicle is loaded, the combined center of gravity shifts toward line B-C. Theoretically the max load will result in the CG at the line B-C. In actual practice, the combined CG should never be at line B-C.

2. The addition of additional counterweight will cause the truck CG to shift toward point A and result in a truck that is less stable laterally.

B-4.2. When the line of action of the vehicle or load-vehicle falls within the stability triangle, the vehicle is stable and will not tip over. However, when the line of action of the vehicle or the vehicle/load combination falls outside the stability triangle, the vehicle is unstable and may tip over. (See Figure 2.)

BILLING CODE 4510-26-P

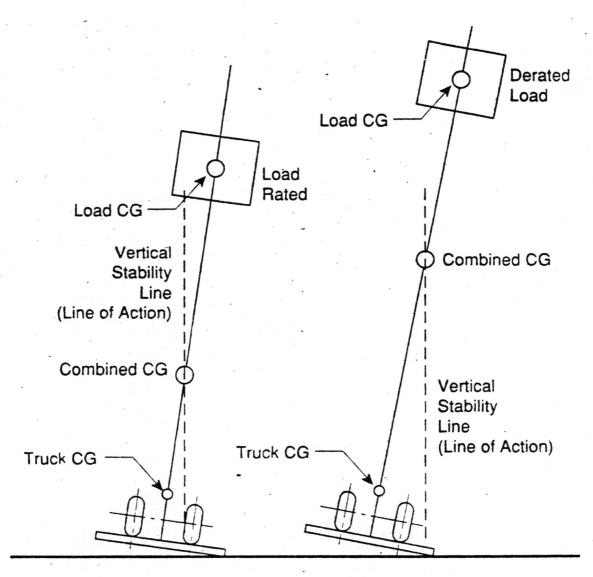

The vehicle is stable

This vehicle is unstable and will continue to tip over

Figure 2.

B–5. Longitudinal Stability

B–5.1. The axis of rotation when a truck tips forward is the point of contact of the front wheels of the vehicle with the pavement. When a powered industrial truck tips forward, it is this line that the truck will rotate about. When a truck is stable the vehicle-moment must exceed the load-moment. As long as the vehicle-moment is equal to or exceeds the load-moment, the vehicle will not tip over. On the other hand, if the load-moment slightly exceeds the vehicle-moment, the truck will begin the tip forward, thereby causing loss of steering control. If the load-moment greatly exceeds the vehicle-moment, the truck will tip forward.

B–5.2. In order to determine the maximum safe load moment, the truck manufacturer normally rates the truck at a maximum load at a given distance from the front face of the forks. The specified distance from the front face of the forks to the line of action of the load is commonly called a load center. Because larger trucks normally handle loads that are physically larger, these vehicles have greater load centers. A truck with a capacity of 30,000 pounds or less capacity is normally rated at a given load weight at a 24 inch load center. For trucks of greater than 30,000 pound capacity, the load center is normally rated at 36 or 48 inch load center distance. In order to safely operate the vehicle, the operator should always check the data plate and determine the maximum allowable weight at the rated load center.

B–5.3. Although the true load moment distance is measured from the front wheels, this distance is greater than the distance from the front face of the forks. Calculation of the maximum allowable load moment using the load center distance always provides a lower load moment than the truck was designed to handle. When handling unusual loads, such as those that are larger than 48 inches long (the center of gravity is greater than 24 inches), with an offset center of gravity, etc., then calculation of a maximum allowable load moment should be undertaken and this value used to determine whether a load can be handled. For example, if an operator is operating a 3,000 pound capacity truck (with a 24 inch load center), the maximum allowable load moment is 72,000 inch-pounds (3,000 times 24). If a probable load is 60 inches long (30 inch load center), then the maximum weight that this load can weigh is 2,400 pounds (72,000 divided by 30).

B–6. Lateral Stability

B–6.1. The lateral stability of a vehicle is determined by the position of the line of action (a vertical line that passes through the combined center of gravity of the vehicle and the load) relative to the stability triangle. When the vehicle is not loaded, the location of the center of gravity of the truck is the only factor to be considered in determining the stability of the truck. As long as the line of action of the combined center of gravity of the vehicle and the load falls within the stability triangle, the truck is stable and will not tip over. However, if the line of action falls outside the stability triangle, the truck is not stable and may tip over.

B–6.2. Factors that affect the lateral stability of a vehicle include the placement of the load on the truck, the height of the load above the surface on which the vehicle is operating, and the degree of lean of the vehicle.

B–7. Dynamic Stability

B–7.1. Up to this point, we have covered stability of a powered industrial truck without consideration of the dynamic forces that result when the vehicle and load are put into motion. The transfer of weight and the resultant shift in the center of gravity due to the dynamic forces created when the machine is moving, braking, cornering, lifting, tilting, and lowering loads, etc., are important stability considerations.

B–7.2. When determining whether a load can be safely handled, the operator should exercise extra caution when handling loads that cause the vehicle to approach its maximum design characteristics. For example, if an operator must handle a maximum load, the load should be carried at the lowest position possible, the truck should be accelerated slowly and evenly, and the forks should be tilted forward cautiously. However, no precise rules can be formulated to cover all of these eventualities.

PART 1915—OCCUPATIONAL SAFETY AND HEALTH STANDARDS FOR SHIPYARD EMPLOYMENT

3. The authority citation for part 1915 would be revised to read as follows:

Authority: Section 41, Longshore and Harbor Workers' Compensation Act (33 U.S.C. 941); secs. 4, 6, 8, Occupational Safety and Health Act of 1970 (29 U.S.C. 653, 655, 657); Secretary of Labor's Order No. 12-71 (36 FR 8754), 8-76 (41 FR 25059), 9-83 (48 FR 35736) or 1-90 (55 FR 9033), as applicable.

Sections 1915.120 and 1915.152 also issued under 29 CFR part 1911.

4. A new § 1915.120 with appendices A and B would be added to subpart G to read as follows:

§ 1915.120 Powered industrial trucks.

(a) *Operator training.* (1) Operator qualifications. (i) The employer shall ensure that each potential operator of a powered industrial truck is capable of performing the duties that are required of the job.

(ii) In determining operator qualifications, the employer shall ensure that each potential operator has received the training required by this paragraph, that each potential operator has been evaluated by a designated person while performing the required duties, and that each potential operator performs those operations competently.

(2) *Training program implementation.*

(i) The employer shall implement a training program and ensure that only trained drivers who have successfully completed the training program are allowed to operate powered industrial trucks. Exception: Trainees under the direct supervision of a designated person shall be allowed to operate a powered industrial truck provided the operation of the vehicle is conducted in an area where other employees are not near and the operation of the truck is under controlled conditions.

(ii) Training shall consist of a combination of classroom instruction (Lecture, discussion, video tapes, and/or conference) and practical training (demonstrations and practical exercises by the trainee).

(iii) All training and evaluation shall be conducted by a designated person who has the requisite knowledge, training and experience to train powered industrial truck operators and judge their competency.

(3) *Training program content.* Powered industrial truck operator trainees shall be trained in the following topics unless the employer can demonstrate that some of the topics are not needed for safe operation.

(i) Truck related topics.

(A) All operating instructions, warnings and precautions for the types of trucks the operator will be authorized to operate;

(B) Similarities to and differences from the automobile;

(C) Controls and instrumentation: location, what they do and how they work;

(D) Power plant operation and maintenance;

(E) Steering and maneuvering;

(F) Visibility (including restrictions due to loading);

(G) Fork and attachment adaption, operation and limitations of their utilization;

(H) Vehicle capacity;
(I) Vehicle stability;
(J) Vehicle inspection and maintenance;
(K) Refueling or charging, recharging batteries;
(L) Operating limitations; and
(M) Any other operating instruction, warning or precaution listed in the operator's manual for the type vehicle which the employee is being trained to operate.

(ii) Workplace related topics.
(A) Surface conditions where the vehicle will be operated;
(B) Composition of probable loads and load stability;
(C) Load manipulation, stacking, unstacking;
(D) Pedestrian traffic;
(E) Narrow aisles and other restricted places of operation;
(F) Operating in hazardous classified locations;
(G) Operating the truck on ramps and other sloped surfaces that could affect the stability of the vehicle;
(H) Other unique or potentially hazardous environmental conditions that exist or may exist in the workplace; and
(I) Operating the vehicle in closed environments and other areas where insufficient ventilation could cause a buildup of carbon monoxide or diesel exhaust.

(iii) The requirements of this section.

(4) *Evaluation and refresher or remedial training.*
(i) Sufficient evaluation and remedial training shall be conducted so that the employee retains and uses the knowledge, skills and ability needed to operate the powered industrial truck safely.
(ii) An evaluation of the performance of each powered industrial truck operator shall be conducted at least annually by a designated person.
(iii) Refresher or remedial training shall be provided when there is reason to believe that there has been unsafe operation, when an accident or a near-miss occurs or when an evaluation indicates that the operator is not capable of performing the assigned duties.

(5) *Certification.*
(i) The employer shall certify that each operator has received the training, has been evaluated as required by this paragraph, and has demonstrated competency in the performance of the operator's duties. The certification shall include the name of the trainee, the date of training, and the signature of the person performing the training and evaluation.
(ii) The employer shall retain the current training materials and course outline or the name and address of the person who conducted the training if it was conducted by an outside trainer.

(6) *Avoidance of duplicative training.*
(i) Each current truck operator who has received training in any of the elements specified in paragraph (a)(3) of this section for the types of trucks the employee is authorized to operate and the type workplace that the trucks are being operated in need not be retrained in those elements if the employer certifies in accordance with paragraph (a)(5)(i) of this section that the operator has been evaluated to be competent to perform those duties.
(ii) Each new truck operator who has received training in any of the elements specified in paragraph (a)(3) of this section for the types of trucks the employee will be authorized to operate and the type of workplace in which the trucks will be operated need not be retrained in those elements before initial assignment in the workplace if the employer has written documentation of the training and if the employee is evaluated pursuant to paragraph (a)(4) of this section to be competent.

(b) [Reserved]

Appendixes to § 1915.120

Appendix A—Training of Powered Industrial Truck Operators

(Non-mandatory appendix to paragraph (a) of this section)

A–1. Operator Selection

A–1.1. Prospective operators of powered industrial trucks should be identified based upon their ability to be trained and accommodated to perform job functions that are essential to the operation of a powered industrial truck. Determination of the capabilities of a prospective operator to fulfill the demands of the job should be based upon the tasks that the job demands.

A–1.2. The employer should identify all the aspects of the job that the employee must meet/perform when doing his or her job. These aspects could include the level at which the employee must see and hear, the physical demands of the job, and the environmental extremes of the job.

A–1.3. One factor to be considered is the ability of the candidate to see and hear within reasonably acceptable limits. Included in the vision requirements are the ability to see at distance and peripherally. In certain instances, there also is a requirement for the candidate to discern different colors, primarily red, yellow and green.

A–1.4. The environmental extremes that might be demanded of a potential powered industrial truck operator include that ability of the person to work in areas of excessive cold or heat.

A–1.5. After an employee has been trained and appropriate accommodations have been made, the employer needs to determine whether the employee can safely perform the job.

A–2. The Method(s) of Training

A–2.1. Among the many methods of training are the lecture, conference, demonstration, test (written and/or oral) and the practical exercise. In most instances, a combination of these methods have been successfully used to train employees in the knowledge, skills and abilities that are essential to perform the job function that the employee is being trained to perform. To enhance the training and to make the training more understandable to the employee, employers and other trainers have used movies, slides, video tapes and other visual presentations. Making the presentation more understandable has several advantages including:

(1) The employees being trained remain more attentive during the presentation if graphical presentation are used, thereby increasing the effectiveness of the training;
(2) The use of visual presentations allows the trainer to ensure that the necessary information is covered during the training;
(3) The use of graphics makes better utilization of the training time by decreasing the need for the instructor to carry on long discussions about the instructional material; and
(4) The use of graphics during instruction provides greater retention by the trainees.

A–3. Training Program Content

A–3.1. Because each type (make and model) powered industrial truck has different operating characteristics, limitations and other unique features, an optimum employee training program for powered industrial truck operators must be based upon the type vehicles that the employee will be trained and authorized to operate. The training must also emphasize the features of the workplace which will affect the manner in which the vehicle must be operated. Finally, the training must include the general safety rules applicable to the operation of all powered industrial trucks.

A–3.2. Selection of the methods of training the operators has been left to the reasonable determination of the employer. Whereas some employees can assimilate instructional material while seated in a classroom, other employees may learn best by observing the conduct of operations (demonstration) and/or by having to personally conduct the operations (practical exercise). In some instances, an employee can receive valuable instruction through the use of electronic mediums, such as the use of video tapes and movies. In most instances, a combination of the different training methods may provide the mechanism for providing the best training in the least amount of time. OSHA has specified at paragraph (a)(2)(ii) of this section that the training must consist of a combination classroom instruction and practical exercise. The use of both these modes of instruction is the only way of assuring that the trainee has received and comprehended the instruction and can utilize the information to safely operate a powered industrial truck.

A-4. Initial Training

A-4.1. The following is an outline of a generalized forklift operator training program:

(1) Characteristics of the powered industrial truck(s) the employee will be allowed to operate:
  (a) Similarities to and differences from the automobile;
  (b) Controls and instrumentation: location, what they do and how they work;
  (c) Power plant operation and maintenance;
  (d) Steering and maneuvering;
  (e) Visibility;
  (f) Fork and/or attachment adaption, operation and limitations of their utilization;
  (g) Vehicle capacity;
  (h) Vehicle stability;
  (i) Vehicle inspection and maintenance;
  (j) Refueling or charging, recharging batteries;
  (k) Operating limitations;
  (l) Any other operating instruction, warning or precaution listed in the operator's manual for the type vehicle which the employee is being trained to operate.

(2) The operating environment:
  (a) Floor surfaces and/or ground conditions where the vehicle will be operated;
  (b) Composition of probable loads and load stability;
  (c) Load manipulation, stacking, unstacking;
  (d) Pedestrian traffic;
  (e) Narrow aisle and restricted place operation;
  (f) Operating in classified hazardous locations;
  (g) Operating the truck on ramps and other sloped surfaces which would affect the stability of the vehicle;
  (h) Other unique or potentially hazardous environmental conditions which exist or may exist in the workplace;
  (i) Operating the vehicle in closed environments and other areas where insufficient ventilation could cause a buildup of carbon monoxide or diesel exhaust.

(3) The requirements of this OSHA Standard.

A-5. Trainee Evaluation

A-5.1. The provisions of these proposed requirements specify that an employee evaluation be conducted both as part of the training and after completion of the training. The initial evaluation is useful for many reasons, including:

(1) the employer can determine what methods of instruction will produce a proficient truck operator with the minimum of time and effort;
(2) the employer can gain insight into the previous training that the trainee has received; and
(3) a determination can be made as to whether the trainee will be able to successfully operate a powered industrial truck. This initial evaluation can be completed by having the employee fill out a questionnaire, by an oral interview, or by a combination of these mechanisms. In many cases, answers received by the employee can be substantiated by contact with other employees or previous employers.

A-6. Refresher or Remedial Training

A-6.1. (The type information listed at paragraph A-6.2 of this appendix would be used when the training is more than an on-the-spot correction being made by a supervisor or when there have been multiple instances of on-the-spot corrections having to be made.) When an on-the-spot correction is used, the person making the correction should point out the incorrect manner of operation of the truck or other unsafe act being conducted, tell the employee how to do the operation correctly, and then ensure that the employee does the operation correctly.

A-6.2. The following items may be used when a more general, structured retraining program is utilized to train employees and eliminate unsafe operation of the vehicle:

(1) Common unsafe situations encountered in the workplace;
(2) Unsafe methods of operating observed or known to be used;
(3) The need for constant attentiveness to the vehicle, the workplace conditions and the manner in which the vehicle is operated.

A-6.3. Details about the above subject areas need to be expanded upon so that the operator receives all the information which is necessary for the safe operation of the vehicle. Insight into some of the specifics of the above subject areas may be obtained from the vehicle manufacturers' literature, the national consensus standards [e.g. the ANSI B56 series of standards (current revisions)] and this OSHA Standard.

Appendix B—Stability of Powered Industrial Trucks

(Non-mandatory appendix to paragraph (a) of this section)

B-1. Definitions

To understand the principle of stability, understanding definitions of the following is necessary:

Center of Gravity is that point of an object at which all of the weight of an object can be considered to be concentrated.

Counterweight is the weight that is a part of the basic structure of a truck that is used to offset the weight of a load and to maximize the resistance of the vehicle to tipping over.

Fulcrum is the axis of rotation of the truck when it tips over.

Grade is the slope of any surface that is usually measured as the number of feet or rise of fall over a hundred foot horizontal distance (this measurement is designated as a percent).

Lateral stability is the resistance of a truck to tipping over sideways.

Line of action is a imaginary vertical line through the center of gravity of an object.

Load center is the horizontal distance from the edge of the load (or the vertical face of the forks or other attachment) to the line of action through the center of gravity of the load.

Longitudinal stability is the resistance of a truck to overturning forward or rearward.

Moment is the product of the weight of the object times the distance from a fixed point. In the case of a powered industrial truck, the distance is measured from the point that the truck will tip over to the line of action of the object. The distance is always measured perpendicular to the line of action.

Track is the distance between wheels on the same axle of a vehicle.

Wheelbase is the distance between the centerline of the front and rear wheels of a vehicle.

B-2. General

B-2.1. Stability determination for a powered industrial truck is not complicated once a few basic principles are understood. There are many factors that influence vehicle stability. Vehicle wheelbase, track, height and weight distribution of the load, and the location of the counterweights of the vehicle (if the vehicle is so equipped), all contribute to the stability of the vehicle.

B–2.2. The "stability triangle", used in most discussions of stability, is not mysterious but is used to demonstrate truck stability in rather simple fashion.

B–3. Basic Principles

B–3.1. The determination of whether an object is stable is dependent on the moment of a system being greater than, equal to or smaller than the moment of an object at the other end of that system. This is the same principle on which a see saw or teeter-totter works, that is, if the product of the load and distance from the fulcrum (moment) is equal to the moment at the other end of the device, the device is balanced and it will not move. However, if there is a greater moment at one end of the device, the device will try to move downward at the end with the greater moment.

B–3.2. Longitudinal stability of a counterbalanced powered industrial truck is dependent on the moment of the vehicle and the moment of the load. In other words, if the mathematic product of the load moment (the distance is from the front wheels, the point about which the vehicle would tip forward) the system is balanced and will not tip forward. However, if the load-moment is greater than the vehicle-moment, the greater load-moment will force the truck to tip forward.

B–4. The Stability Triangle

B–4.1. Almost all counterbalanced powered industrial trucks have a three point suspension system, that is, the vehicle is supported at three points. This is true even if it has four wheels. The steer axle of most trucks is attached to the truck by means of a pivot pin in the center of the axle. This three point support forms a triangle called the stability triangle when the points are connected with imaginary lines Figure 1 depicts the stability triangle.

BILLING CODE 4510–26–P

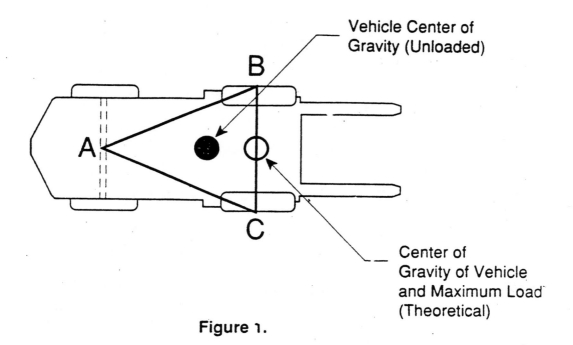

Figure 1.

NOTES:

1. When the vehicle is loaded, the combined center of gravity shifts toward line B-C. Theoretically the max load will result in the CG at the line B-C. In actual practice, the combined CG should never be at line B-C.

2. The addition of additional counterweight will cause the truck CG to shift toward point A and result in a truck that is less stable laterally.

B-4 2. When the line of action of the vehicle or load-vehicle falls within the stability triangle, the vehicle is stable and will not tip over. However, when the line of action of the vehicle or the vehicle/load combination falls outside the stability triangle, the vehicle is unstable and may tip over. (See Figure 2.)

BILLING CODE 4510-26-P

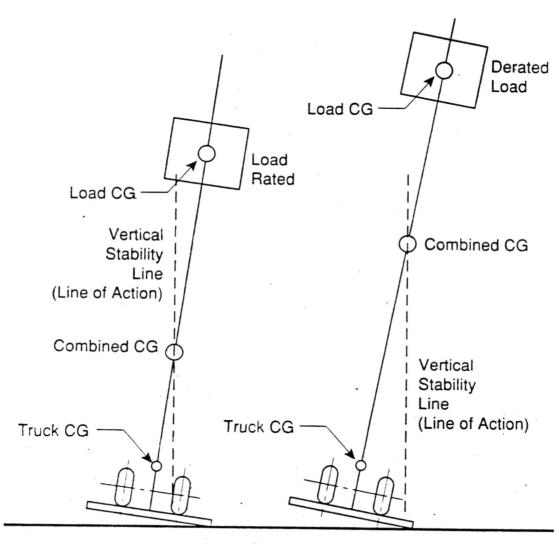

Figure 2.

B-5. Longitudinal Stability

B-5.1. The axis of rotation when a truck tips forward is the point of contact of the front wheels of the vehicle with the pavement. When a powered industrial truck tips forward, it is this line that the truck will rotate about. When a truck is stable the vehicle-moment must exceed the load-moment. As long as the vehicle-moment is equal to or exceeds the load-moment, the vehicle will not tip over. On the other hand, if the load-moment slightly exceeds the vehicle-moment, the truck will begin the tip forward, thereby causing loss of steering control. If the load-moment greatly exceeds the vehicle-moment, the truck will tip forward.

B-5.2. In order to determine the maximum safe load moment, the truck manufacturer normally rates the truck at a maximum load at a given distance from the front face of the forks. The specified distance from the front face of the forks to the line of action of the load is commonly called a load center. Because larger trucks normally handle loads that are physically larger, these vehicles have greater load centers. A truck with a capacity of 30,000 pounds or less capacity is normally rated at a given load weight at a 24 inch load center. For trucks of greater than 30,000 pound capacity, the load center is normally rated at 36 or 48 inch load center distance. In order to safely operate the vehicle, the operator should always check the data plate and determine the maximum allowable weight at the rated load center.

B-5.3. Although the true load moment distance is measured from the front wheels, this distance is greater than the distance from the front face of the forks. Calculation of the maximum allowable load moment using the load center distance always provides a lower load moment than the truck was designed to handle. When handling unusual loads, such as those that are larger than 48 inches long (the center of gravity is greater than 24 inches), with an offset center of gravity, etc., then calculation of a maximum allowable load moment should be undertaken and this value used to determine whether a load can be handled. For example, if an operator is operating a 3000 pound capacity truck (with a 24 inch load center), the maximum allowable load moment is 72,000 inch-pounds (3,000 times 24). If a probable load is 60 inches long (30 inch load center), then the maximum weight that this load can weigh is 2,400 pounds (72,000 divided by 30).

B-6. Lateral Stability

B-6.1. The lateral stability of a vehicle is determined by the position of the line of action (a vertical line that passes through the combined center of gravity of the vehicle and the load) relative to the stability triangle. When the vehicle is not loaded, the location of the center of gravity of the truck is the only factor to be considered in determining the stability of the truck. As long as the line of action of the combined center of gravity of the vehicle and the load falls within the stability triangle, the truck is stable and will not tip over. However, if the line of action falls outside the stability triangle, the truck is not stable and may tip over.

B-6.2. Factors that affect the lateral stability of a vehicle include the placement of the load on the truck, the height of the load above the surface on which the vehicle is operating, and the degree of lean of the vehicle.

B-7. Dynamic Stability

B-7.1. Up to this point, we have covered stability of a powered industrial truck without consideration of the dynamic forces that result when the vehicle and load are put into motion. The transfer of weight and the resultant shift in the center of gravity due to the dynamic forces created when the machine is moving, braking, cornering, lifting, tilting, and lowering loads, etc., are important stability considerations.

B-7.2. When determining whether a load can be safely handled, the operator should exercise extra caution when handling loads that cause the vehicle to approach its maximum design characteristics. For example, if an operator must handle a maximum load, the load should be carried at the lowest position possible, the truck should be accelerated slowly and evenly, and the forks should be tilted forward cautiously. However, no precise rules can be formulated to cover all of these eventualities.

PART 1917—MARINE TERMINALS

5. The authority citation for part 1917 would be revised to read as follows:

Authority: Section 41, Longshore and Harbor Workers' Compensation Act (33 U.S.C. 941); secs. 4, 6, 8, Occupational Safety and Health Act of 1970 (29 U.S.C. 653, 655, 657); Secretary of Labor's Order No. 12-71 (36 FR 8754), 8-76 (41 FR 25059), 9-83 (48 FR 35736) or 1-90 (55 FR 9033), as applicable.

Section 1917.43 also issued under 29 CFR part 1911.

6. Section 1917.43 would be amended by adding a new paragraph (i) and by adding appendices A and B at the end of the section to read as follows:

§ 1917.43 Powered industrial trucks.

* * * * *

(i) *Operator training.*

(1) *Operator qualifications.* (i) The employer shall ensure that each potential operator of a powered industrial truck is capable of performing the duties that are required of the job.

(ii) In determining operator qualifications, the employer shall ensure that each potential operator has received the training required by this paragraph, that each potential operator has been evaluated by a designated person while performing the required duties, and that each potential operator performs those operations competently

(2) *Training program implementation.*

(i) The employer shall implement a training program and ensure that only trained drivers who have successfully completed the training program are allowed to operate powered industrial trucks. Exception: Trainees under the direct supervision of a designated person shall be allowed to operate a powered industrial truck provided the operation of the vehicle is conducted in an area where other employees are not near and the operation of the truck is under controlled conditions.

(ii) Training shall consist of a combination of classroom instruction (Lecture, discussion, video tapes, and/or conference) and practical training (demonstrations and practical exercises by the trainee).

(iii) All training and evaluation shall be conducted by a designated person who has the requisite knowledge, training and experience to train powered industrial truck operators and judge their competency.

(3) *Training program content.* Powered industrial truck operator trainees shall be trained in the following topics unless the employer can demonstrate that some of the topics are not needed for safe operation.

(i) Truck related topics.

(A) All operating instructions, warnings and precautions for the types of trucks the operator will be authorized to operate;

(B) Similarities to and differences from the automobile;

(C) Controls and instrumentation: location, what they do and how they work;

(D) Power plant operation and maintenance;

(E) Steering and maneuvering;

(F) Visibility (including restrictions due to loading);

(G) Fork and attachment adaption, operation and limitations of their utilization;

(H) Vehicle capacity;
(I) Vehicle stability;
(J) Vehicle inspection and maintenance;
(K) Refueling or charging, recharging batteries;
(L) Operating limitations; and
(M) Any other operating instruction, warning or precaution listed in the operator's manual for the type vehicle which the employee is being trained to operate.

(ii) Workplace related topics.
(A) Surface conditions where the vehicle will be operated;
(B) Composition of probable loads and load stability;
(C) Load manipulation, stacking, unstacking;
(D) Pedestrian traffic;
(E) Narrow aisles and other restricted places of operation;
(F) Operating in hazardous classified locations;
(G) Operating the truck on ramps and other sloped surfaces that could affect the stability of the vehicle;
(H) Other unique or potentially hazardous environmental conditions that exist or may exist in the workplace; and
(I) Operating the vehicle in closed environments and other areas where insufficient ventilation could cause a buildup of carbon monoxide or diesel exhaust.

(iii) The requirements of this section.

(4) *Evaluation and refresher or remedial training.* (i) Sufficient evaluation and remedial training shall be conducted so that the employee retains and uses the knowledge, skills and ability needed to operate the powered industrial truck safely.

(ii) An evaluation of the performance of each powered industrial truck operator shall be conducted at least annually by a designated person.

(iii) Refresher or remedial training shall be provided when there is reason to believe that there has been unsafe operation, when an accident or a near-miss occurs or when an evaluation indicates that the operator is not capable of performing the assigned duties.

(5) *Certification.*
(i) The employer shall certify that each operator has received the training, has been evaluated as required by this paragraph, and has demonstrated competency in the performance of the operator's duties. The certification shall include the name of the trainee, the date of training, and the signature of the person performing the training and evaluation.

(ii) The employer shall retain the current training materials and course outline or the name and address of the person who conducted the training if it was conducted by an outside trainer.

(6) *Avoidance of duplicative training.*
(i) Each current truck operator who has received training in any of the elements specified in paragraph (i)(3) of this section for the types of trucks the employee is authorized to operate and the type workplace that the trucks are being operated in need not be retrained in those elements if the employer certifies in accordance with paragraph (i)(5)(i) of this section that the operator has been evaluated to be competent to perform those duties.

(ii) Each new truck operator who has received training in any of the elements specified in paragraph (i)(3) of this section for the types of trucks the employee will be authorized to operate and the type of workplace in which the trucks will be operated need not be retrained in those elements before initial assignment in the workplace if the employer has written documentation of the training and if the employee is evaluated pursuant to paragraph (i)(4) of this section to be competent.

Note to paragraph (i): Appendices A and B provide non-mandatory guidance to assist employers in implementing this paragraph (i).

Appendices to § 1917.43

Appendix A—Training of Powered Industrial Truck Operators

(Non-mandatory appendix to paragraph (i) of this section)

A-1. Operator Selection

A-1.1. Prospective operators of powered industrial trucks should be identified based upon their ability to be trained and accommodated to perform job functions that are essential to the operation of a powered industrial truck. Determination of the capabilities of a prospective operator to fulfill the demands of the job should be based upon the tasks that the job demands.

A-1.2. The employer should identify all the aspects of the job that the employee must meet/perform when doing his or her job. These aspects could include the level at which the employee must see and hear, the physical demands of the job, and the environmental extremes of the job.

A-1.3. One factor to be considered is the ability of the candidate to see and hear within reasonably acceptable limits. Included in the vision requirements are the ability to see at distance and peripherally. In certain instances, there also is a requirement for the candidate to discern different colors, primarily red, yellow and green.

A-1.4. The environmental extremes that might be demanded of a potential powered industrial truck operator include that ability of the person to work in areas of excessive cold or heat.

A-1.5. After an employee has been trained and appropriate accommodations have been made, the employer needs to determine whether the employee can safely perform the job.

A-2. The Method(s) of Training

A-2.1. Among the many methods of training are the lecture, conference, demonstration, test (written and/or oral) and the practical exercise. In most instances, a combination of these methods have been successfully used to train employees in the knowledge, skills and abilities that are essential to perform the job function that the employee is being trained to perform. To enhance the training and to make the training more understandable to the employee, employers and other trainers have used movies, slides, video tapes and other visual presentations. Making the presentation more understandable has several advantages including:

(1) The employees being trained remain more attentive during the presentation if graphical presentation are used, thereby increasing the effectiveness of the training;

(2) The use of visual presentations allows the trainer to ensure that the necessary information is covered during the training;

(3) The use of graphics makes better utilization of the training time by decreasing the need for the instructor to carry on long discussions about the instructional material; and

(4) The use of graphics during instruction provides greater retention by the trainees.

A-3. Training Program Content

A-3.1. Because each type (make and model) powered industrial truck has different operating characteristics, limitations and other unique features, an optimum employee training program for powered industrial truck operators must be based upon the type vehicles that the employee will be trained and authorized to operate. The training must also emphasize the features of the workplace which will affect the manner in which the vehicle must be operated. Finally, the training must include the general safety rules applicable to the operation of all powered industrial trucks.

A-3.2. Selection of the methods of training the operators has been left to the reasonable determination of the employer. Whereas some employees can assimilate instructional material while seated in a classroom, other employees may learn best by observing the conduct of operations (demonstration) and/or by having to personally conduct the operations (practical exercise). In some instances, an employee can receive valuable instruction through the use of electronic mediums, such as the use of video tapes and movies. In most instances, a combination of the different training methods may provide the mechanism for providing the best training in the least amount of time. OSHA has specified at paragraph (i)(2)(ii) of this section that the training must consist of a combination classroom instruction and practical exercise. The use of both these modes of instruction is the only way of assuring that the trainee has received and comprehended the instruction and can utilize the information to safely operate a powered industrial truck.

A–4. Initial Training

A–4.1. The following is an outline of a generalized forklift operator training program:

(1) Characteristics of the powered industrial truck(s) the employee will be allowed to operate:

(a) Similarities to and differences from the automobile;
(b) Controls and instrumentation: location, what they do and how they work;
(c) Power plant operation and maintenance;
(d) Steering and maneuvering;
(e) Visibility;
(f) Fork and/or attachment adaption, operation and limitations of their utilization;
(g) Vehicle capacity;
(h) Vehicle stability;
(i) Vehicle inspection and maintenance;
(j) Refueling or charging, recharging batteries.
(k) Operating limitations.
(l) Any other operating instruction, warning or precaution listed in the operator's manual for the type vehicle which the employee is being trained to operate.

(2) The operating environment:

(a) Floor surfaces and/or ground conditions where the vehicle will be operated;
(b) Composition of probable loads and load stability;
(c) Load manipulation, stacking, unstacking;
(d) Pedestrian traffic;
(e) Narrow aisle and restricted place operation;
(f) Operating in classified hazardous locations;
(g) Operating the truck on ramps and other sloped surfaces which would affect the stability of the vehicle;
(h) Other unique or potentially hazardous environmental conditions which exist or may exist in the workplace.
(i) Operating the vehicle in closed environments and other areas where insufficient ventilation could cause a buildup of carbon monoxide or diesel exhaust.

(3) The requirements of this OSHA Standard.

A–5. Trainee Evaluation

A–5.1. The provisions of these proposed requirements specify that an employee evaluation be conducted both as part of the training and after completion of the training. The initial evaluation is useful for many reasons, including:

(1) the employer can determine what methods of instruction will produce a proficient truck operator with the minimum of time and effort;
(2) the employer can gain insight into the previous training that the trainee has received; and
(3) a determination can be made as to whether the trainee will be able to successfully operate a powered industrial truck. This initial evaluation can be completed by having the employee fill out a questionnaire, by an oral interview, or by a combination of these mechanisms. In many cases, answers received by the employee can be substantiated by contact with other employees or previous employers.

A–6. Refresher or Remedial Training

A–6.1. (The type information listed in paragraph A–6.2 of this appendix would be used when the training is more than an on-the-spot correction being made by a supervisor or when there have been multiple instances of on-the-spot corrections having to be made.) When an on-the-spot correction is used, the person making the correction should point out the incorrect manner of operation of the truck or other unsafe act being conducted, tell the employee how to do the operation correctly, and then ensure that the employee does the operation correctly.

A–6.2. The following items may be used when a more general, structured retraining program is utilized to train employees and eliminate unsafe operation of the vehicle:

(1) Common unsafe situations encountered in the workplace;
(2) Unsafe methods of operating observed or known to be used;
(3) The need for constant attentiveness to the vehicle, the workplace conditions and the manner in which the vehicle is operated.

A–6.3. Details about the above subject areas need to be expanded upon so that the operator receives all the information which is necessary for the safe operation of the vehicle. Insight into some of the specifics of the above subject areas may be obtained from the vehicle manufacturers' literature, the national consensus standards [e.g. the ANSI B56 series of standards (current revisions)] and this OSHA Standard.

Appendix B—Stability of Powered Industrial Trucks

(Non-mandatory appendix to paragraph (i) of this section)

B–1. Definitions

To understand the principle of stability, understanding definitions of the following is necessary:

Center of Gravity is that point of an object at which all of the weight of an object can be considered to be concentrated.

Counterweight is the weight that is a part of the basic structure of a truck that is used to offset the weight of a load and to maximize the resistance of the vehicle to tipping over.

Fulcrum is the axis of rotation of the truck when it tips over.

Grade is the slope of any surface that is usually measured as the number of feet of rise of fall over a hundred foot horizontal distance (this measurement is designated as a percent).

Lateral stability is the resistance of a truck to tipping over sideways.

Line of action is a imaginary vertical line through the center of gravity of an object.

Load center is the horizontal distance from the edge of the load (or the vertical face of the forks or other attachment) to the line of action through the center of gravity of the load.

Longitudinal stability is the resistance of a truck to overturning forward or rearward.

Moment is the product of the weight of the object times the distance from a fixed point. In the case of a powered industrial truck, the distance is measured from the point that the truck will tip over to the line of action of the object. The distance is always measured perpendicular to the line of action.

Track is the distance between wheels on the same axle of a vehicle.

Wheelbase is the distance between the centerline of the front and rear wheels of a vehicle.

B–2. General

B–2.1. Stability determination for a powered industrial truck is not complicated once a few basic principles are understood. There are many factors that influence vehicle stability. Vehicle wheelbase, track, height and weight distribution of the load, and the location of the counterweights of the vehicle (if the vehicle is so equipped), all contribute to the stability of the vehicle.

B-2.2. The "stability triangle", used in most discussions of stability, is not mysterious but is used to demonstrate truck stability in rather simple fashion.

B-3. Basic Principles

B-3.1. The determination of whether an object is stable is dependent on the moment of an object at one end of a system being greater than, equal to or smaller than the moment of an object at the other end of that system. This is the same principle on which a see saw or teeter-totter works, that is, if the product of the load and distance from the fulcrum (moment) is equal to the moment at the other end of the device, the device is balanced and it will not move. However, if there is a greater moment at one end of the device, the device will try to move downward at the end with the greater moment.

B-3.2. Longitudinal stability of a counterbalanced powered industrial truck is dependent on the moment of the vehicle and the moment of the load. In other words, if the mathematic product of the load moment (the distance is from the front wheels, the point about which the vehicle would tip forward) the system is balanced and will not tip forward. However, if the load-moment is greater than the vehicle-moment, the greater load-moment will force the truck to tip forward.

B-4. The Stability Triangle

B-4.1. Almost all counterbalanced powered industrial trucks have a three point suspension system, that is, the vehicle is supported at three points. This is true even if it has four wheels. The steer axle of most trucks is attached to the truck by means of a pivot pin in the center of the axle. This three point support forms a triangle called the stability triangle when the points are connected with imaginary lines. Figure 1 depicts the stability triangle.

BILLING CODE 4510-26-P

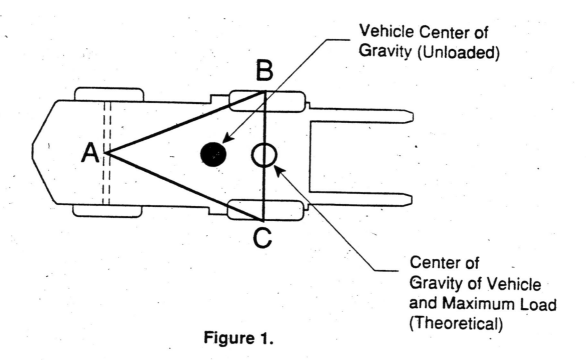

Figure 1.

NOTES:

1. When the vehicle is loaded, the combined center of gravity shifts toward line B-C. Theoretically the max load will result in the CG at the line B-C. In actual practice, the combined CG should never be at line B-C.

2. The addition of additional counterweight will cause the truck CG to shift toward point A and result in a truck that is less stable laterally.

B–4.2. When the line of action of the vehicle or load-vehicle falls within the stability triangle, the vehicle is stable and will not tip over. However, when the line of action of the vehicle or the vehicle/load combination falls outside the stability triangle, the vehicle is unstable and may tip over. (See Figure 2.)

BILLING CODE 4510-26-P

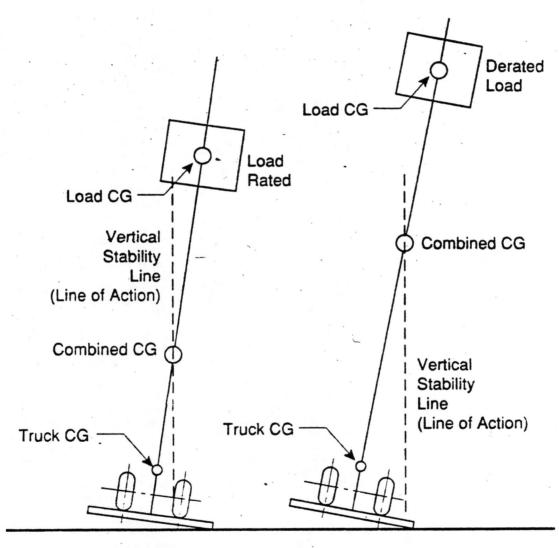

Figure 2.

B-5. Longitudinal Stability

B-5.1. The axis of rotation when a truck tips forward is the point of contact of the front wheels of the vehicle with the pavement. When a powered industrial truck tips forward, it is this line that the truck will rotate about. When a truck is stable the vehicle-moment must exceed the load-moment. As long as the vehicle-moment is equal to or exceeds the load-moment, the vehicle will not tip over. On the other hand, if the load-moment slightly exceeds the vehicle-moment, the truck will begin the tip forward, thereby causing loss of steering control. If the load-moment greatly exceeds the vehicle-moment, the truck will tip forward.

B-5.2. In order to determine the maximum safe load moment, the truck manufacturer normally rates the truck at a maximum load at a given distance from the front face of the forks. The specified distance from the front face of the forks to the line of action of the load is commonly called a load center. Because larger trucks normally handle loads that are physically larger, these vehicles have greater load centers. A truck with a capacity of 30,000 pounds or less capacity is normally rated at a given load weight at a 24 inch load center. For trucks of greater than 30,000 pound capacity, the load center is normally rated at 36 or 48 inch load center distance. In order to safely operate the vehicle, the operator should always check the data plate and determine the maximum allowable weight at the rated load center.

B-5.3. Although the true load moment distance is measured from the front wheels, this distance is greater than the distance from the front face of the forks. Calculation of the maximum allowable load moment using the load center distance always provides a lower load moment than the truck was designed to handle. When handling unusual loads, such as those that are larger than 48 inches long (the center of gravity is greater than 24 inches), with an offset center of gravity, etc., then calculation of a maximum allowable load moment should be undertaken and this value used to determine whether a load can be handled. For example, if an operator is operating a 3,000 pound capacity truck (with a 24 inch load center), the maximum allowable load moment is 72,000 inch-pounds (3,000 times 24). If a probable load is 60 inches long (30 inch load center), then the maximum weight that this load can weigh is 2,400 pounds (72,000 divided by 30).

B-6. Lateral Stability

B-6.1. The lateral stability of a vehicle is determined by the position of the line of action (a vertical line that passes through the combined center of gravity of the vehicle and the load) relative to the stability triangle. When the vehicle is not loaded, the location of the center of gravity of the truck is the only factor to be considered in determining the stability of the truck. As long as the line of action of the combined center of gravity of the vehicle and the load falls within the stability triangle, the truck is stable and will not tip over. However, if the line of action falls outside the stability triangle, the truck is not stable and may tip over.

B-6.2. Factors that affect the lateral stability of a vehicle include the placement of the load on the truck, the height of the load above the surface on which the vehicle is operating, and the degree of lean of the vehicle.

B-7. Dynamic Stability

B-7.1. Up to this point, we have covered stability of a powered industrial truck without consideration of the dynamic forces that result when the vehicle and load are put into motion. The transfer of weight and the resultant shift in the center of gravity due to the dynamic forces created when the machine is moving, braking, cornering, lifting, tilting, and lowering loads, etc., are important stability considerations.

B-7.2. When determining whether a load can be safely handled, the operator should exercise extra caution when handling loads that cause the vehicle to approach its maximum design characteristics. For example, if an operator must handle a maximum load, the load should be carried at the lowest position possible, the truck should be accelerated slowly and evenly, and the forks should be tilted forward cautiously. However, no precise rules can be formulated to cover all of these eventualities.

PART 1918—SAFETY AND HEALTH REGULATIONS FOR LONGSHORING

7. The authority citation for part 1918 would be revised to read as follows:

Authority: Section 41, Longshore and Harbor Workers' Compensation Act (33 U.S.C. 941); secs. 4, 6, 8, Occupational Safety and Health Act of 1970 (29 U.S.C. 653, 655, 657); Secretary of Labor's Order No. 12-71 (36 FR 8754), 8-76 (41 FR 25059), 9-83 (48 FR 35736) or 1-90 (55 FR 9033), as applicable. Section 1918.77 also issued under 29 CFR part 1911.

8. A new § 1918.77 with appendices A and B would be added to subpart G to read as follows:

§ 1918.77 Powered Industrial Trucks.

(a) *Operator training.*

(1) *Operator qualifications.* (i) The employer shall ensure that each potential operator of a powered industrial truck is capable of performing the duties that are required of the job.

(ii) In determining operator qualifications, the employer shall ensure that each potential operator has received the training required by this paragraph, that each potential operator has been evaluated by a designated person while performing the required duties, and that each potential operator performs those operations competently.

(2) *Training program implementation.*

(i) The employer shall implement a training program and ensure that only trained drivers who have successfully completed the training program are allowed to operate powered industrial trucks. Exception: Trainees under the direct supervision of a designated person shall be allowed to operate a powered industrial truck provided the operation of the vehicle is conducted in an area where other employees are not near and the operation of the truck is under controlled conditions.

(ii) Training shall consist of a combination of classroom instruction (Lecture, discussion, video tapes, and/or conference) and practical training (demonstrations and practical exercises by the trainee).

(iii) All training and evaluation shall be conducted by a designated person who has the requisite knowledge, training and experience to train powered industrial truck operators and judge their competency.

(3) *Training program content.* Powered industrial truck operator trainees shall be trained in the following topics unless the employer can demonstrate that some of the topics are not needed for safe operation.

(i) Truck related topics.

(A) All operating instructions, warnings and precautions for the types of trucks the operator will be authorized to operate;

(B) Similarities to and differences from the automobile;

(C) Controls and instrumentation: location, what they do and how they work;

(D) Power plant operation and maintenance;

(E) Steering and maneuvering;

(F) Visibility (including restrictions due to loading);

(G) Fork and attachment adaption, operation and limitations of their utilization;

(H) Vehicle capacity;

(I) Vehicle stability;

(J) Vehicle inspection and maintenance;

(K) Refueling or charging, recharging batteries;
(L) Operating limitations; and
(M) Any other operating instruction, warning or precaution listed in the operator's manual for the type vehicle which the employee is being trained to operate.

(ii) Workplace related topics.
(A) Surface conditions where the vehicle will be operated;
(B) Composition of probable loads and load stability;
(C) Load manipulation, stacking, unstacking;
(D) Pedestrian traffic;
(E) Narrow aisles and other restricted places of operation;
(F) Operating in hazardous classified locations;
(G) Operating the truck on ramps and other sloped surfaces that could affect the stability of the vehicle;
(H) Other unique or potentially hazardous environmental conditions that exist or may exist in the workplace; and
(I) Operating the vehicle in closed environments and other areas where insufficient ventilation could cause a buildup of carbon monoxide or diesel exhaust.

(iii) The requirements of this section.
(4) *Evaluation and refresher or remedial training.*
(i) Sufficient evaluation and remedial training shall be conducted so that the employee retains and uses the knowledge, skills and ability needed to operate the powered industrial truck safely.
(ii) An evaluation of the performance of each powered industrial truck operator shall be conducted at least annually by a designated person.
(iii) Refresher or remedial training shall be provided when there is reason to believe that there has been unsafe operation, when an accident or a near-miss occurs or when an evaluation indicates that the operator is not capable of performing the assigned duties.
(5) *Certification.*
(i) The employer shall certify that each operator has received the training, has been evaluated as required by this paragraph, and has demonstrated competency in the performance of the operator's duties. The certification shall include the name of the trainee, the date of training, and the signature of the person performing the training and evaluation.
(ii) The employer shall retain the current training materials and course outline or the name and address of the person who conducted the training if it was conducted by an outside trainer.
(6) *Avoidance of Duplicative Training.*

(i) Each current truck operator who has received training in any of the elements specified in paragraph (a)(3) of this section for the types of trucks the employee is authorized to operate and the type workplace that the trucks are being operated in need not be retrained in those elements if the employer certifies in accordance with paragraph (a)(5)(i) of this section that the operator has been evaluated to be competent to perform those duties.
(ii) Each new truck operator who has received training in any of the elements specified in paragraph (a)(3) of this section for the types of trucks the employee will be authorized to operate and the type of workplace in which the trucks will be operated need not be retrained in those elements before initial assignment in the workplace if the employer has written documentation of the training and if the employee is evaluated pursuant to paragraph (a)(4) of this section to be competent.
(b) [Reserved]

Appendixes to § 1918.77

Appendix A—Training of Powered Industrial Truck Operators

(Non-mandatory appendix to paragraph (a) of this section)

A–1. Operator Selection

A–1.1. Prospective operators of powered industrial trucks should be identified based upon their ability to be trained and accommodated to perform job functions that are essential to the operation of a powered industrial truck. Determination of the capabilities of a prospective operator to fulfill the demands of the job should be based upon the tasks that the job demands.

A–1.2. The employer should identify all the aspects of the job that the employee must meet/perform when doing his or her job. These aspects could include the level at which the employee must see and hear, the physical demands of the job, and the environmental extremes of the job.

A–1.3. One factor to be considered is the ability of the candidate to see and hear within reasonably acceptable limits. Included in the vision requirements are the ability to see at distance and peripherally. In certain instances, there also is a requirement for the candidate to discern different colors, primarily red, yellow and green.

A–1.4. The environmental extremes that might be demanded of a potential powered industrial truck operator include that ability of the person to work in areas of excessive cold or heat.

A–1.5. After an employee has been trained and appropriate accommodations have been made, the employer needs to determine whether the employee can safely perform the job.

A–2. The Method(s) of Training

A–2.1. Among the many methods of training are the lecture, conference, demonstration, test (written and/or oral) and the practical exercise. In most instances, a combination of these methods have been successfully used to train employees in the knowledge, skills and abilities that are essential to perform the job function that the employee is being trained to perform. To enhance the training and to make the training more understandable to the employee, employers and other trainers have used movies, slides, video tapes and other visual presentations. Making the presentation more understandable has several advantages including:

(1) The employees being trained remain more attentive during the presentation if graphical presentation are used, thereby increasing the effectiveness of the training;
(2) The use of visual presentations allows the trainer to ensure that the necessary information is covered during the training;
(3) The use of graphics makes better utilization of the training time by decreasing the need for the instructor to carry on long discussions about the instructional material; and
(4) The use of graphics during instruction provides greater retention by the trainees.

A–3. Training Program Content

A–3.1. Because each type (make and model) powered industrial truck has different operating characteristics, limitations and other unique features, an optimum employee training program for powered industrial truck operators must be based upon the type vehicles that the employee will be trained and authorized to operate. The training must also emphasize the features of the workplace which will affect the manner in which the vehicle must be operated. Finally, the training must include the general safety rules applicable to the operation of all powered industrial trucks.

A–3.2. Selection of the methods of training the operators has been left to the reasonable determination of the employer. Whereas some employees can assimilate instructional material while seated in a classroom, other employees may learn best by observing the conduct of operations (demonstration) and/or by having to personally conduct the operations (practical exercise). In some instances, an employee can receive valuable instruction through the use of electronic mediums, such as the use of video tapes and movies. In most instances, a combination of the different training methods may provide the mechanism for providing the best training in the least amount of time. OSHA has specified at paragraph (a)(2)(ii) of this section that the training must consist of a combination classroom instruction and practical exercise. The use of both these modes of instruction is the only way of assuring that the trainee has received and comprehended the instruction and can utilize the information to safely operate a powered industrial truck.

A-4. Initial Training

A-4.1. The following is an outline of a generalized forklift operator training program:

(1) Characteristics of the powered industrial truck(s) the employee will be allowed to operate:
  (a) Similarities to and differences from the automobile;
  (b) Controls and instrumentation: location, what they do and how they work;
  (c) Power plant operation and maintenance;
  (d) Steering and maneuvering;
  (e) Visibility;
  (f) Fork and/or attachment adaption, operation and limitations of their utilization;
  (g) Vehicle capacity;
  (h) Vehicle stability;
  (i) Vehicle inspection and maintenance;
  (j) Refueling or charging, recharging batteries.
  (k) Operating limitations.
  (l) Any other operating instruction, warning or precaution listed in the operator's manual for the type vehicle which the employee is being trained to operate.

(2) The operating environment:
  (a) Floor surfaces and/or ground conditions where the vehicle will be operated;
  (b) Composition of probable loads and load stability;
  (c) Load manipulation, stacking, unstacking;
  (d) Pedestrian traffic;
  (e) Narrow aisle and restricted place operation;
  (f) Operating in classified hazardous locations;
  (g) Operating the truck on ramps and other sloped surfaces which would affect the stability of the vehicle;
  (h) Other unique or potentially hazardous environmental conditions which exist or may exist in the workplace.
  (i) Operating the vehicle in closed environments and other areas where insufficient ventilation could cause a buildup of carbon monoxide or diesel exhaust.

(3) The requirements of this OSHA Standard.

A-5. Trainee Evaluation

A-5.1. The provisions of these proposed requirements specify that an employee evaluation be conducted both as part of the training and after completion of the training. The initial evaluation is useful for many reasons, including:

(1) the employer can determine what methods of instruction will produce a proficient truck operator with the minimum of time and effort;
(2) the employer can gain insight into the previous training that the trainee has received; and
(3) a determination can be made as to whether the trainee will be able to successfully operate a powered industrial truck. This initial evaluation can be completed by having the employee fill out a questionnaire, by an oral interview, or by a combination of these mechanisms. In many cases, answers received by the employee can be substantiated by contact with other employees or previous employers.

A-6. Refresher or Remedial Training

A-6.1. (The type information listed at paragraph A-6.2 of this appendix would be used when the training is more than an on-the-spot correction being made by a supervisor or when there have been multiple instances of on-the-spot corrections having to be made.) When an on-the-spot correction is used, the person making the correction should point out the incorrect manner of operation of the truck or other unsafe act being conducted, tell the employee how to do the operation correctly, and then ensure that the employee does the operation correctly.

A-6.2. The following items may be used when a more general, structured retraining program is utilized to train employees and eliminate unsafe operation of the vehicle:
(1) Common unsafe situations encountered in the workplace;
(2) Unsafe methods of operating observed or known to be used;
(3) The need for constant attentiveness to the vehicle, the workplace conditions and the manner in which the vehicle is operated.

A-6.3. Details about the above subject areas need to be expanded upon so that the operator receives all the information which is necessary for the safe operation of the vehicle. Insight into some of the specifics of the above subject areas may be obtained from the vehicle manufacturers' literature, the national consensus standards [e.g. the ANSI B56 series of standards (current revisions)] and this OSHA Standard.

Appendix B—Stability of Powered Industrial Trucks

(Non-mandatory appendix to paragraph (a) of this section)

B-1. Definitions

To understand the principle of stability, understanding definitions of the following is necessary:

Center of Gravity is that point of an object at which all of the weight of an object can be considered to be concentrated.

Counterweight is the weight that is a part of the basic structure of a truck that is used to offset the weight of a load and to maximize the resistance of the vehicle to tipping over.

Fulcrum is the axis of rotation of the truck when it tips over.

Grade is the slope of any surface that is usually measured as the number of feet of rise of fall over a hundred foot horizontal distance (this measurement is designated as a percent).

Lateral stability is the resistance of a truck to tipping over sideways.

Line of action is a imaginary vertical line through the center of gravity of an object.

Load center is the horizontal distance from the edge of the load (or the vertical face of the forks or other attachment) to the line of action through the center of gravity of the load.

Longitudinal stability is the resistance of a truck to overturning forward or rearward.

Moment is the product of the weight of the object times the distance from a fixed point. In the case of a powered industrial truck, the distance is measured from the point that the truck will tip over to the line of action of the object. The distance is always measured perpendicular to the line of action.

Track is the distance between wheels on the same axle of a vehicle.

Wheelbase is the distance between the centerline of the front and rear wheels of a vehicle.

B-2. General

B-2.1. Stability determination for a powered industrial truck is not complicated once a few basic principles are understood. There are many factors that influence vehicle stability. Vehicle wheelbase, track, height and weight distribution of the load, and the location of the counterweights of the vehicle (if the vehicle is so equipped), all contribute to the stability of the vehicle.

B-2.2. The "stability triangle", used in most discussions of stability, is not mysterious but is used to demonstrate truck stability in rather simple fashion.

B-3. Basic Principles

B-3.1. The determination of whether an object is stable is dependent on the moment of an object at one end of a system being greater than, equal to or smaller than the moment of an object at the other end of that system. This is the same principle on which a seesaw or teeter-totter works, that is, if the product of the load and distance from the fulcrum (moment) is equal to the moment at the other end of the device, the device is balanced and it will not move. However, if there is a greater moment at one end of the device, the device will try to move downward at the end with the greater moment.

B-3.2. Longitudinal stability of a counterbalanced powered industrial truck is dependent on the moment of the vehicle and the moment of the load. In other words, if the mathematic product of the load moment (the distance is from the front wheels, the point about which the vehicle would tip forward) the system is balanced and will not tip forward. However, if the load-moment is greater than the vehicle-moment, the greater load-moment will force the truck to tip forward.

B-4. The Stability Triangle

B-4.1. Almost all counterbalanced powered industrial trucks have a three-point suspension system, that is, the vehicle is supported at three points. This is true even if it has four wheels. The steer axle of most trucks is attached to the truck by means of a pivot pin in the center of the axle. This three-point support forms a triangle called the stability triangle when the points are connected with imaginary lines. Figure 1 depicts the stability triangle.

BILLING CODE 4510-26-P

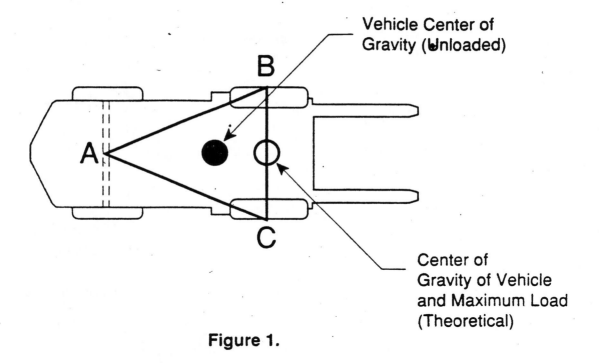

Figure 1.

NOTES:

1. When the vehicle is loaded, the combined center of gravity shifts toward line B-C. Theoretically the max load will result in the CG at the line B-C. In actual practice, the combined CG should never be at line B-C.

2. The addition of additional counterweight will cause the truck CG to shift toward point A and result in a truck that is less stable laterally.

B-4.2. When the line of action of the vehicle or load-vehicle falls within the stability triangle, the vehicle is stable and will not tip over. However, when the line of action of the vehicle or the vehicle/load combination falls outside the stability triangle, the vehicle is unstable and may tip over. (See Figure 2.)

BILLING CODE 4510-26-P

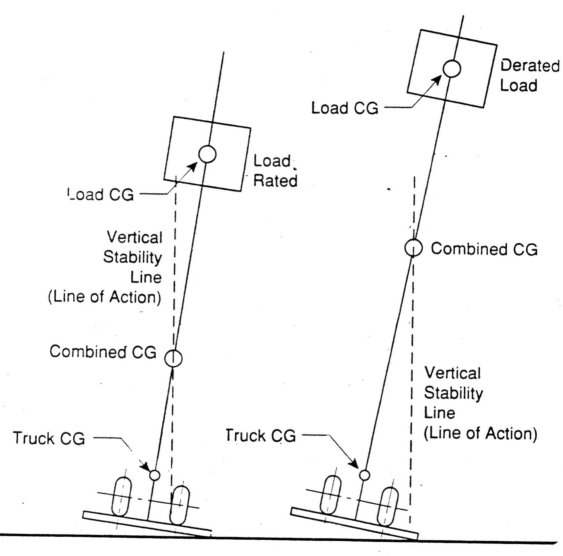

The vehicle is stable

This vehicle is unstable and will continue to tip over

Figure 2.

B-5. Longitudinal Stability

B-5.1. The axis of rotation when a truck tips forward is the point of contact of the front wheels of the vehicle with the pavement. When a powered industrial truck tips forward, it is this line that the truck will rotate about. When a truck is stable the vehicle-moment must exceed the load-moment. As long as the vehicle-moment is equal to or exceeds the load-moment, the vehicle will not tip over. On the other hand, if the load-moment slightly exceeds the vehicle-moment, the truck will begin the tip forward, thereby causing loss of steering control. If the load-moment greatly exceeds the vehicle-moment, the truck will tip forward.

B-5.2. In order to determine the maximum safe load moment, the truck manufacturer normally rates the truck at a maximum load at a given distance from the front face of the forks. The specified distance from the front face of the forks to the line of action of the load is commonly called a load center. Because larger trucks normally handle loads that are physically larger, these vehicles have greater load centers. A truck with a capacity of 30,000 pounds or less capacity is normally rated at a given load weight at a 24 inch load center. For trucks of greater than 30,000 pound capacity, the load center is normally rated at 36 or 48 inch load center distance. In order to safely operate the vehicle, the operator should always check the data plate and determine the maximum allowable weight at the rated load center.

B-5.3. Although the true load moment distance is measured from the front wheels, this distance is greater than the distance from the front face of the forks. Calculation of the maximum allowable load moment using the load center distance always provides a lower load moment than the truck was designed to handle. When handling unusual loads, such as those that are larger than 48 inches long (the center of gravity is greater than 24 inches), with an offset center of gravity, etc., then calculation of a maximum allowable load moment should be undertaken and this value used to determine whether a load can be handled. For example, if an operator is operating a 3000 pound capacity truck (with a 24 inch load center), the maximum allowable load moment is 72,000 inch-pounds (3,000 times 24). If a probable load is 60 inches long (30 inch load center), then the maximum weight that this load can weigh is 2,400 pounds (72,000 divided by 30).

B-6. Lateral Stability

B-6.1. The lateral stability of a vehicle is determined by the position of the line of action (a vertical line that passes through the combined center of gravity of the vehicle and the load) relative to the stability triangle. When the vehicle is not loaded, the location of the center of gravity of the truck is the only factor to be considered in determining the stability of the truck. As long as the line of action of the combined center of gravity of the vehicle and the load falls within the stability triangle, the truck is stable and will not tip over. However, if the line of action falls outside the stability triangle, the truck is not stable and may tip over.

B-6.2. Factors that affect the lateral stability of a vehicle include the placement of the load on the truck, the height of the load above the surface on which the vehicle is operating, and the degree of lean of the vehicle.

B-7. Dynamic Stability

B-7.1. Up to this point, we have covered stability of a powered industrial truck without consideration of the dynamic forces that result when the vehicle and load are put into motion. The transfer of weight and the resultant shift in the center of gravity due to the dynamic forces created when the machine is moving, braking, cornering, lifting, tilting, and lowering loads, etc., are important stability considerations.

B-7.2. When determining whether a load can be safely handled, the operator should exercise extra caution when handling loads that cause the vehicle to approach its maximum design characteristics. For example, if an operator must handle a maximum load, the load should be carried at the lowest position possible, the truck should be accelerated slowly and evenly, and the forks should be tilted forward cautiously. However, no precise rules can be formulated to cover all of these eventualities.

[FR Doc. 95-5826 Filed 3-13-95; 8.45 am]
BILLING CODE 4510-26-P

APPENDIX B

Quiz Questions for Forklift Operators

Powered Walkie/Pallet Trucks

True/False Questions

Check true or false as your correct answer.

		True	False
1.	Manufacturers' models of walkie trucks could have different features and operating controls.	☐	☐
2.	If a defect or problem is discovered on your vehicle during the shift, wait until the end of the shift to report it.	☐	☐
3.	Alert employees near you when you going to set a load down.	☐	☐
4.	Walkie trucks are designed to be overloaded.	☐	☐
5.	When turning on a ramp, do it very slowly.	☐	☐
6.	When traveling up a ramp, have the load on the high side.	☐	☐
7.	An empty walkie or pallet truck can tip over.	☐	☐
8.	It is important to move slowly near the edge of a dock.	☐	☐
9.	Taking short cuts is an indication of an experienced operator.	☐	☐

		True	False
10.	When entering a trailer at the dock, wheel chocks are not important because the pallet jack is light in weight.	☐	☐
11.	Pallet trucks make wide swings in aisles.	☐	☐
12.	It is a good idea to drive slowly over dock boards and bridge plates.	☐	☐
13.	When the operating handle is in the vertical or horizontal position, the brake is on.	☐	☐
14.	Report slippery floors or surfaces to your supervisor immediately.	☐	☐
15.	It is all right to move an improperly stacked pallet as long as it is moved slowly.	☐	☐
16.	Always face the direction of travel.	☐	☐
17.	When walking alongside of a walkie unit, keep your feet to the side and away from the bottom of the unit.	☐	☐
18.	Never operate the walkie truck at speeds that are faster than you walk.	☐	☐
19.	When turning with the truck behind you, allow for plenty of room for the load and forks.	☐	☐
20.	Always come to a complete stop before changing directions.	☐	☐
21.	Keep parked units clear of exits, doors, and emergency equipment.	☐	☐
22.	Always yield the right of way to pedestrians.	☐	☐
23.	Keep loads against the load back rest while being transported.	☐	☐
24.	While traveling, avoid bumps. blocks of wood or holes in the floor.	☐	☐
25.	Operators should perform an inspection of their pallet trucks every once in a while.	☐	☐

Answers to True/False Questions

Powered Walkie Trucks

1. T	10. F	19. T
2. F	11. T	20. T
3. T	12. T	21. T
4. F	13. T	22. T
5. F	14. T	23. T
6. F	15. F	24. T
7. T	16. T	25. F
8. T	17. T	
9. F	18. T	

Lift Truck Operator

True / False Questions

Check true or false as your correct answer.

		True	False
1.	If an operator has any mechanical problems with the truck, they should be reported to a supervisor immediately.	☐	☐
2.	Forks should be used for picking up loads, not for pushing, ramming, or shoving.	☐	☐
3.	Loads should be centered on the forks and back against the mast.	☐	☐
4.	Wheel chocks placed against truck trailer tires at loading docks are mandatory.	☐	☐
5.	Pedestrians always have the right of way.	☐	☐
6.	Only use your emergency brake when parking at the end of your shift.	☐	☐
7.	A driver should not be concerned about the rearend swing of the truck when turning a corner.	☐	☐
8.	Sometimes it is okay to carry loads which are rated higher than the truck's capacity.	☐	☐
9.	Propane tanks are dangerous and have as much explosive power as dynamite.	☐	☐
10.	You don't have to check out your brakes, steering, horn, or other equipment at the start of each shift; once a week is fine.	☐	☐
11.	A lift truck turns in the same manner as an automobile.	☐	☐
12.	It is not necessary to report slippery conditions in your driving area.	☐	☐
13.	Always drive up a ramp with the load in front of you.	☐	☐
14.	Back down a ramp with the load in front of you; load on the high side.	☐	☐
15.	It is okay to lift anyone on your forks to a higher elevation, just as long as they hold on to keep from falling.	☐	☐
16.	You are required to report all accidents immediately.	☐	☐
17.	It is okay to transport fellow employees as long as you don't speed.	☐	☐

		True	False
18.	Travel with forks low at all times.	☐	☐
19.	You can drive in reverse when forward visibility is blocked or limited.	☐	☐
20.	When parking the truck, leave the forks about 6-10 inches from the floor.	☐	☐
21.	Levers, pedals, and controls are the same on *all* lift trucks.	☐	☐
22.	The average cost of a lift truck is usually more than two times that of the average automobile.	☐	☐
23.	When your lift truck is empty, operate it as if it were carrying a load.	☐	☐
24.	When leaving a lift truck equipped with liquid propane fuel overnight, be sure to close the shut-off valve.	☐	☐
25.	Operators should wipe off their machines sometime during their working shift when the machines are not in operation.	☐	☐
26.	Operators should check fuel, oil, water, and any gauges at the beginning of each shift.	☐	☐
27.	Operators should not have to watch for pedestrians and workers when the lift truck is in motion.	☐	☐
28.	Operators may park on railroad tracks when no railroad cars are spotted in the area.	☐	☐
29.	Power trucks should not be left unattended with the motor running.	☐	☐
30.	Stunt driving is a good way to show that the operator is a good operator.	☐	☐
31.	Operators must immediately report any mechanical deficiency.	☐	☐
32.	It is safe to fill the gas tank of the vehicle inside a facility.	☐	☐
33.	To stop their machine, operators are permitted to use reverse rather than the brake.	☐	☐
34.	You may have people stand or add weight to the back of the truck in order to lift more with the forks.	☐	☐
35.	Bent or damaged loading plates must never be used.	☐	☐
36.	The operator should travel on the left side of an aisle whenever possible.	☐	☐
37.	It is a safe practice to permit people to ride on the forks for short periods of time when doing work at various heights.	☐	☐

	True	False
38. Operators must face in the direction in which the machine is being operated.	☐	☐
39. It is okay for an operator to ride the clutch pedal.	☐	☐
40. Operator should pump the foot accelerator while the machine is running.	☐	☐
41. Trucks should be operated at full speed at all times.	☐	☐
42. Propane tanks must be stored outdoors when not on the forklift.	☐	☐
43. It is permissible to start, turn, or stop a truck abruptly.	☐	☐
44. A good operator uses common sense.	☐	☐
45. Overhead obstacles are never a hazard to experienced operators.	☐	☐
46. When picking up a load, you can save time by backing up and turning while the load is coming down.	☐	☐
47. A good practice is to hit the dock plate at a high rate of speed to avoid bumps.	☐	☐
48. When turning in reverse, allow plenty of room for forks and loads.	☐	☐
49. Quick, snappy moves are a sign of a good operator.	☐	☐
50. When turning on a ramp, be sure that your load is secure.	☐	☐
51. A driver is responsible for ensuring that no one passes under the elevated forks.	☐	☐
52. Overloading your forklift will have no effect on the steering, because rear wheels have sufficient counterbalance weight.	☐	☐
53. When parking your truck, make sure forks are left high enough to clear obstructions.	☐	☐
54. A good driver will always make turns smoothly and slowly.	☐	☐
55. Keep the engine oil level at the full mark on the dipstick.	☐	☐
56. The best test for hydraulic sump tank oil leakage is to drop the fork carriage fast and look for spillage.	☐	☐
57. On many electric trucks, leaving the seat *automatically* brakes the vehicle.	☐	☐

	True	False

58. It is okay to elevate a person on a regular pallet if you raise the pallet slowly. ☐ ☐

59. Lift capacities of all forklifts are the same. ☐ ☐

60. A lift truck steers with its rear wheels and hence, its rearend swings wide on turns. ☐ ☐

61. It's okay to smoke when refueling as long as you're in a restricted area. ☐ ☐

62. Start in low gear with every load. ☐ ☐

63. For greater efficiency, place a loaded pallet within three to six inches of the desired spot and *carefully* nudge it into place with your forks. ☐ ☐

64. Loads, when lifted high, reduce a truck's stability. ☐ ☐

65. Lift trucks are designed to be operated in the same manner whether the loads are in the elevated or lowered position. ☐ ☐

66. Standing under, or walking under, elevated forks or an elevated load is an acceptable procedure, as long as the operator exercises proper caution. ☐ ☐

67. Traveling with the load elevated is not good practice under any circumstances. ☐ ☐

68. When operating a truck with a tilting mast, the safe capacity is the same, regardless of the position of the tilt. ☐ ☐

69. When picking up a load, the forks should be pre-positioned about 12 inches apart. ☐ ☐

70. Pushing loads that are too heavy can damage the lift truck. ☐ ☐

71. Double-tiered loads may be handled only if proper caution is exercised. ☐ ☐

72. Tractor-trailer and boxcar floor conditions affect the safe operation of lift trucks. ☐ ☐

73. Oil puddles on the floor, have no effect on the safe operation of your lift truck, as long as the tires are maintained in good operating condition. ☐ ☐

74. If you take special care, it's okay to let bystanders stand beneath loads when you're high-stacking. ☐ ☐

		True	False
75.	A careful driver follows the "rules of the road" and backs up when a bulky load obstructs forward vision.	☐	☐
76.	A careful driver may hang a leg outside a machine when it's protected by a wide load.	☐	☐
77.	A careful driver drives slowly on slippery, wet floors.	☐	☐
78.	An efficient operator checks loads for stability before moving them.	☐	☐
79.	Skid marks are generally a sign of unsafe driving.	☐	☐
80.	There should be a designated storage area for your lift truck where you pick it up at the beginning of your shift and leave it when you finish.	☐	☐
81.	Always test parking and service brakes before starting.	☐	☐
82.	Hydraulic brake fluid may be substituted for hydraulic oil.	☐	☐
83.	The forks of your lift truck can be used to jack up another truck that needs repairs.	☐	☐
84.	It is normal for lift trucks to use large amounts of hydraulic oil.	☐	☐
85.	A load should always be carried high so the driver can see under it.	☐	☐
86.	The forks are to be used only for handling freight and never to be used as an elevated platform to lift workers when replacing burned-out bulbs or removing trailer tarps.	☐	☐
87.	A good operator drives and maneuvers carefully in time to anticipate traffic hazards and avoid collision or panic stops.	☐	☐
88.	In tight quarters, it is often necessary to use fork blades to push or position loads in trailers.	☐	☐
89.	When working inside a truck or boxcar, it is important to watch overhead clearance, particularly when lifting.	☐	☐
90.	It is the responsibility of pedestrians and workers to be in their own traffic lanes, and out of the way of your truck.	☐	☐
91.	When driving in and out of trailers or boxcars with loads, it's a good rule to operate your sit-down lift truck in low gear.	☐	☐
92.	Parking on an incline without putting chocks under wheels is allowed.	☐	☐

		True	False
93.	It is permissible to tow railroad cars or large trucks.	☐	☐
94.	Loads that stick out past the truck need a flag.	☐	☐
95.	The freight car brakes should be set before driving in.	☐	☐
96.	Steering wheel knobs may swing and break fingers.	☐	☐
97.	Bent or damaged loading plates must never be used.	☐	☐
98.	Operators must face in the direction in which their machines are moving.	☐	☐
99.	It is not permissible for operators to let their foot ride the clutch pedal or inching pedal.	☐	☐
100.	When you are lifting a load by hand off the pallet on your forks, the forks should be elevated about two feet off the floor to make lifting easier.	☐	☐
101.	When a crane or hoist is lifting a heavy load off the pallet on your forks, it is important that the forks are lowered to the floor.	☐	☐
102.	Driving in or out of a building into the bright sunlight can be dangerous.	☐	☐
103.	It saves time to position the height of a load while traveling to the stacking area.	☐	☐
104.	Do not place heavier freight on top of lighter freight or it will collapse under the load.	☐	☐
105.	Tilt the mast back to cradle drums or barrels and keep them from rolling off the forks.	☐	☐
106.	Tilt the mast forward to lift or lower long containers off and on low dollies.	☐	☐
107.	A good operator makes certain loads will ride safely before moving them.	☐	☐
108.	It is not necessary to slow down at a blind corner if the driver has a good horn, and uses it for 20 feet before entering the intersection.	☐	☐
109.	It's a good idea to use reverse gear as a brake, to save wear on the brake bands.	☐	☐
110.	A sit-down lift truck turns the same way an automobile does.	☐	☐

		True	False
111.	It is permissible to carry an extra operator or passenger on your truck, if you have your supervisor's permission.	☐	☐
112.	It is all right to move an improperly made-up load if it doesn't extend above the carriage of the lift mechanism.	☐	☐
113.	It is permissible to bump a load into position on a stack, if the stack is not over two feet high.	☐	☐
114.	The same weight can be lifted with the end of the forks as when the forks are run all the way under the load.	☐	☐
115.	It is all right to start your sit-down truck lift in high gear when carrying a load up a grade.	☐	☐
116.	If the bridge plate into a rail car is not secure, a fast-moving truck can spin the plate right out from under its wheels.	☐	☐
117.	You don't have to worry about floor strengths, because most floors are strong enough to support a lift truck and its load.	☐	☐
118.	It is mandatory that overhead guards be used on all lift trucks whenever freight will be stacked higher than ten feet.	☐	☐
119.	If your lift truck is making a "funny" noise, the correct procedure is to tell your supervisor at the end of the shift.	☐	☐
120.	The faster you drive, the more work can be completed in your workday.	☐	☐
121.	A safe speed in a plant or warehouse is about 35 miles per hour.	☐	☐
122.	Wheel chocks are only necessary if you feel that the trailer may be a dangerous one to enter.	☐	☐
123.	The stacking of empty pallets in storage must never exceed six feet in height.	☐	☐
124.	The "no smoking" rule would apply to propane fuel use and storage and not the battery recharge areas.	☐	☐
125.	Batteries produce hydrogen gas which can explode if ignited by a spark, cigarette, or other source.	☐	☐
126.	Driving in reverse is not allowed.	☐	☐

		True	False

127. If a propane tank runs out of fuel, shut off your engine, grasp the metal coupling on the fuel lines with your bare hands, and turn counterclockwise to release. ☐ ☐

128. Driving too fast is the number one cause of forklift accidents. ☐ ☐

129. A forklift rated at two thousand pounds can actually lift three thousand pounds. ☐ ☐

130. When you're approaching a blind intersection, sound your horn loudly and continue on through. ☐ ☐

131. The load center is different on each forklift and on each load being lifted. ☐ ☐

132. When approaching an aisle that does not require you to stop, you must still be prepared to stop. ☐ ☐

133. Overturning a forklift can still take place with the load down and the forklift going less than full speed. ☐ ☐

134. When parking your lift truck, place forks flat on the floor, put on the emergency brake, and shut off the engine. ☐ ☐

135. Always cut off the engine when changing LP cylinders on a lift truck. ☐ ☐

136. The best method of determining the amount of gas in a cylinder is by using a scale. ☐ ☐

137. When installing an LP cylinder on a forklift, keep the relief valve vent upward and the dowel pin the hole in the collar to prevent the tank from rotating. ☐ ☐

138. To turn on the LP gas feed line, turn clockwise to the stop position, then make a counterclockwise quarter-turn. ☐ ☐

139. LP gas can cause frostbite on the hands when connecting the fuel line to the cylinder. ☐ ☐

140. Watch out for escaped propane, which accumulates in confined areas. It is extremely dangerous. ☐ ☐

141. LP in liquid form can give second degree burns. ☐ ☐

142. When the LP cylinder is installed on the lift truck, insert the relief valve in a face-down position. ☐ ☐

		True	False
143.	Sparks caused by throwing LP cylinders on concrete or against steel cannot ignite the LP fuel.	☐	☐
144.	Liquid propane cylinder valves should be closed when not in use.	☐	☐
145.	When transporting truck lifts to have LP cylinders installed, it is not necessary to close the feeder valve.	☐	☐
146.	When refueling an LP cylinder mounted on a lift truck, the truck lift should have the engine cut off and the lift truck should also be grounded.	☐	☐
147.	It is permissible to use an LP cylinder if the foot ring is missing.	☐	☐
148.	Heat from a blower-type heater that is more than three feet away will not affect LP gas.	☐	☐
149.	Raw LP fuel and odor contaminants cannot harm your respiratory system.	☐	☐
150.	It is a good idea to wear gloves when handling propane cylinders.	☐	☐
151.	Lifting propane cylinders can injure your back.	☐	☐
152.	If any of the fuel lines show wear, contact maintenance.	☐	☐
153.	When filling a LP cylinder, it is unsafe to depend solely upon the pressure relief valve for an indication the cylinder is full.	☐	☐
154.	It is better to allow an LP fire to continue burning when the fire is controlled and you have no source of turning off the fuel supply.	☐	☐
155.	The dry chemical fire extinguisher is not effective against fighting LP fires.	☐	☐
156.	Trained professionals should fight an LP fire.	☐	☐
157.	As long as you sit beside them, it is okay to let unauthorized people operate your lift truck.	☐	☐
158.	An accident usually results from or includes an emotional upset influencing the judgment of the operator.	☐	☐
159.	You should always back down a ramp whether you are carrying a load or are empty.	☐	☐
160.	It's a good idea to always rely on others to see the hazards or danger for you.	☐	☐

		True	False
161.	When you raise the forks and the load you are carrying, the center of gravity changes on the forklift.	☐	☐
162.	Every forklift has a built-in margin of safety. Lifting too much weight, for example, may be possible for the forklift, but the steering becomes ineffective during an overload.	☐	☐
163.	When crossing railroad tracks, you should always drive slowly and cross at an angle even if your truck is not loaded.	☐	☐
164.	When leaving or entering a building, you must stop at the building entrance or exit before proceeding.	☐	☐
165.	Even with perfect steering, no load, and a good driving surface, a forklift can still be dangerous.	☐	☐
166.	Both clamps must be used to hold down a propane tank on a forklift.	☐	☐
167.	It is important to check dock plates before loading, and during the loading and unloading operations of a trailer or boxcar.	☐	☐
168.	Extending your feet, legs, or arms outside the confines of the truck, or placing them between the uprights of the mast, is considered an acceptable operating procedure as long as you are careful.	☐	☐
169.	Quick starting and stopping are not considered acceptable operating procedures.	☐	☐
170.	While it is an accepted procedure to come to a full stop before reversing direction of travel when carrying a load, this practice can be ignored when the forks are empty.	☐	☐
171.	It is important to keep fire aisles, access to stairways, and fire equipment areas clear.	☐	☐
172.	Cross aisles, doorways, and corners are considered extremely hazardous in the operation of lift trucks.	☐	☐
173.	Materials can be placed within three feet of electrical panels.	☐	☐

	True	False
174. It is permissible to use a match to check the fluid level in a battery.	☐	☐
175. Floor capacity of a trailer is the supervisor's responsibility, not the operator's.	☐	☐

Lift Truck Operator
True/False Quiz Answers

1.	T	39.	F	77.	T	115.	F
2.	T	40.	F	78.	T	116.	T
3.	T	41.	F	79.	T	117.	F
4.	T	42.	T	80.	T	118.	T
5.	T	43.	F	81.	T	119.	F
6.	F	44.	T	82.	F	120.	F
7.	F	45.	F	83.	F	121.	F
8.	F	46.	F	84.	T	122.	F
9.	T	47.	F	85.	F	123.	T
10.	F	48.	T	86.	T	124.	F
11.	F	49.	F	87.	T	125.	T
12.	F	50.	F	88.	T	126.	F
13.	T	51.	T	89.	T	127.	F
14.	T	52.	F	90.	T	128.	T
15.	F	53.	F	91.	T	129.	F
16.	T	54.	T	92.	F	130.	F
17.	F	55.	T	93.	F	131.	T
18.	T	56.	F	94.	T	132.	T
19.	T	57.	T	95.	T	133.	T
20.	F	58.	F	96.	T	134.	T
21.	F	59.	F	97.	T	135.	T
22.	T	60.	T	98.	T	136.	T
23.	T	61.	F	99.	T	137.	T
24.	T	62.	T	100.	T	138.	T
25.	T	63.	F	101.	T	139.	T
26.	T	64.	T	102.	T	140.	T
27.	F	65.	F	103.	F	141.	T
28.	F	66.	F	104.	T	142.	T
29.	T	67.	T	105.	T	143.	F
30.	F	68.	F	106.	T	144.	T
31.	T	69.	F	107.	T	145.	F
32.	F	70.	T	108.	F	146.	T
33.	F	71.	T	109.	F	147.	F
34.	F	72.	T	110.	F	148.	F
35.	T	73.	F	111.	F	149.	F
36.	F	74.	F	112.	F	150.	T
37.	F	75.	T	113.	F	151.	T
38.	T	76.	F	114.	F	152.	T

153.	F	159.	T	165.	T	171.	T
154.	T	160.	F	166.	T	172.	T
155.	F	161.	T	167.	T	173.	F
156.	T	162.	T	168.	F	174.	F
157.	F	163.	T	169.	T	175.	F
158.	T	164.	T	170.	F		

Lift Truck Operator

Multiple Choice Questions

Check true or false as your correct answer.

1. When crossing a railroad, you should:

    a. Proceed slowly and cross at an angle.
    b. Proceed slowly and go straight across.
    c. Continue at traveling speed, railroad crossings will not affect the lift truck.

2. When transporting a load with the lift truck, keep the load:

    a. As high as possible to keep from hitting anyone.
    b. As low as possible so you can see where you're going.
    c. Only about an inch high for good floor clearance.

3. When transporting a load, the operator should:

    a. Look in the direction of travel.
    b. Watch the load.
    c. Have someone walk along in front of the load and tell the operator which direction to go.

4. If a load is unstable, you should:

    a. Pick it up and transport it very carefully.
    b. Have someone stand on the forks and hold the load.
    c. Have it stacked correctly.

5. It is all right to carry a load that exceeds the limit of the lift truck:

    a. When it is transported slowly and carefully.
    b. Anytime, because the lift trucks are underrated for safety.
    c. Never.

6. While operating a lift truck, you should watch for overhead obstructions:

    a. When you're transporting a load.
    b. All the time.
    c. When you're unloading.

308 / Forklift Safety

7. The drive wheels of a lift truck:

   a. Get better traction when the lift truck is empty.
   b. Get better traction when it is loaded.
   c. Get the same amount of traction whether the lift load is loaded or empty.

8. When operating a lift truck on a wet surface, you should:

   a. Slow down, because the tires get less traction.
   b. Proceed as normal, because a wet surface doesn't affect traction.
   c. Speed up, in order to get through the wet spot or area quicker.

9. When loading a truck trailer, you should:

   a. Check the trailer tires to make sure they are blocked.
   b. Check to make sure the tractor is attached.
   c. Load the left side of the trailer first.

10. A typical forklift costs about:

    a. $9,000.
    b. $15,000.
    c. $25,000.

11. When you leave your lift truck, you should:

    a. Set the brake so the lift truck won't roll.
    b. Set the brake and turn the key switch off.
    c. Set the brake, turn the key switch off, remove the key, and lower the forks to the floor.

12. When going up an incline:

    a. You should have the load leading.
    b. You should have the load trailing.
    c. It doesn't make any difference which way you have the load.

13. When going down an incline:

    a. You should have the load leading.
    b. You should have the load trailing.
    c. It doesn't make any difference which way you have the load.

14. Before loading or unloading a boxcar or trailer, you should:

    a. Check the floor of the boxcar or trailer to be sure that it is safe.
    b. Check the sides and top of the boxcar or trailer to be sure that they are solid.
    c. Neither of the above.

15. When changing your direction of travel:

    a. Slow down before shifting.
    b. Speed doesn't affect you when changing directions.
    c. Come to a complete stop before shifting.

16. The cost of a typical used wood pallet is:

    a. About $5 each.
    b. About $20 each.
    c. About $35 each.

17. Who has the right of way?

    a. A pedestrian.
    b. A lift truck.
    c. The one on the right.

18. When following another lift truck, you should be:

    a. As close as possible.
    b. About 4 lift truck lengths behind.
    c. About 10 lift truck lengths behind.

19. If you meet another lift truck approaching from the opposite direction, you should:

    a. Pass it on your right.
    b. Pass it on your left.
    c. Wait and see on which side it will pass you.

20. The capacity of your lift truck is:

    a. How much weight the lift truck can safety lift at the far ends of the forks, away from the carriage.
    b. Determined by the load center.
    c. How much weight the lift truck can lift at the close ends of the forks, next to the carriage.

310 / Forklift Safety

21. An operator is normally responsible for:

   a. Performing maintenance checks on the lift truck.
   b. Performing all of the maintenance on the lift truck.
   c. Only operating the lift truck.

22. When leaving a building:

   a. You should slow down.
   b. Stop, and then drive out.
   c. Stop, sound horn, and then drive out.

23. When parking on an incline, you should:

   a. Set the brake.
   b. Block the wheels.
   c. Set the brake and block the wheels.

24. When working with loads on pallets, the pallet should be:

   a. Against the carriage.
   b. About 6 inches from the carriage.
   c. About 12 inches from the carriage.

25. Who is responsible for checking floors for safety where a lift truck is to be used?

   a. The lift truck operator.
   b. The supervisor of the area.
   c. The Maintenance Department.

26. When crossing a railroad, or similar obstacle:

   a. Hold the steering wheel or handle securely.
   b. Continue as normal.
   c. Release the steering wheel or handle.

27. If any mechanical problems occur, the lift truck operator should:

   a. Correct the problem personally.
   b. Report the problem to a supervisor.
   c. Ignore the problem; leave it for the next shift.

28. When lifting a load, be sure:

    a. That the load is centered on the forks or platform.
    b. The weight of the load is centered on the forks or platform.
    c. The load is parallel to the mast.

29. To move a load that weighs about 300 pounds more than the capacity of the lift truck:

    a. Push it.
    b. If possible, divide the load and make two trips.
    c. Have a couple of workers stand on the back of the lift truck.

30. Gaseous areas where paint vapors, gasoline vapors, or others are present:

    a. Must be avoided.
    b. Can be driven through about every other trip.
    c. Do not present any danger to the lift truck operator.

31. Speed has:

    a. No effect on the operator's control of the lift truck.
    b. Little effect on the operator's control of the lift truck.
    c. Great effect on the operator's control of the lift truck.

32. When a lift truck is not carrying a load, you should:

    a. Operate the lift truck as though it was loaded.
    b. Speed up, because the lift truck has a low center of gravity and maneuvers better.
    c. Carry plant workers to increase productivity.

33. The speed at which a lift truck should be operated is:

    a. Determined by the operator.
    b. Determined by the load and its condition.
    c. Determined by signs posted along aisles and passageways.

34. It is okay to pull boxcars with a lift truck:

    a. Any time.
    b. Occasionally.
    c. Never.

35. When maneuvering a lift truck in a confined space, you should:

    a. Raise the load for better visibility.
    b. Be especially careful of how far the rear of the lift truck swings.
    c. Raise the load and then tilt it back for better visibility.

36. When parking the lift truck:

    a. Raise the forks as high as possible.
    b. Lower the forks as low as possible with the tips down.
    c. Lower them about 2 inches above the floor.

37. It is okay to place material in the aisle ways:

    a. Any time.
    b. Never.
    c. Makes no difference.

38. Loads should be raised:

    a. While the mast is tilted forward.
    b. In a quick jerking motion to prevent strain on the lift mechanism.
    c. While the mast is in a vertical position.

39. When lowering the load, a quick stop could damage:

    a. The load.
    b. The hydraulic system.
    c. The load and the hydraulic system.

40. When stacking palletized loads, the pallets:

    a. Should be leaned toward the aisle for easy loading.
    b. Should be staggered to make a more stable stack.
    c. Should be stacked as straight as possible.

41. The condition of a pallet:

    a. Affects load stability.
    b. Affects load center.
    c. Neither of the above.

42. Stopping a loaded lift truck abruptly could:

    a. Cause the load to pull.
    b. Damage the lift mechanism.
    c. Bend the forks.

43. When operating a lift truck in an aisle on which doors open from both sides, you should:

    a. Drive the lift truck on the left side of the aisle.
    b. Drive the lift truck in the middle of the aisle.
    c. Drive the lift truck on the right side of the aisle.

44. You can use your lift truck as a jack:

    a. Whenever necessary.
    b. Occasionally.
    c. Never.

45. When explaining a mechanical malfunction to mechanics:

    a. Have them ride with you for a short time during operation.
    b. You ride with them while they drive.
    c. Let them drive the lift truck alone.

46. When approaching a blind corner, you should:

    a. Stop and then proceed.
    b. Slow down and sound the horn.
    c. Stop, sound the horn, and then proceed.

47. The gases from a battery are:

    a. Dangerous, and could explode if a match or cigarette is too close.
    b. Dangerous from certain types of batteries.
    c. Harmless.

48. If a wet spot is noticed where lift trucks are being used, you should:

    a. Report it to your supervisor, and steer around it.
    b. Report it to your relief on the next shift.
    c. Steer clear of it.

49. Corrosion on the battery posts and cable terminals:

    a. Could cause loss of electrical power.
    b. Indicates a strong battery.
    c. Indicates a weak battery.

50. While proceeding up an incline, if the way is blocked and you must take another route, you should:

    a. Turn around.
    b. Stop and then back down.
    c. Stop and then turn around.

51. Most forklift accidents take place because of:

    a. The manner or way in which the forklift is operated.
    b. Bad brakes.
    c. Wet and oily floor surfaces.
    d. Safety rules which are not known by the operator.

52. A skilled, safe operator:

    a. Drives at the speed limit or just a little above it.
    b. Knows what a stop sign represents.
    c. Looks for workplace hazards and prepares for trouble in advance.

53. One of the reasons a forklift can tip over on its side is due to:

    a. Too much speed.
    b. Going too slow up a ramp.
    c. Operator applying the brakes too often.
    d. Operator using too much speed and carrying a load too high.

54. Unmarked hazards or dangers can be made safe by:

    a. Discussing them with the safety committee.
    b. Being aware of them and preparing for emergencies.
    c. Making a hazard more obvious by identifying it.
    d. All of the above.

55. It is recommended that you make a safety check on your forklift:

    a. After lunch.
    b. At the end of the shift.
    c. At the beginning of the shift.
    d. Never; maintenance is responsible for checking it out.

56. Which is the most important item for dock safety:

    a. A secure dock plate.
    b. A non-slippery deck.
    c. Wheel chocks.
    d. All of the above.

57. When parking the forklift:

    a. Lower forks to the floor.
    b. Park in an out-of-the-way spot.
    c. Park near a door or entrance.
    d. Park on an incline, but be sure to chock your wheels.

58. The best time to recognize the apparent workplace hazards:

    a. Is after an accident.
    b. Is before an accident.
    c. Is never; this is a job for the safety committee.

59. When transporting baskets or tubs/containers:

    a. Carry three at a time.
    b. Carry four at a time, but travel in reverse.
    c. Carry two at a time.
    d. Consider the skill of the operator in regard to how many can be carried.

60. When a repair is required on the lift truck, the operator should report the problem *directly* to:

    a. The supervisor.
    b. The maintenance department.
    c. The personnel department.

61. Forks used for picking up loads can also be used for:

   a. Pushing or shoving.
   b. Ramming.
   c. Neither a nor b.

62. When checking your battery for water level, or gas tank for fuel consumption and leakage, the prescribed source of light comes from a:

   a. Match.
   b. Flashlight.
   c. Cigarette lighter.

63. The *acceptable* number of passengers who can be safely transported on a lift truck, in addition to the operator are:

   a. A maximum of one.
   b. A maximum of two.
   c. No passengers at all.

64. The proper direction of travel while transporting a load down a ramp during a lift truck operation is:

   a. To put the load first, with the mast tilted back.
   b. To back down, in low gear.
   c. Either a or b, depending on the size of the load.

65. When transporting a load that is obstructing the operator's vision, the correct safety procedure is to:

   a. Drive backwards, and look in the direction of travel.
   b. Place your head beyond the side limits of the truck and attempt to look around the load.
   c. Raise the load high enough to permit unobstructed vision underneath.

66. The one person who is primarily responsible for knowing the location of crane hooks, chainfalls, sprinkler heads, hose reels, or any other construction which may cause overhead clearance problems *during* lift truck operations. This person is the:

   a. Superintendent.
   b. Lift truck operator.
   c. Foreman.

67. Lift truck accidents are prevented when operators:

   a. Check out their equipment before operating.
   b. Observe safety rules and regulations.
   c. Both a and b.

68. Report all accidents regarding persons and property *directly* to:

   a. The nurse on duty.
   b. Your immediate supervisor.
   c. The personnel department.

69. Rough, wet, and slippery floors are hazardous to:

   a. Pedestrians.
   b. Fork lift operation.
   c. Both a and b.

70. Professional operators look at "stunt driving" and horseplay as:

   a. A true test of courage and driving ability.
   b. Acts that are hazardous and unsafe.
   c. Good clean fun.

71. The primary responsibility for pedestrian safety during the operation of lift trucks rests with the:

   a. Pedestrian.
   b. Safety supervisor.
   c. Lift truck operator.

72. The characterization most representative of a safety-minded, rule-observing lift truck operator is the:

   a. "Cowboy."
   b. "Know it all."
   c. "Professional."

73. The *fact* that most lift trucks have rear wheel steering is:

   a. A vital safety factor.
   b. A provision for easier steering and better mobility.
   c. Neither a nor b.

74. When traveling, the maximum number of inches which the forks or loads should usually be raised off of the floor is:

   a. 2 to 4.
   b. 6 to 8.
   c. 12.

75. When operating your truck, you find the power steering system squeals as soon as you turn it. You then:

   a. Add fluid to the reservoir.
   b. Notify your supervisor.
   c. Ignore it until the end of the shift.

76. Your engine hour meter should be checked:

   a. Daily.
   b. Weekly.
   c. Every month, to be sure it's running.

77. In operating your lift truck, it is *most* important to:

   a. Listen for unusual noises.
   b. Keep it clean.
   c. Shut off the engine while you're waiting for work.

78. When changing directions:

   a. Always stop.
   b. Slow down to about one mile per hour.
   c. Don't worry about speed; a lift truck is built to take the shock.

79. Never exceed the rated load capacity of your truck:

   a. Unless you add additional counterweighting.
   b. Unless your lift is under 84 inches high.
   c. Never.

80. The *greatest* cause of accidents among *new* lift truck drivers is:

   a. Driving on the wrong side of the aisle.
   b. Daydreaming.
   c. Forgetting to watch overhead obstructions when lifting.

81. When taking a load down a ramp:

    a. Drive backwards.
    b. Drive forward.
    c. Reduce speed by zigzagging.

82. If you cannot start your truck immediately, you should:

    a. Fix it yourself.
    b. Call the mechanic.
    c. Report the trouble to your supervisor.

83. If mechanical deficiency develops, you should:

    a. Fix it yourself.
    b. Keep on driving it.
    c. Report the trouble to your supervisor.

84. Forks on an empty parked truck must always be:

    a. Two inches from the floor.
    b. Four inches from the floor.
    c. Resting on the floor.

85. Compressed air or gas cylinders:

    a. Can be laid on the forks for transporting.
    b. Can be lifted by the forks if kept low.
    c. Must be transported upright with caps in place in a cage.

86. Operators should drive their trucks:

    a. Two truck lengths behind other vehicles.
    b. Four truck lengths behind other vehicles.
    c. As fast as they wish.

87. When entering or leaving a building, you should:

    a. Stop and drive out.
    b. Drive right out.
    c. Stop, sound horn, and go out slowly.
    d. Slow down.

88. If a load hangs up, you should:

   a. Reach through the mast uprights and loosen it.
   b. Get off the machine and free it.
   c. Jerk it loose by driving the lift truck back and forth.

89. Forks on a moving, empty, or loaded truck must always be:

   a. One inch from the floor.
   b. Four inches from the floor.
   c. About a foot high to miss any floor obstruction.

90. Suppose you are approaching a blind intersection. You can't hear a truck coming the other way and you're in a hurry to get your load to its destination. What should you do?

   a. Since you don't hear anything approaching, accelerate slightly and race across before anything comes.
   b. Slow down, sound horn, and proceed with caution.
   c. Keep going. Since your truck is loaded, you have the right-of-way over pedestrians and any unloaded vehicles.

91. The horn on your forklift:

   a. Can only be used outdoors.
   b. Should be tested only at the beginning of the shift.
   c. Must always be functional.

92. When using an elevator:

   a. Drive quickly but carefully through doors before elevator is called to another floor.
   b. With all types of lift trucks, especially hand-powered trucks, drive forward, perpendicular to the ledge.
   c. Align truck at a proper angle five feet from car entrance; drive slowly into car.

93. Suppose you are driving your lift truck on a loading dock. You suddenly realize you are getting too close to the edge. What should you do?

   a. Turn away forward.
   b. Stop, get off, and study the situation.
   c. Back as close to the edge as possible, then pull forward away from the aisle.

94. Battery condition is indicated by the:

    a. Ammeter gauge.
    b. Warning light.
    c. Voltmeter.
    d. None of the above.

95. If you *have* to park on a ramp:

    a. Leave the machine in gear.
    b. Set the brakes and block the wheels.
    c. Set the load down, and rest the machine against it.

96. As a driver, it is:

    a. Your responsibility to watch for pedestrians.
    b. Their responsibility to watch for you.
    c. Management's responsibility to keep them out of your work area.

97. On a lift truck, your horn:

    a. Makes a good device to catch the foreman's attention.
    b. Should be sounded at intersections.
    c. Should be sounded when you are racing another lift truck.

98. When refueling, the truck motor must be:

    a. Idling.
    b. Shut off.
    c. Inspected.

99. What is peculiar about lift truck steering?

    a. It does not allow sharp turns.
    b. It steers with rear wheels.
    c. It steers with front wheels.

100. Who makes the best fork lift operator?

    a. An employee seeking a job that won't require his best service.
    b. A temperamental employee.
    c. A mature employee with a good driving record.

Lift Truck Operator
Multiple Choice Quiz Answers

1. a	26. a	51. a	76. a
2. b	27. b	52. c	77. a
3. a	28. b	53. d	78. a
4. c	29. b	54. d	79. c
5. c	30. a	55. c	80. c
6. b	31. c	56. c	81. a
7. b	32. a	57. a	82. c
8. a	33. c	58. b	83. c
9. a	34. c	59. c	84. c
10. c	35. b	60. a	85. c
11. c	36. b	61. a	86. b
12. a	37. b	62. b	87. c
13. b	38. c	63. c	88. b
14. a	39. c	64. b	89. b
15. c	40. c	65. a	90. b
16. a	41. a	66. b	91. c
17. a	42. a	67. c	92. b
18. b	43. b	68. b	93. a
19. b	44. c	69. c	94. c
20. b	45. c	70. b	95. b
21. a	46. c	71. a	96. a
22. c	47. a	72. c	97. b
23. c	48. a	73. b	98. b
24. a	49. a	74. a	99. b
25. a	50. b	75. b	100. c

APPENDIX C

Sources of Help for Forklift and Powered Truck Training

Equipment Manufacturers and Programs

Clark Material Handling Company
   749 W. Short Street
   Lexington, KY 40508
   (606) 288-1200

Crown Equipment
   40 South Washington Street
   New Bremen, OH 45869
   (419) 629-2311

Hyster Company
   P.O. Box 847
   Danville, IL 61834
   (217) 446-4888

Kalmar AC, Inc.
   555 Metro Place North, Suite 250
   Dublin, OH 43017
   (614) 798-3600

Komatsu Forklift Inc.
   Komatsu Canada Ltd.
   1725 Sismet Road
   Mississauga, Ontario L4W 1P9
   Canada
   (905) 625-6292

Komatsu Forklift USA, Inc.
5595 Fresca Drive
LaPalma, CA 90623
(714) 228-3877

Linde/Baker Material Handling Company
P.O. Box 2400
Summerville, SC 29483
(803) 875-8000

Mitsubishi Caterpillar Forklift America, Inc.
Caterpillar Lift Truck Group
2011 West Sam Houston Parkway, N.
Houston, TX 77043
(713) 365-1818

Nissan
  Canada
    Nissan Canada Inc., Forklift Division
    5290 Orbitor Drive
    P.O. Box 1709, Station B
    Mississauga, Ontario L4Y 4H6
    Canada
    (905) 629-2888

  Mexico
    Nissan Mexico (MARSA)
    Av. Ceylan No. 5
    Col. la Joya Ixtacala
    54160 Tlalnepantla
    Estado, de Mexico
    011 525 388 1515

  USA
    Nissan Forklift Corp., North America
    240 North Prospect Street
    Marengo, IL 60152
    (815) 568-0061

The Raymond Corporation
P.O. Box 130
South Canal Street
Greene, NY 13778-0130
(607) 656-2311

Toyota
 Canada
  Toyota Canada Inc.
  Industrial Equipment Division
  One Toyota Place
  Scarborough, Ontario M1H 1H9
  Canada
  (416) 438-6320

 Mexico
  Toyota Tsusho
  Paseo de la Reforma 403-703
  Mexico, D.F.C.P. 06500
  011 525 208 51 98

 USA
  Toyota Industrial Equipment/TMS
  1900 South Western Avenue
  Torrance, CA 90509
  (310) 618-8600

Yale Materials Handling Corporation
 15 Junction Road
 Flemington, NJ 08822-9499
 (908) 788-3100

(This list is reprinted from the Industrial Truck Association's handbook)

Forklift Training Programs: Audiovisuals, literature, manuals, courses, computer programs

BNA Communications Inc.
9439 Key West Ave
Rockville, MD 20850
(800) 217-2338

Broner Glove & Safety Co.
1750 Hammon Road
Auburn Hills, MI 48326
(810) 391-5000

Brown & Root Environmental
4100 Clinton Drive
Houston, TX 77020
(713) 676-4993

Coastal Video Communications
3083 Brickhouse Court
Virginia Beach, VA 23452
(800) 767-7703

Comprehensive Loss Management
15800 32nd Ave N, Suite 106
Minneapolis, MN 55447
(612) 551-1022

Conney Safety Products
3202 Latham Drive
Madison, WI 53713
(800) 356-9100

Direct Safety Company
7815 S. 46th Street
Phoenix, AZ 85044
(602) 968-6241

DuPont Co. Safety and Environmental Management Services
1007 North Market Street
Wilmington, DE 19898
(800) 532-SAFE

ERM Group
   855 Springdale Drive
   Exton, PA 19341
   (800) 544-3117

FLI Learning Systems, Inc.
   P. O. Box 2233
   Princeton, NJ 08543
   (609) 466-9000

Government Institutes, Inc.
   4 Research Place
   Rockville, MD 20850
   (301) 921-2345

IDESCO Corp.
   37 W. 26th St.
   New York, NY 10010
   (800) 336-1383

Industrial Accident Prevention Association
   250 Yonge Street 28th Foor
   Toronto, Ontario
   Canada M5E 2N4
   (416) 506-8888

Industrial Training, Inc.
   5989 Tahoe Drive SE
   Grand Rapids, MI 49546
   (800) 253-4623

Injury Prevention Technology
   2732 Woodstock Road
   Los Alamitos, CA 90720
   (310) 430-5646

Interactive Media Communications
   204 Second Avenue
   Waltham, MA 02154
   (617) 890-7707

J.J.Keller and Associates, Inc.
  3003 W. Bridgewood Lane
  P. O. Box 368
  Neenah, WI 54956-0368
  (800) 843-3174

Labelmaster
  5724 North Pulaski Road
  Chicago, IL 60646
  (800) 621-5808

Marcom Group, Ltd.
  20 Creek Parkway
  Boothwyn, PA 19061
  (800) 654-2448

Marshall Productions
  529 South Clinton Avenue
  Trenton, NJ 08611
  (800) 619-1901

MPC Promotions
  2026 Shepherdsville Road
  Louisville, KY 40218
  (800) 331-0989

Multi-Media Access (M-Max), Inc.
  Barrington, IL
  (800) 742-6629

National Safety Council
  1121 Spring Lake Drive
  Itasca, IL 60143-3201
  (630) 285-1121

NJ & Associates, Inc.
  Two Bent Tree Tower
  16479 Dallas Parkway Suite 700
  Dallas, TX 75248
  (800) 622-0177

OmniTrain/Safety Shorts
   2960 North 23rd St.
   LaPorte, TX 77571
   (713) 470-9999

Ray Jewell Productions
   139 Loma Media Road
   Santa Barbara, CA 93103
   (805) 568-1184

Seton Name Plate Company
   20 Thompson Road
   Branford, CT 06405
   (203) 488-8059

Shepherd's Industrial Training Systems
   P. O. Box 341033
   Bartlett, TN 38184
   (901) 382-5507

Vallen Safety Supply Company
   13333 Northwest Freeway
   Houston, TX 77040
   (800) 372-3389

Magazines

Material Handling Product News
   Gordon Publications
   301 Gilbralter Drive, Box 650
   Morris Plains, NJ 07950-0650
   (201) 292-5100

Material Handling Engineering
   Penton Publishing Compnay
   1100 Superior Avenue
   Cleveland, OH 44114-2543

Modern Materials Handling
   Cahners Publishing Company
   P. O. Box 7563
   Highlands Ranch, CO 80163-7563

Other

American National Standards Institute
  11 W. 42nd Street, 13th floor
  New York, NY 10036
  (212) 642-4900

American Society of Safety Engineers
  1800 East Oakton
  Des Plaines, IL 60018-2187
  (847) 699-2929

Canadian Standards Association
  178 Rexdale Blvd.
  Rexdale, Toronto
  Ontario, Canada M9W 1R3
  (B 335-94 Industrial Lift Truck Operator Training)

Industrial Truck Association
  1750 K. Street NW, Suite 460
  Washington, DC 20006
  (202) 296-9880

Kenhar Products, Inc.
  P.O. 1508
  Guelph, Ontario
  Canada N1H 6N9
  (Fork Inspections)

National Fire Protection Association
  1 Batterymarch Park
  Quincy, MA 02269-9101
  (800) 344-3555

APPENDIX D

OSHA 1910.178 Guidelines for Powered Industrial Trucks

General requirements. This section contains safety requirements relating to fire protection, design, maintenance, and use of fork trucks, tractors, platform lift trucks, motorized hand trucks, and other specialized industrial trucks powered by electric motors or internal combustion engines. This section does not apply to compressed air or nonflammable compressed gas-operated industrial trucks, nor to farm vehicles, nor to vehicles intended primarily for earth moving or over-the-road hauling.

All new powered industrial trucks acquired and used by an employer after the effective date specified in paragraph (b) of 1910.182 shall meet the design and construction requirements for powered industrial trucks established in the "American National Standard for Powered Industrial Trucks, Part II, ANSI B56.1-1969", except for vehicles intended primarily for earth moving or over-the-road hauling.

Approved trucks shall bear a label or some other identifying mark indicating approval by the testing laboratory. See paragraph (a)(7) of this section and paragraph 405 of "American National Standard for Powered Industrial Trucks, Part II, ANSI B56.1-1969", which is incorporated by reference in paragraph (a)(2) of this section and which provides that if the powered industrial truck is accepted by a nationally recognized testing laboratory it should be so marked.

Modifications and additions which affect capacity and safe operation shall not be performed by the customer or user without manufacturers prior written approval. Capacity, operation, and maintenance instruction plates, tags, or decals shall be changed accordingly.

If the truck is equipped with front-end attachments other than factory installed attachments, the user shall request that the truck be marked to identify the attachments and show the approximate weight of the truck and attachment combination at maximum elevation with load laterally centered.

The user shall see that all nameplates and markings are in place and are maintained in a legible condition.

As used in this section, the term, "approved truck" or "approved industrial truck" means a truck that is listed or approved for fire safety purposes for the intended use by a nationally recognized testing laboratory, using nationally recognized testing standards. Refer to 1910.155(c)(3)(iv)(A) for definition of nationally recognized testing laboratory.

Designations. For the purpose of this standard there are eleven different designations of industrial trucks or tractors as follows: D, DS, DY, E, ES, EE, EX, G, GS, LP, and LPS.

The D designated units are units similar to the G units except that they are diesel engine powered instead of gasoline engine powered.

The DS designated units are diesel powered units that are provided with additional safeguards to the exhaust, fuel and electrical systems. They may be used in some locations where a D unit may not be considered suitable.

The DY designated units are diesel powered units that have all the safeguards of the DS units and in addition do not have any electrical equipment including the ignition and are equipped with temperature limitation features.

The E designated units are electrically powered units that have minimum acceptable safeguards against inherent fire hazards.

The ES designated units are electrically powered units that, in addition to all of the requirements for the E units, are provided with additional safeguards to the electrical system to prevent emission of hazardous sparks and to limit surface temperatures. They may be used in some locations where the use of an E unit may not be considered suitable.

The EE designated units are electrically powered units that have, in addition to all of the requirements for the E and ES units, the electric motors and all other electrical equipment completely enclosed. In certain locations the EE unit may be used where the use of an E and ES unit may not be considered suitable.

The EX designated units are electrically powered units that differ from the E, ES, or EE units in that the electrical fittings and equipment are so designed, constructed and assembled that the units may be used in certain atmospheres containing flammable vapors or dusts.

The G designated units are gasoline powered units having minimum acceptable safeguards against inherent fire hazards.

The GS designated units are gasoline powered units that are provided with additional safeguards to the exhaust, fuel, and electrical systems. They may be used in some locations where the use of a G unit may not be considered suitable.

The LP designated unit is similar to the G unit except that liquefied petroleum gas is used for fuel instead of gasoline.

The LPS designated units are liquefied petroleum gas powered units that are provided with additional safeguards to the exhaust, fuel, and electrical systems. They may be used in some locations where the use of an LP unit may not be considered suitable.

The atmosphere or location shall have been classified as to whether it is hazardous or nonhazardous prior to the consideration of industrial trucks being used therein and the type of industrial truck required shall be as provided in paragraph (d) of this section for such location.

Designated locations. The industrial trucks specified under subparagraph (2) of this paragraph are the minimum types required but industrial trucks having greater safeguards may be used if desired.

For specific areas of use see Table N-1 which tabulates the information contained in this section. References are to the corresponding classification as used in subpart S of this part.

Power-operated industrial trucks shall not be used in atmospheres containing hazardous concentration of acetylene, butadiene, ethylene oxide, hydrogen (or gases or vapors equivalent in hazard to hydrogen, such as manufactured gas), propylene oxide, acetaldehyde, cyclopropane, diethyl ether, ethylene, isoprene, or unsymmetrical dimethyl hydrazine (UDMH).

Power-operated industrial trucks shall not be used in atmospheres containing hazardous concentrations of metal dust, including aluminum, magnesium, and their commercial alloys, other metals of similarly hazardous characteristics, or in atmospheres containing carbon black, coal or coke dust except approved power-operated industrial trucks designated as EX may be used in such atmospheres.

In atmospheres where dust of magnesium, aluminum or aluminum bronze may be present, fuses, switches, motor controllers, and circuit breakers of trucks shall have enclosures specifically approved for such locations.

Only approved power-operated industrial trucks designated as EX may be used in atmospheres containing acetone, acrylonitrile, alcohol, ammonia, benzine, benzol, butane, ethylene dichloride, gasoline, hexane, lacquer solvent vapors, naphtha, natural gas, propane, propylene, styrene, vinyl acetate, vinyl chloride, or xylenes in quantities sufficient to produce explosive or ignitable mixtures and where such concentrations of these gases or vapors exist continuously, intermittently or periodically under normal operating conditions or may exist frequently because of repair, maintenance operations, leakage, breakdown or faulty operation of equipment.

Power-operated industrial trucks designated as DY, EE, or EX may be used in locations where volatile flammable liquids or flammable gases are handled, processed or used, but in which the hazardous liquids, vapors or gases will normally be confined within closed containers or closed systems from which they can escape only in case of accidental rupture or breakdown of such containers or systems, or in the case of abnormal operation of equipment; also in locations in which hazardous concentrations of gases or vapors are normally prevented by positive mechanical ventilation but which might become hazardous through failure or abnormal operation of the ventilating equipment; or in locations which are adjacent to Class I, Division 1 locations, and to which hazardous concentrations of gases or vapors might occasionally be communicated unless such communication is prevented by adequate positive-pressure ventilation from a source of clear air, and effective safeguards against ventilation failure are provided.

TABLE N-1. -- SUMMARY TABLE ON USE OF INDUSTRIAL TRUCKS IN VARIOUS LOCATIONS

Classes	Unclassified	Class I locations	Class II locations	Class III locations
Description of classes.	Locations not possessing atmospheres as described in other columns.	Locations in which flammable gases or vapors are, or may be, present in the air in quantities sufficient to produce explosive or ignitable mixtures.	Locations which are hazardous because of the presence of combustible dust.	Locations where easily ignitable fibers or flyings are present but not likely to be in suspension in quantities sufficient to produce ignitable mixtures.
Groups in classes	None	A B	C	D
Examples of locations or atmospheres in classes and groups.	Piers and wharves inside and outside general storage, general industrial or commercial properties.	Acetylene Hydrogen	Ethyl ether	Gasoline Naphtha Alcohols Acetone Lacquer solvent Benzene

(Continued)

E	F	G	None
Metal dust	Carbon black coal dust, coke dust	Grain dust, flour dust, starch dust, organic dust.	Baled waste, cocoa fiber, cotton, excelsior, hemp, istle, jute, kapok, oakum, sisal, Spanish moss, synthetic fibers, tow.
Divisions (nature of hazardous conditions)	None	Above condition exists continuously, intermittently, or periodically under normal operating conditions.	Above condition may occur accidentally as due to a puncture of a storage drum.

(Continued)

1	2	1	2
Explosive mixture may be present under normal operating conditions, or where failure of equipment may cause the condition to exist simultaneously with arcing or sparking of electrical equipment, or where dusts of an electrically conducting nature may be present.	Explosive mixture not normally present, but where deposits of dust may cause heat rise in electrical equipment, or where such deposits may be ignited by arcs or sparks from electrical equipment.	Locations in which easily ignitable fibers or materials producing combustible flyings are handled, manufactured, or used.	Locations in which easily ignitable fibers are stored or handled (except in the process of manufacture).

Authorized uses of trucks by types in groups of classes and divisions

Groups in classes	None	A	B	C	D	A	B	C
Type of truck authorized:								
Diesel:								
Type D	D**	....	....	....	....	....	....	....
Type DS	......	....	....	....	....	....	....	....
Type DY	......	....	....	....	....	....	....	....
Electric:								
Type E	E**	....	....	....	....	....	....	....
Type ES	......	....	....	....	....	....	....	....
Type EE	......	....	....	....	....	....	....	....
Type EX	......	....	....	....	EX	....	....	....
Gasoline:								
Type G	G**	....	....	....	....	....	....	....
Type GS	......	....	....	....	....	....	....	....
LP-Gas:								
Type LP	LP**	....	....	....	....	....	....	....
Type LPS	......	....	....	....	....	....	....	....
Paragraph Ref. in No. 505.	210.211		201 (a)		203 (a)		209 (a)	
Diesel:								
Type D	....	....	....	....	....	....	....	....
Type DS	DS	....	....	....	....	DS	....	DS
Type DY	DY	....	....	....	....	DY	DY	DY
Electric:								
Type E	....	....	....	....	....	....	....	E
Type ES	ES	....	....	....	....	ES	....	ES
Type EE	EE	....	....	....	....	EE	EE	EE
Type EX	EX	EX	EX	....	....	EX	EX	EX
Gasoline:								
Type G	....	....	....	....	....	....	....	....
Type GS	GS	....	....	....	....	GS	....	GS
LP-Gas:								
Type LP	....	....	....	....	....	....	....	....
Type LPS	LPS	....	....	....	....	LPS	....	LPS
Paragraph Ref. in No. 505.	204 (a), (b)	202 (a)	205 (a)		209 (a)	206 (a), (b)	207 (a)	208 (a)

** Trucks conforming to these types may also be used -- see subdivision (c)(2)(x) and (c)(2)(xii) of this section.

In locations used for the storage of hazardous liquids in sealed containers or liquified or compressed gases in containers, approved power-operated industrial trucks designated as DS, ES, GS, or LPS may be used. This classification includes locations where volatile flammable liquids or flammable gases or vapors are used, but which, would become hazardous only in case of an accident or of some unusual operating condition. The quantity of hazardous material that might escape in case of accident, the adequacy of ventilating equipment, the total area involved, and the record of the industry or business with respect to explosions or fires are all factors that should receive consideration in determining whether or not the DS or DY, ES, EE, GS, LPS designated truck possesses sufficient safeguards for the location. Piping without valves, checks, meters and similar devices would not ordinarily be deemed to introduce a hazardous condition even though used for hazardous liquids or gases. Locations used for the storage of hazardous liquids or of liquified or compressed gases in sealed containers would not normally be considered hazardous unless subject to other hazardous conditions also.

Only approved power operated industrial trucks designated as EX shall be used in atmospheres in which combustible dust is or may be in suspension continuously, intermittently, or periodically under normal operating conditions, in quantities sufficient to produce explosive or ignitable mixtures, or where mechanical failure or abnormal operation of machinery or equipment might cause such mixtures to be produced.

The EX classification usually includes the working areas of grain handling and storage plants, room containing grinders or pulverizers, cleaners, graders, scalpers, open conveyors or spouts, open bins or hoppers, mixers, or blenders, automatic or hopper scales, packing machinery, elevator heads and boots, stock distributors, dust and stock collectors (except all-metal collectors vented to the outside), and all similar dust producing machinery and equipment in grain processing plants, starch plants, sugar pulverizing plants, malting plants, hay grinding plants, and other occupancies of similar nature; coal pulverizing plants (except where the pulverizing equipment is essentially dust tight); all working areas where metal dusts and powders are produced, processed, handled, packed, or stored (except in tight containers); and other similar locations where combustible dust may, under normal operating conditions, be present in the air in quantities sufficient to produce explosive or ignitable mixtures.

Only approved power-operated industrial trucks designated as DY, EE, or EX shall be used in atmospheres in which combustible dust will not normally be in suspension in the air or will not be likely to be thrown into suspension by the normal operation of equipment or apparatus in quantities sufficient to produce explosive or ignitable mixtures but where deposits or accumulations of such dust may be ignited by arcs or sparks originating in the truck.

Only approved power-operated industrial trucks designated as DY, EE, or EX shall be used in locations which are hazardous because of the presence of easily ignitable fibers or flyings but in which such fibers or flyings are not likely to be in suspension in the air in quantities sufficient to produce ignitable mixtures.

Only approved power-operated industrial trucks designated as DS, DY, ES, EE, EX, GS, or LPS shall be used in locations where easily ignitable fibers are stored or handled, including outside storage, but are not being processed or manufactured. Industrial trucks designated as E, which have been previously used in these locations may be continued in use.

On piers and wharves handling general cargo, any approved power-operated industrial truck designated as Type D, E, G, or LP may be used, or trucks which conform to the requirements for these types may be used.

If storage warehouses and outside storage locations are hazardous only the approved power-operated industrial truck specified for such locations in this paragraph (c) (2) shall be used. If not classified as hazardous, any approved power-operated industrial truck designated as Type D, E, G, or LP may be used, or trucks which conform to the requirements for these types may be used.

If general industrial or commercial properties are hazardous, only approved power-operated industrial trucks specified for such locations in this paragraph (c) (2) shall be used. If not classified as hazardous, any approved power-operated industrial truck designated as Type D, E, G, or LP may be used, or trucks which conform to the requirements of these types may be used.

Converted industrial trucks. Power-operated industrial trucks that have been originally approved for the use of gasoline for fuel, when converted to the use of liquefied petroleum gas fuel in accordance with paragraph (q) of this section, may be used in those locations where G, GS or LP, and LPS designated trucks have been specified in the preceding paragraphs.

Safety guards. High Lift Rider trucks shall be fitted with an overhead guard manufactured in accordance with paragraph (a) (2) of this section, unless operating conditions do not permit.

If the type of load presents a hazard, the user shall equip fork trucks with a vertical load backrest extension manufactured in accordance with paragraph (a) (2) of this section.

Fuel handling and storage. The storage and handling of liquid fuels such as gasoline and diesel fuel shall be in accordance with NFPA Flammable and Combustible Liquids Code (NFPA No. 30-1969).

The storage and handling of liquefied petroleum gas fuel shall be in accordance with NFPA Storage and Handling of Liquefied Petroleum Gases (NFPA No. 58-1969).

Changing and charging storage batteries. Battery charging installations shall be located in areas designated for that purpose.

Facilities shall be provided for flushing and neutralizing spilled electrolyte, for fire protection, for protecting charging apparatus from damage by trucks, and for adequate ventilation for dispersal of fumes from gassing batteries.

[Reserved]

A conveyor, overhead hoist, or equivalent material handling equipment shall be provided for handling batteries.

Reinstalled batteries shall be properly positioned and secured in the truck.

A carboy tilter or siphon shall be provided for handling electrolyte.

When charging batteries, acid shall be poured into water; water shall not be poured into acid.

Trucks shall be properly positioned and brake applied before attempting to change or charge batteries.

Care shall be taken to assure that vent caps are functioning. The battery (or compartment) cover(s) shall be open to dissipate heat.

Smoking shall be prohibited in the charging area.

Precautions shall be taken to prevent open flames, sparks, or electric arcs in battery charging areas.

Tools and other metallic objects shall be kept away from the top of uncovered batteries.

Lighting for operating areas. [Reserved]

Where general lighting is less than 2 lumens per square foot, auxiliary directional lighting shall be provided on the truck.

Control of noxious gases and fumes. Concentration levels of carbon monoxide gas created by powered industrial truck operations shall not exceed the levels specified in 1910.1000.

Dockboards (bridge plates). See 1910.30(a).

Trucks and railroad cars. The brakes of highway trucks shall be set and wheel chocks placed under the rear wheels to prevent the trucks from rolling while they are boarded with powered industrial trucks.

Wheel stops or other recognized positive protection shall be provided to prevent railroad cars from moving during loading or unloading operations.

Fixed jacks may be necessary to support a semitrailer and prevent upending during the loading or unloading when the trailer is not coupled to a tractor.

Positive protection shall be provided to prevent railroad cars from being moved while dockboards or bridge plates are in position.

Operator training. Only trained and authorized operators shall be permitted to operate a powered industrial truck. Methods shall be devised to train operators in the safe operation of powered industrial trucks.

Truck operations. Trucks shall not be driven up to anyone standing in front of a bench or other fixed object.

No person shall be allowed to stand or pass under the elevated portion of any truck, whether loaded or empty.

Unauthorized personnel shall not be permitted to ride on powered industrial trucks. A safe place to ride shall be provided where riding of trucks is authorized.

The employer shall prohibit arms or legs from being placed between the uprights of the mast or outside the running lines of the truck.

When a powered industrial truck is left unattended, load engaging means shall be fully lowered, controls shall be neutralized, power shall be shut off, and brakes set. Wheels shall be blocked if the truck is parked on an incline.

A powered industrial truck is unattended when the operator is 25 ft. or more away from the vehicle which remains in his view, or whenever the operator leaves the vehicle and it is not in his view.

When the operator of an industrial truck is dismounted and within 25 ft. of the truck still in his view, the load engaging means shall be fully lowered, controls neutralized, and the brakes set to prevent movement.

A safe distance shall be maintained from the edge of ramps or platforms while on any elevated dock, or platform or freight car. Trucks shall not be used for opening or closing freight doors.

Brakes shall be set and wheel blocks shall be in place to prevent movement of trucks, trailers, or railroad cars while loading or unloading. Fixed jacks may be necessary to support a semitrailer during loading or unloading when the trailer is not coupled to a tractor. The flooring of trucks, trailers, and railroad cars shall be checked for breaks and weakness before they are driven onto.

There shall be sufficient headroom under overhead installations, lights, pipes, sprinkler system, etc.

An overhead guard shall be used as protection against falling objects. It should be noted that an overhead guard is intended to offer protection from the impact of small packages, boxes, bagged material, etc., representative of the job application, but not to withstand the impact of a falling capacity load.

A load backrest extension shall be used whenever necessary to minimize the possibility of the load or part of it from falling rearward.

Only approved industrial trucks shall be used in hazardous locations.

Whenever a truck is equipped with vertical only, or vertical and horizontal controls elevatable with the lifting carriage or forks for lifting personnel, the following additional precautions shall be taken for the protection of personnel being elevated.

Use of a safety platform firmly secured to the lifting carriage and/or forks.

Means shall be provided whereby personnel on the platform can shut off power to the truck.

Such protection from falling objects as indicated necessary by the operating conditions shall be provided.

[Reserved]

Fire aisles, access to stairways, and fire equipment shall be kept clear.

Traveling. All traffic regulations shall be observed, including authorized plant speed limits. A safe distance shall be maintained approximately three truck lengths from the truck ahead, and the truck shall be kept under control at all times.

The right of way shall be yielded to ambulances, fire trucks, or other vehicles in emergency situations.

Other trucks traveling in the same direction at intersections, blind spots, or other dangerous locations shall not be passed.

The driver shall be required to slow down and sound the horn at cross aisles and other locations where vision is obstructed. If the load being carried obstructs forward view, the driver shall be required to travel with the load trailing.

Railroad tracks shall be crossed diagonally wherever possible. Parking closer than 8 feet from the center of railroad tracks is prohibited.

The driver shall be required to look in the direction of, and keep a clear view of the path of travel.

Grades shall be ascended or descended slowly.

When ascending or descending grades in excess of 10 percent, loaded trucks shall be driven with the load upgrade.

[Reserved]

On all grades the load and load engaging means shall be tilted back if applicable, and raised only as far as necessary to clear the road surface.

Under all travel conditions the truck shall be operated at a speed that will permit it to be brought to a stop in a safe manner.

Stunt driving and horseplay shall not be permitted.

The driver shall be required to slow down for wet and slippery floors.

Dockboard or bridgeplates, shall be properly secured before they are driven over. Dockboard or bridgeplates shall be driven over carefully and slowly and their rated capacity never exceeded.

Elevators shall be approached slowly, and then entered squarely after the elevator car is properly leveled. Once on the elevator, the controls shall be neutralized, power shut off, and the brakes set.

Motorized hand trucks must enter elevator or other confined areas with load end forward.

Running over loose objects on the roadway surface shall be avoided.

While negotiating turns, speed shall be reduced to a safe level by means of turning the hand steering wheel in a smooth, sweeping motion. Except when maneuvering at a very low speed, the hand steering wheel shall be turned at a moderate, even rate.

Loading. Only stable or safely arranged loads shall be handled. Caution shall be exercised when handling off-center loads which cannot be centered.

Only loads within the rated capacity of the truck shall be handled.

The long or high (including multiple-tiered) loads which may affect capacity shall be adjusted.

Trucks equipped with attachments shall be operated as partially loaded trucks when not handling a load.

A load engaging means shall be placed under the load as far as possible; the mast shall be carefully tilted backward to stabilize the load.

Extreme care shall be used when tilting the load forward or backward, particularly when high tiering. Tilting forward with load engaging means elevated shall be prohibited except to pick up a load. An elevated load shall not be tilted forward except when the load is in a deposit position over a rack or stack. When stacking or tiering, only enough backward tilt to stabilize the load shall be used.

Operation of the truck. If at any time a powered industrial truck is found to be in need of repair, defective, or in any way unsafe, the truck shall be taken out of service until it has been restored to safe operating condition.

Fuel tanks shall not be filled while the engine is running. Spillage shall be avoided.

Spillage of oil or fuel shall be carefully washed away or completely evaporated and the fuel tank cap replaced before restarting engine.

No truck shall be operated with a leak in the fuel system until the leak has been corrected.

Open flames shall not be used for checking electrolyte level in storage batteries or gasoline level in fuel tanks.

Maintenance of industrial trucks. Any power-operated industrial truck not in safe operating condition shall be removed from service. All repairs shall be made by authorized personnel.

No repairs shall be made in Class I, II, and III locations.

Those repairs to the fuel and ignition systems of industrial trucks which involve fire hazards shall be conducted only in locations designated for such repairs.

Trucks in need of repairs to the electrical system shall have the battery disconnected prior to such repairs.

All parts of any such industrial truck requiring replacement shall be replaced only by parts equivalent as to safety with those used in the original design.

Industrial trucks shall not be altered so that the relative positions of the various parts are different from what they were when originally received from the manufacturer, nor shall they be altered either by the addition of extra parts not provided by the manufacturer or by the elimination of any parts, except as provided in paragraph (q)(12) of this section. Additional counterweighting of fork trucks shall not be done unless approved by the truck manufacturer.

Industrial trucks shall be examined before being placed in service, and shall not be placed in service if the examination shows any condition adversely affecting the safety of the vehicle. Such examination shall be made at least daily. Where industrial trucks are used on a round-the-clock basis, they shall be examined after each shift. Defects when found shall be immediately reported and corrected.

Water mufflers shall be filled daily or as frequently as is necessary to prevent depletion of the supply of water below 75 percent of the filled capacity. Vehicles with mufflers having screens or other parts that may become clogged shall not be operated while such screens or parts are clogged. Any vehicle that emits hazardous sparks or flames from the exhaust system shall immediately be removed from service, and not returned to service until the cause for the emission of such sparks and flames has been eliminated.

When the temperature of any part of any truck is found to be in excess of its normal operating temperature, thus creating a hazardous condition, the vehicle shall be removed from service and not returned to service until the cause for such overheating has been eliminated.

Industrial trucks shall be kept in a clean condition, free of lint, excess oil, and grease. Noncombustible agents should be used for cleaning trucks. Low flash point (below 100 deg. F.) solvents shall not be used. High flash point (at or above 100 deg. F.) solvents may be used. Precautions regarding toxicity, ventilation, and fire hazard shall be consonant with the agent or solvent used.

[Reserved]

Industrial trucks originally approved for the use of gasoline for fuel may be converted to liquefied petroleum gas fuel provided the complete conversion results in a truck which embodies the features specified for LP or LPS designated trucks. Such conversion equipment shall be approved. The description of the component parts of this conversion system and the recommended method of installation on specific trucks are contained in the "Listed by Report."

[39 FR 23502, June 27, 1974, as amended at 40 FR 23073, May 28, 1975; 43 FR 49749, Oct. 24, 1978; 49 FR 5322, Feb. 10, 1984; 53 FR 12122, Apr. 12, 1988; 55 FR 32015, Aug. 6, 1990]

GLOSSARY

A

ABC Extinguisher: A multi-purpose dry chemical extinguisher for fighting class A, B, and C fires.

Accident: An unplanned event, not necessarily injurious or damaging to property, interrupting the activity in process.

ACGIH: American Conference of Governmental Industrial Hygienists. An association with membership open to anyone who is engaged in the practice of industrial hygiene or occupational and environmental health and safety.

Acute: Health effects that show up a short length of time after exposure. An acute exposure runs a comparatively short course.

ADA: Americans with Disabilities Act. A 1991 federal law prohibiting discrimination against people with disabilities in most public activities, including the workplace.

Alpha particle: A small, positively charged particle made up of two neutrons and two protons, moving at very high velocity, thrown off by many radioactive materials, including uranium and radium.

Alveoli: Tiny air sacs of the lungs, formed at the ends of bronchioles; through the thin walls of the alveoli, the blood takes in oxygen and gives up carbon dioxide in respiration.

ANSI: American National Standards Institute. A voluntary membership organization (operated with private funding) that develops consensus standards nationally for a wide variety of devices and procedures.

Asbestos: A hydrated magnesium silicate in fibrous form.

ASME: American Society of Mechanical Engineers.

Attachment: A device other than conventional forks or load backrest extension, mounted permanently or removably on the levating mechanism of a lift truck for handling the load, e.g. fork extensions, clamps, rotating devices, side shifters, load stabilizers, rams, etc.

Audiogram: A record of hearing loss or hearing level measured at several different frequencies - usually 500 to 6,000 Hz. The audiogram may be presented graphically or numerically. Hearing level is shown as a function of frequency.

B

Beta particle (beta radiation): A small electrically charged particle thrown off by many radioactive materials; identical to the electron. Beta particles emerge from radioactive materials at high speeds.

BLS: Bureau of Labor Statistics

C

Carbon dioxide (CO_2): A nonflammable (inert) liquified, industrial gas. Used in CO_2 fire extinguishers mainly to fight electrical fires. Can be used to fight flammable liquid fires but will not smother the flame or liquid; will only remove the oxygen. Although nontoxic, carbon dioxide can cause asphyxiation due to the displacement of air.

Carbon monoxide: A colorless, odorless, toxic gas produced by any process that involves the incomplete combustion of carbon-containing substances. It is emitted through the exhaust of gasoline-powered vehicles.

Carboxyhemoglobin: The reversible combination of carbon monoxide (CO) with hemoglobin.

Chronic: Persistent, prolonged, repeated.

Class A Fires: Fires in ordinary combustible materials, such as wood, cloth, paper, rubber and many plastics, that require the heat absorbing, cooling effects of water to be extinguished.

Class B Fires: Fires in flammable liquids, flammable gases, greases, and similar materials that must be extinguished by excluding air (oxygen), by inhibiting the release of combustible vapors with AFFF or FFFP agents, or by interrupting the combustion chain reaction.

Class C Fires: Fires in live electrical equipment. For operator safety requires the use of electrically nonconductive extinguishing agents, such as dry chemical or halon. Once the electrical equipment is deenergized, extinguishers for Class A or B fires may be used.

Class D Fires: Fires in certain combustible metals such as magnesium, titanium, zirconium, sodium, and potassium, that require a heat-absorbing extinguishing medium that does not react with the burning metals.

Combustible liquids: Combustible liquids are those having a flash point at or above 100°F (37.8°C).

Corrosive: A substance that causes visible destruction or permanent changes in human skin tissue at the site of contact.

D

Decibel (dB): A unit used to express sound power level (Lw) and sound-pressure level (Lp). Sound power is the total acoustic output of a sound source in watts (W). By definition, sound-power level, in decibels, is : $Lw = 10 \log W/Wo$, where W is the sound power of the source and Wo is the reference sound power of 10-12. Because the decibel is also used to describe other physical quantities, such as electrical current and electrical voltage, the correct reference must be specified.

Dock/bridge plate: A portable or fixed device for spanning the gap, or compensating for differences in levels, between a loading platform and carriers.

Dusts: Solid particles generated by handling, crushing, grinding, rapid impact, detonation, and decrepitation of organic or inorganic materials, such as rock, ore, metal, coal, wood, and grain. Dusts do not tend to flocculate, except under electrostatic forces, they do not diffuse in air but settle under the influence of gravity.

E

Ergonomics: The study of human characteristics for the appropriate design of living and work environments.

EPA: Environmental Protection Agency.

F

FDA: The U.S. Food and Drug Administration. The FDA establishes requirements for the labeling of foods and drugs to protect consumers from misbranded, unwholesome, ineffective, and hazardous products. FDA also regulates materials for food contact service and the conditions under which such materials are approved.

Filter, HEPA: High-efficiency particulate air filter that is at least 99.97% efficient in removing thermally generated monodisperse dioctylphthalate smoke particles with a diameter of 0.3 microns.

Flammable: Any substance that is easily ignited, burns intensely, or has a rapid rate of flame spread. Flammable and inflammable are identical in meaning; however, the prefix "in" indicates negative in many words and can cause confusion. Flammable, therefore, is the preferred term.

Flammable liquid: Any liquid having a flash point below 100°F. (37.8°C).

Flammable range: The difference between the lower and upper flammable limits, expressed in terms of percentage of vapor or gas in air by volume, and is also often referred to as the "explosive range."

Fulcrum: The axis of rotation of the truck when it tips over.

Fume: Airborne particulates formed by the condensation of solid particles from the gaseous state. Usually, fumes are generated after initial volatilization from a combustion process, or from a melting process (such as metal fume emitted during welding). Usually measure less than 1 micron in diameter.

G

Gas: A state of matter in which the material has very low density and viscosity, can expand and contract greatly in response to changes in temperature and pressure, easily diffused into other gases, and readily and uniformly distributes itself throughout any container. A gas can be changed to the liquid or solid state only by the combined effect of increased pressure and decreased temperature (below the critical temperature).

Geiger counter: A gas-filled electrical device that counts the presence of an atomic particle or ray by detecting the ions produced. Sometimes referred to as a Geiger-Müller counter.

Grab sample: A sample taken, within a very short time period to determine the constituents at a specific time.

H

Hazardous Materials Identification System (HMIS): A labeling system to identify health, flammability, and reactivity in chemicals. This system uses horizontal bars to identify the color associated with the hazard. A numerical rating system, using a 0-4, identifies the degree of hazards of the chemical.

Hemoglobin: The red coloring matter of the blood that carries the oxygen.

I

ICC: U.S. Interstate Commerce Commission.

Incident: An undesired event that may cause personal harm or other damage. In the United States, OSHA defines the criteria for recordkeeping purposes. For this book, incidents involve close calls, near misses, property damage, equipment damage, the discovery of unsafe behavior and conditions, and minor first aid cases.

J

Job Hazard Analysis (JHA): A method for studying a job in order to a) identify hazards or potential accidents associated with each step or task, and b) develop solutions that will eliminate, nullify, or prevent such hazards or accidents. Sometimes called Job Safety Analysis.

Job Safety Observation (JSO): A program in which employees are observed by a supervisor, or another person, that studies the work in progress to determine if the employee is behaving in a safe manner and is completing the task as required. This activity is then properly recorded.

L

Laser: Light amplification by stimulated emission of radiation. Lasers may operate in either pulsed or continuous mode.

Lateral stability: The resistance of a truck to tipping over sideways.

Liquefied petroleum gas (LP gas): A flammable liquefied (including cryogenic) fuel. Used as a domestic, commercial, agricultural, and industrial fuel. The NFPA rates this gas as a "4" on the 704 label system.

Load center: The position of the load's center of gravity in the horizontal plane only, relative to the vertical load-engaging face of the forks (or the equivalent for other load-engaging means). It constitutes the standard base for rating the load capacity of the lift truck.

Longitudinal stability: The resistance of a truck to overturning forward or rearward.

M

Mast: The vertical arms of a forklift that allow the forks and load to travel a given distance for raising or lowering. The mast may also be tilted forward or backward with operator controls.

Micron (micrometer): A unit of length equal to 10-4 cm, approximately 1/25,000 of an inch.

Moment: The product of the weight of the object times the distance from a fixed point. In the case of a powered industrial truck, the distance is measured from the point that the truck will tip over to the line of action of the object. The distance is always measured perpendicular to the line of action.

MSDS: Material Safety Data Sheet. A document prepared by a chemical manufacturer, describing the properties and hazards of the chemical.

N

NFPA: The National Fire Protection Association. A voluntary membership organization with the aim to promote and improve fire protection and prevention. The NFPA publishes the National Fire Codes.

NFPA 704: A labeling system to identify the fire and other hazards of materials. This identification scheme, based on the "704 diamond," can visually present information on flammability health and self-reactivity hazards, as well as special information associated with the hazards of the materials being identified.

NIOSH: The National Institute for Occupational Safety and Health. A federal agency that conducts research on health and safety concerns, tests and certifies respirators, and trains occupational health and safety professionals.

NRR: Noise reduction rating. As applied to ear protection, the number of dB that the device reduces in transmission to the ear.

O

OSHA: The U.S. Occupational Safety and Health Administration of the Department of Labor. A federal agency with safety and health regulatory and enforcement authorities for general U.S. industry and business.

Overhead guard: A framework mounted on the lift truck and positioned above the head of the riding operator to protect against falling objects but not a capacity load.

P

Permissible Exposure Limit (PEL): The legally enforced exposure limit for a substance established by the U.S. OSHA. The PEL indicates the permissible concentration of air contaminants to which nearly all workers may be repeatedly exposed for eight (8) hours a day, forty (40) hours a week, over a working lifetime (30 years) without adverse health effects.

Powered industrial truck: A mobile, power-driven vehicle used to carry, push, pull, lift, stack, or tier material. This includes fork lift trucks, straddle carriers, pallet or platform lifts, hoisters, stackers, personnel carriers and lift platforms, and industrial tractors. The definition does not include earthmoving equipment, or excavating or manipulating machinery. It also does not include material handling equipment confined to a fixed system such as a stacker crane conveyor system.

PPE: Personal Protective Equipment. Devices worn by the worker to protect against hazards in the environment. Respirators, gloves, glasses, hard hats, and hearing protectors are some examples.

PPM: Parts per million part of air by volume of vapor or gas or other contaminant.

R

Radiation (nuclear): The emission of atomic particles or electromagnetic radiation from the nucleus of an atom.

RCRA: Resource Conservation and Recovery Act. Environmental legislation administered by the EPA aimed at controlling the generation, treating, storage, transportation, and disposal of hazardous wastes.

Route of entry: The path by which chemicals can enter the body, primarily inhalation, ingestion, and skin absorption.

S

Safety belt: A strap, webbing, or similar device including all necessary buckles and fasteners, designed to secure a person, and to mitigate the possible injuries in case of an accident.

Sprinkler head: A device that allows water to flow on a fire or heat source after a fuseable link melts on the head.

Stability triangle: The triangle that has one point at the center of each front wheel and one point at the center of the lift truck's steering wheel axle. The relative location, inside or outside the triangle, of the truck's and load's combined center of gravity determines the stability of the truck.

T

Time Weighted Average Concentration (TWA): Refers to concentrations of airborne toxic materials weighted for a certain time duration, usually 8 hours.

TLV: Threshold limit value. A term used by ACGIH to express the airborne concentration of a material to which *nearly* all persons can be exposed day after day, without adverse effects. ACGIH expresses TLVs in three ways:

TLV-C: The ceiling limit. The concentration that should not be exceeded even instantaneously.

TLV-STEL: The short-term exposure limit, or maximum concentration for a continuous 15-minute exposure period (maximum of four such periods per day, with at least 60 minutes between exposure periods, and provided that the daily TLV-TWA is not exceeded).

TLV-TWA: The allowable time weighted average concentration for a normal 8-hour work day or 40-hour work week.

Toxic substance: Any substance that can cause acute or chronic injury to the human body, or which is suspected of being able to cause diseases or injury under some conditions.

V

Vapors: The gaseous form of substances that are normally in the solid or liquid state (at room temperature and pressure). The vapor can be changed back to the solid or liquid state either by increasing the pressure or decreasing the temperature alone. Vapors also diffuse. Evaporation is the process by which a liquid is changed to the vapor state and mixed with the surrounding air. Solvents with low boiling points volatize readily.

W

Wheel chock: A block or wedge placed under a wheel to keep the wheel from moving.

X

X-rays: Highly penetrating radiation similar to gamma rays. Unlike gamma rays, X-rays do not come from the nucleus of the atom but from the surrounding electrons. They are produced by electron bombardment. When these rays pass through an object they give a shadow picture of the denser portions.

INDEX

A

Accident costs, 7, 135, 183
    back injuries, 153
    eye injuries, 135
    fires, 7
    material handling, 178
Accident investigation, 173, 179, 185
    form, 184
Acid, 92, 94, 128
Aisles, 22, 54, 55, 65, 117, 118, 121, 123, 124, 125
Alarms, 40, 92, 123, 125, 145, 155, 220
Americans with Disabilities Act, 10
Anchorage point, 133
ANSI Standards, 8, 21, 22, 61, 82, 98, 134, 135, 136, 137, 222
Audits - Forklift training, 239, 240
Awards, 11, 215

B

Back safety, 153-155
Back-up alarms, 70, 72, 121, 145
Battery, 48, 148
    acid, 61, 92
    charging, 92, 93
Belly button switch, 60
Body harness, 38
Brakes, 71, 72, 177

C

Carbon monoxide, 82, 92, 141-144, 170, 233
Catalytic converter, 41
Chemicals, 128, 146, 147, 148
Cold weather clothing, 134
Compressed natural gas, 141-144, 232
Computer based training, 196-197
Controls, 72
Coolant level, 72
Cost savings, 9, 14
Counterbalance walkie truck, 54
Counterbalance, 48

D

Diesel fuel, 48, 51, 96, 232
Direct costs of accidents, 10, 183
Dock inspection form, 110-113
Dock plates, 92, 107, 108
Dock safety, 90-92, 211
Dock slope, 102
Docks, 101, 102
Dust masks, 114, 138
Dynamic stability, 35

E

Electric battery, 232
Electric lift trucks, 29, 48
Elevator, 85
Emergency spills, 92
Employee training - PPE, 140
Engines, 41
Environmental Protection Agency, 41, 150, 233, 234
Environmental, 231

Ergonomics, 151-155
Eye and face protection, 136, 137
Eye wash station, 94

F

Factory mutual, 7
Fire extinguishers, 7, 72, 89, 94, 95, 114, 236, 237, 238
Fire losses, 7
Flip charts, 196
Forklifts:
    causes of accidents, 4
    citations, 4, 5
    damage, 219
    fatalities, 1
    fatalities classifications, 2
    quantity, 43
    rodeos, 199, 207
    stability, 11, 30
    tip over, 17, 40
Forks, 73, 220-225
Front end attachments, 98
Fuel level, 74
Full protection device, 132-133

G

Gasoline, 48, 96, 232
Gauges, 65, 70, 74

H

Hand protection 134, 139
Hard hats, 134
Hazard assessment, 61, 139
Hazardous areas, 29
Hazardous waste, 233
Hearing protection 138, 139, 145
Heat curtains, 119
Heinrich, Wm., 175, 176, 187
High lift trucks, 48

HMIS label, 150
Horns, 71, 122
Hybrid forklift models, 52
Hydraulic oil leaks, 74

I

ICC Bar, 91, 104, 105, 106
Incident investigation form, 180-181
Incident investigation, 176-179
Indirect costs of accidents, 10, 183
Industrial Truck Association, 222
Inspection forms, 70
Inspections, 69
Instructors check list, 193

J

Job Hazard Analysis, 159
    form #1, 162, 163
    form #2, 162 - 167
    form #3, 162 - 168, 172
    master list, 161
    overview, 160
Job Safety Observation, 240-244
    form, 244
Judges Guidelines - Rodeo, 213, 214, 216
Judges scoring sheets, 206

L

Lanyard, 38, 133
Lasers, 132
Lateral stability, 34
Learning, 188
Lift cage 96, 228-231
Lift truck models classification, 55-56
Lift truck training log, 194
Lift trucks inspection form, 76-78
Lifting, 107, 153-154
Lights, 72, 105, 118, 120, 155
Liquid natural gas, 29, 48

Liquid propane gas, 29, 48, 51, 74, 94, 95, 141, 143, 144, 152, 232
Listening skills, 190, 191
Load center, 30
Longitudinal stability, 35
Low lift trucks, 43, 48

M

Maintenance, 96
Management training, 171
Manually propelled, 52
Material safety data sheet, 146, 148, 233
Metatarsal shoes, 138
Mirrors, 118, 121, 122, 155
Models of fork lift trucks, 43
Moment, 11, 30, 31, 35

N

Name plates, 98
Narrow aisle lift trucks, 50
National Fire Protection Association, 94
NFPA 704 Label, 149
NFPA Standards, 29, 82, 94, 96
NIOSH dust masks, 138
Noise, 138, 139, 145

O

Oil filters, 235
Operating characteristics, 36
Operating handle - walkie truck, 64
Operator training, 21, 22
Operators, 99
Order pickers, 50
OSHA standards, 82
OSHA's proposed training requirements, 24
Overhead guard, 75, 98
Overhead projector, 196

P

Pallet trucks, 48
Pallets, 215, 227, 228
Pedals, 153
Pedestrian safety, 11, 24, 38, 40, 65, 83, 88, 101, 117, 121, 123, 124, 220
Personal protective equipment, 61, 93, 94, 96, 127, 129, 138, 145, 148, 150, 154, 156, 162, 207
Pinch point, 60
Platform trucks, 48, 53
PPE risk assesment survey, 128, 131, 132, 133
Preventing property damage, 8, 9
Propane tank, 75

Q

Quizzes, 195

R

Radiation, 129, 151
Railroad tracks, 99
Ramps, 67, 85, 86, 90
Resource Conservation and Recovery Act, 233
Respirators, 129, 138
Rider reach truck, 50

S

Safety committee, 12
Seat belts, 19, 20, 25, 72, 75
SHARP Evaluation Program, 239
Shoes, 59, 129, 135, 136, 138
Side loaders, 51
Signs, 118, 120
Skills evaluation form, 208
Skills for training, 196
Skills testing form, 205
Skills testing, 201, 205
Slides for training, 196

Slopes, 108
Sources of training, 23
Stability triangle, 31-33
Sulfuric acid, 94
Supervisors training, 191

T

Tip over, 17, 18, 19, 20, 25, 32, 33, 34, 35, 36, 37, 38, 40, 64, 65, 86, 90, 117, 124
Tire safety, 97, 98
Tires, 33, 35, 51, 73, 235
Trailers, 101, 102
Training form - PPE, 130
Training plan, 191, 192
Training program, which are successful, 12-15
Training tips, 189
Traveling with power equipment, 83, 90
Truck restraints, 103, 104, 106
Turret trucks, 52

U

Unattended trucks, 99
Unsafe behavior, 182
Unsafe conditions, 182, 183

V

Video training 150, 195
Vision requirements, 11
Voluntary Protection Program, 12, 15, 238, 239

W

Walkie inspection form, 66
Walkie reach trucks, 54
Walkie straddle trucks, 54
Walkie trucks, 53, 59, 62-65
Wheel chocks, 18, 22, 61, 65, 90, 91, 92, 99, 103-105, 115, 161, 170
Winged seats, 19, 25, 40

 # GOVERNMENT INSTITUTES MINI-CATALOG

PC #	ENVIRONMENTAL TITLES	Pub Date	Price
585	Book of Lists for Regulated Hazardous Substances, 8th Edition	1997	$79
4088	CFR Chemical Lists on CD ROM, 1997 Edition	1997	$125
4089	Chemical Data for Workplace Sampling & Analysis, Single User	1997	$125
512	Clean Water Handbook, 2nd Edition	1996	$89
581	EH&S Auditing Made Easy	1997	$79
587	E H & S CFR Training Requirements, 3rd Edition	1997	$89
4082	EMMI-Envl Monitoring Methods Index for Windows-Network	1997	$537
4082	EMMI-Envl Monitoring Methods Index for Windows-Single User	1997	$179
525	Environmental Audits, 7th Edition	1996	$79
548	Environmental Engineering and Science: An Introduction	1997	$79
578	Environmental Guide to the Internet, 3rd Edition	1997	$59
560	Environmental Law Handbook, 14th Edition	1997	$79
353	Environmental Regulatory Glossary, 6th Edition	1993	$79
625	Environmental Statutes, 1998 Edition	1998	$69
4098	Environmental Statutes Book/Disk Package, 1998 Edition	1997	$208
4994	Environmental Statutes on Disk for Windows-Network	1997	$405
4994	Environmental Statutes on Disk for Windows-Single User	1997	$139
570	Environmentalism at the Crossroads	1995	$39
536	ESAs Made Easy	1996	$59
515	Industrial Environmental Management: A Practical Approach	1996	$79
4078	IRIS Database-Network	1997	$1,485
4078	IRIS Database-Single User	1997	$495
510	ISO 14000: Understanding Environmental Standards	1996	$69
551	ISO 14001: An Executive Repoert	1996	$55
518	Lead Regulation Handbook	1996	$79
478	Principles of EH&S Management	1995	$69
554	Property Rights: Understanding Government Takings	1997	$79
582	Recycling & Waste Mgmt Guide to the Internet	1997	$49
603	Superfund Manual, 6th Edition	1997	$115
566	TSCA Handbook, 3rd Edition	1997	$95
534	Wetland Mitigation: Mitigation Banking and Other Strategies	1997	$75

PC #	SAFETY AND HEALTH TITLES	Pub Date	Price
547	Construction Safety Handbook	1996	$79
553	Cumulative Trauma Disorders	1997	$59
559	Forklift Safety	1997	$65
539	Fundamentals of Occupational Safety & Health	1996	$49
535	Making Sense of OSHA Compliance	1997	$59
563	Managing Change for Safety and Health Professionals	1997	$59
589	Managing Fatigue in Transportation, *ATA Conference*	1997	$75
4086	OSHA Technical Manual, Electronic Edition	1997	$99
598	Project Mgmt for E H & S Professionals	1997	$59
552	Safety & Health in Agriculture, Forestry and Fisheries	1997	$125
613	Safety & Health on the Internet, 2nd Edition	1998	$49
597	Safety Is A People Business	1997	$49
463	Safety Made Easy	1995	$49
590	Your Company Safety and Health Manual	1997	$79

Electronic Product available on CD-ROM or Floppy Disk

PLEASE CALL OUR CUSTOMER SERVICE DEPARTMENT AT (301) 921-2323 FOR A FREE PUBLICATIONS CATALOG.

Government Institutes
4 Research Place, Suite 200 • Rockville, MD 20850-3226
Tel. (301) 921-2323 • FAX (301) 921-0264
E mail: giinfo@govinst.com • Internet: http://www.govinst.com

GOVERNMENT INSTITUTES ORDER FORM

4 Research Place, Suite 200 • Rockville, MD 20850-3226 • Tel (301) 921-2323 • Fax (301) 921-0264
Internet: *http://www.govinst.com* • E-mail: *giinfo@govinst.com*

3 EASY WAYS TO ORDER

1. **Phone:** **(301) 921-2323**
   Have your credit card ready when you call.

2. **Fax:** **(301) 921-0264**
   Fax this completed order form with your company purchase order or credit card information.

3. **Mail:** **Government Institutes**
   4 Research Place, Suite 200
   Rockville, MD 20850-3226
   USA
   Mail this completed order form with a check, company purchase order, or credit card information.

PAYMENT OPTIONS

❑ **Check** (*payable to Government Institutes in US dollars*)

❑ **Purchase Order** (this order form must be attached to your company P.O. Note: All International orders must be pre-paid.)

❑ **Credit Card** ❑ ❑ ❑

Exp. ___/___

Credit Card No. _____

Signature _____
Government Institutes' Federal I.D.# is 52-0994196

CUSTOMER INFORMATION

Ship To: (Please attach your Purchase Order)

Name: _____
GI Account# (*7 digits on mailing label*): _____
Company/Institution: _____
Address: _____
(please supply street address for UPS shipping)
_____
City: _____ State/Province: _____
Zip/Postal Code: _____ Country: _____
Tel: ( ) _____
Fax: ( ) _____
E-mail Address: _____

Bill To: (if different than ship to address)

Name: _____
Title/Position: _____
Company/Institution: _____
Address: _____
(please supply street address for UPS shipping)
_____
City: _____ State/Province: _____
Zip/Postal Code: _____ Country: _____
Tel: ( ) _____
Fax: ( ) _____
E-mail Address: _____

Qty.	Product Code	Title	Price

❑ **New Edition No Obligation Standing Order Program**
Please enroll me in this program for the products I have ordered. Government Institutes will notify me of new editions by sending me an invoice. I understand that there is no obligation to purchase the product. This invoice is simply my reminder that a new edition has been released.

15 DAY MONEY-BACK GUARANTEE
If you're not completely satisfied with any product, return it undamaged within 15 days for a full and immediate refund on the price of the product.

Subtotal _____
MD Residents add 5% Sales Tax _____
Shipping and Handling (see box below) _____
Total Payment Enclosed _____

Within U.S:
1-4 products: $6/product
5 or more: $3/product

Outside U.S:
Add $15 for each item (Airmail)
Add $10 for each item (Surface)

SOURCE CODE: BP01

Government Institutes • 4 Research Place, Suite 200 • Rockville, MD 20850
Internet: http://www.govinst.com • E-mail: giinfo@govinst.com